Food Safety and Security

Food Safety and Security

Editor: Ellis Simmons

Callisto Reference

www.callistoreference.com

Callisto Reference,
118-35 Queens Blvd., Suite 400,
Forest Hills, NY 11375, USA

Visit us on the World Wide Web at:
www.callistoreference.com

ISBN: 978-1-64116-255-5 (Hardback)

Printed in China.

Cataloging-in-Publication Data

Food safety and security / edited by Ellis Simmons
p. cm.
Includes bibliographical references and index.
ISBN 978-1-64116-255-5
1. Food--Safety measures. 2. Food security. 3. Food contamination--Prevention.
4. Food industry and trade--Safety measures. I. Simmons, Ellis.

TX531 .F66 2020
363.192--dc23

Table of Contents

Preface

A scientific discipline that describes the careful preparation, handling and storage of food to prevent food-borne illness is called food safety. It includes a number of routines that should be followed to avoid any health hazard. Safety of food is ensured between the industry and the market and then between the market and the consumers. Food safety is concerned with the practices related to food labeling, agricultural and animal husbandry, food additives, food manufacturing, food contamination, novel foods and genetically modified foods. A measure of the availability of food and its accessibility to an individual at an affordable price is called food security. Food availability and accessibility are associated with the supply of food through production, exchange and distribution, and its allocation and affordability according to the preferences of individuals and household. The security of food is measured through the calculation of calorie intake per person per day, available on a household budget. This book studies, analyzes and upholds the pillars of food safety and security and their utmost significance in modern times. It elucidates new techniques and their applications in a multidisciplinary manner. Coherent flow of topics, student-friendly language and extensive use of examples make it an invaluable source of knowledge.

This book has been the outcome of endless efforts put in by authors and researchers on various issues and topics within the field. The book is a comprehensive collection of significant researches that are addressed in a variety of chapters. It will surely enhance the knowledge of the field among readers across the globe.

It gives us an immense pleasure to thank our researchers and authors for their efforts to submit their piece of writing before the deadlines. Finally in the end, I would like to thank my family and colleagues who have been a great source of inspiration and support.

Editor

1

The contribution of Productive Safety Net Program for food security of the rural households in the case of Bale Zone, Southeast Ethiopia

Diriba Welteji[1*], Kerime Mohammed[1] and Kedir Hussein[2]

Abstract

Background: Developing countries in general and Ethiopia in particular are affected by a growing problem of food insecurity. Millions of individuals and households especially the rural poor are the main victims of the problem. The government of Ethiopia together with other development partners launched the Productive Safety Net Program (PSNP) in 2005 to help chronically food-insecure households. This study discusses the role of PSNP on the improvement in consumption needs and asset base of the households, effect on community development and determinants of food insecurity level of beneficiaries.

Methodology: Three Woredas have been selected agro-ecologically. Both qualitative and quantitative data-generating techniques were employed. Qualitative data were analyzed thematically and using narration. Frequency distribution table and *t* test were used. Logistic regression was employed to analyze the determinants of food insecurity.

Results and conclusion: Results indicated that PSNP was helping beneficiaries for consumption smoothing, asset accumulation, and development of the local community. The effectiveness of PSNP was significantly determined by age and education level of the household head and occurrence of shocks on the last five consecutive years. The practice of PSNP was challenged by a lack of monitoring and evaluation of structures, low payment and limited awareness of beneficiaries. To assure the positive role of PSNP, culture of savings and accumulation of assets, engagement of beneficiary households in diversified asset building livelihood strategies, targeting and minimizing wrong inclusion and exclusion, and the management as well as monitoring practices of locally constructed community development infrastructures should be enhanced.

Keywords: Asset building, Community development, Food insecurity, Productive Safety Net Program

Background

Achieving food security is a significant and growing challenge in developing countries. People's health and education, their ability to work, human right and equality are impaired by the problem of food insecurity. Women and girls are the most susceptible to the impacts of food insecurity due to their low access and control over resources [1].

*Correspondence: Diriba.welteji@gmail.com
[1] Department of Rural Development and Agricultural Extension, College of Agriculture and Natural Resources, Madda Walabu University, Bale-Robe, Ethiopia
Full list of author information is available at the end of the article

Chronic food insecurity is one of the defining features of poverty in Ethiopia. Those individuals and households whose livelihood are heavily dependent on subsistence agriculture are highly affected by the problem [2]. Combinations of factors have resulted in serious and growing problems of food insecurity in the country. These problems can be classified under environmental, socioeconomic and technology-related problems [3].

The Productive Safety Net is a program initiated by different development partners like World Bank (WB) and other organizations in response to the problem of food insecurity. The government of Ethiopia launched the program in 2005 with the goal of helping chronically food-insecure households to withstand with stresses and shocks, accumulate and create an asset and to make them food self-sufficient. PSNP transfers food, cash or both based on need and season either through direct support or public work activities. The Program also provides credit and technical support to households based on tailored business plans. Beyond the immediate beneficiaries of the program, the PSN has a multiplier development effect through the participation of able-bodied individuals in different development activities, such as land and water resources rehabilitation and developing community infrastructures, including rural road, schools and clinics [4].

A total of 495,995 households were graduated from PSNP between 2008 and 2012. This indicates that it is bringing a considerable number of poor households out of poverty [4]. In 2008, 252,239 individuals were beneficiaries of the PSNP in Bale Zone. These beneficiaries were from the 13 Woredas of the Zone, and the level of food insecurity was severe in pastoral communities. Therefore, the proportion of program beneficiaries from the pastoral communities were high.

Therefore, it is ideal to investigate the contribution of PSNP on food security of the rural households in Bale Zone, Southeast Ethiopia. It would contribute to the adding insights of the issue and draw some pertinent policy ideas through which the contribution of the program can be maximized.

Review of conceptual and empirical literature

The concept of food security was coined following the first World Food Conference in 1974 in Rome. Ever since its definition has been considerably changing and recently reached more than 250. The recently coined and the relatively comprehensive one is put as 'all people, at all times, have physical and economic access to safe and nutritious food to meet their dietary needs and food preferences for an active, and healthy life' [6].

Sufficiency, access, security and time are the four core issues in the definition of food insecurity. Sufficiency of food involves the calories needed for an active and healthy life, while access refers to entitlement to produce, purchase or exchange food or receive it as a gift. Security is the balance between vulnerability, risk, and insurance. Time refers to the temporal situation of food insecurity whereby food insecurity can be either chronic/permanent or temporal/transitory [7].

There are three main concerns and subsequent theoretical shifts on food security. The first shift is from 'global and national' concern to 'household and individual'. The second shift has been from 'food first' to 'livelihood first' approach. The third shift is from 'objective measurements' to 'people's perception' [7].

The last half a century has brought about significant improvements in aggregate food security and diversity of food. However, many people, particularly in developing countries, have not been able to be benefited from these improvements. In 2013, there were an estimated 842 million people (12% of the world population) who were unable to meet their dietary energy requirements necessary to live an active and healthy life. Around one in eight people on our planet are likely to have suffered from chronic food insecurity [7]. Of the 842 million food-insecure people in 2013, 827 million or 98.2% of the people are living in developing countries, with the highest number of undernourished people coming from Southern Asia, followed by Sub-Saharan Africa and Eastern Asia [8].

Food insecurity in Ethiopia is normally understood in terms of recurrent food crises and famines, and responses to food insecurity have conventionally been dominated by emergency food-based interventions. Since 1998, the numbers of food-aid beneficiaries in Ethiopia have fluctuated between 5 and 14 million every year [9]. Poverty and food insecurity are two different concepts, with their correlation varying significantly among countries, depending on the specific national context. However, in the case of Ethiopia, the overlap of the two concepts is greater than in other countries [10]. Statistics on national poverty trends in Ethiopia indicate that there has been a substantial reduction in poverty over the last 15–20 years [5]. Human Development Index (HDI) increased from 0.275 in 2000 to 0.396 in 2012, although it is still ranked among those countries with the lowest HDI throughout the world [5].

The results from the household consumption and expenditure surveys (HCE) conducted in the country in 1995/1996, 1999/2000, 2004/2005 and 2010/2011 showed that the proportion of people living below the nationally defined poverty line (i.e., headcount index) has decreased from 47.5 to 30.4% in rural areas and from 33.2 to 25.7% in the urban centers during the period of 1995/1996–2010/2011. The depth of poverty (i.e., poverty gap index) stood at 7.8% in 2010/2011 nationwide, with 8% in rural areas and 6.9% in the urban centers. The poverty severity index—while substantially declining during the period of 1999/2000–2004/2005 on a national level—increased in rural (17%) and urban areas (5.1%) from 2004/2005 to 2010/2011. In the context of rural Ethiopia, these data suggest that while the proportion of people below the poverty line and the average gap that separates the poor from the poverty line have declined in the past two decades, there has only been poor improvement in the distribution of income among the rural poor, particularly since 2004/2005 [5].

The causes of food insecurity are many and varied. It includes a low rate of agricultural production, low access to food, the limited capacity of infrastructures and local markets, HIV/AIDS, investment power, finance gap, poor health, shortage of water and poor sanitation, environmental degradation, climate change and natural disasters, conflict and persecution.

In order to deal with the problem of food insecurity, governments in developing countries have implemented various social protection instruments with three functions in common: (1) to maintain the basic level of consumption, (2) to facilitate investments in human capital and other productive assets and (3) to strengthen the capacity of those in poverty. The Productive Safety Net Program (PSNP) was launched by the government of Ethiopia, with donor support, in January 2005. Recognizing that a large component of this food insecurity is 'chronic' rather than 'transitory' and that decades of food aid have had no discernible impact on reducing rural poverty and vulnerability, the PSNP represents an innovative attempt to tackle chronic food insecurity and break Ethiopia's dependence on food aid [11].

The Productive Safety Net Program (PSNP) is targeted toward households that are both food insecure and poor. In Ethiopia, as in many other African countries, there is a pressing need to improve household food security. An emerging consensus suggests that this is most easily accomplished through two development strategies with two complementary dimensions: investments that facilitate income generation and asset accumulation (infrastructure development, improved technologies for agriculture, etc.), and interventions that protect the poorest from hunger, prevent asset depletion and provide a platform on which the growth interventions can take place. Food aid targeting in Ethiopia has a long history of relying on community-based targeting systems, which have been seen as effective. The PSNP adopted this system while further refining the targeting criteria to capture chronic food insecurity—defined as a 3-month food gap or more and receiving food aid for three consecutive years [12].

Chronic food insecurity at the Woreda and household level is a defining feature of the eligibility criteria for PSNP participation. The household must have faced continuous food shortages (usually 3 months of food gap or more) in the last 3 years and received food assistance. The other criteria are households that suddenly become more vulnerable as a result of a severe loss of assets and are unable to support themselves and households without family support and other means of social protection and support [11].

The PSNP uses a mix of geographic and community-based targeting to identify chronically food-insecure households in chronically food-insecure Woredas. After determining PSNP eligibility based on these criteria, households are assigned to public works or direct support: Eligible households with able-bodied adults receive transfers for their participation in public works projects, while those households that cannot provide labor or other means of support receive unconditional transfers. Most beneficiary households participate in public works (90% of all PSNP transfers); a much smaller proportion receives direct support [13]. Public work participants received 6 months of food and cash transfer, while direct support beneficiaries received 12 months of unconditional transfer [12].

Methodology: sampling strategy, data collection and analysis

The study was conducted in Bale Zone, Southeast Ethiopia. Three sample Woredas have been selected from the three agro-ecology classifications into relative *dega*,[1] *weina-dega*[2] and *kolla*.[3] Again each Woreda was classified into relative *dega*, *weina-dega* and *kolla* to pick a single kebele from each agro-ecology zone of the Woreda. Both Woredas and kebeles have been selected using a lottery method of simple random sampling. Kothari's (1990) formula (with 0.5 estimated proportions of respondents, 95% confidence interval and 0.05 margin of error) was used with the following formula:

$$n_0 = \frac{p(1-p)z^2}{e^2}$$

where n_0 = sample size, P = estimated proportion of respondents: 0.5, Z = the number of standard error corresponding to 95% CI which is 1.96. e = margin of error: 0.05 margin of error was selected.

$$n_0 = \frac{0.5(1-0.5)1.96^2}{0.05^2} = 384$$

Therefore, using infinite population sample size determination formula the total numbers of samples included in the study were = 384.

Using finite population sample size determination formula:

$$n = \frac{n_o}{1+\frac{(n-1)}{N}} = \frac{384}{1+\frac{(384-1)}{49{,}470}} = 381$$

With the assumption of 10% non-response rate, 39 respondents were added to the calculation of 381*0.01. Therefore, the true sample size was 381 + 39 = 420.

---

[1] Highland climate.

[2] Midland climate.

[3] Lowland climate.

Survey households were selected in each kebele using the lottery method. The sampling frame of beneficiaries of PSNP was obtained from the program coordinator on each kebele. A total number of samples taken from each *kebele* of selected *Woredas* were allocated proportionally to the total number of PSNP beneficiary households of the *Kebele*. The distribution of sample households in the *Woredas* and *kebeles* is presented in Table 1.

The study employed various data collection techniques, namely household surveys, key informant interviews, direct observations and focus group discussions. A structured interview was conducted based on the questionnaire designed and key informant interview guide with elders, Administrator of the Zone and the three selected *Woredas*, Development Agents of the selected *kebeles* in each *Woreda* and coordinator of the Zone and *Woredas* PSNP.

The results were analyzed using both descriptive such as percentage and mean and inferential (paired *t* test and logistic regression). For the purpose of logistic regression, food security level of beneficiaries was taken as the dependent variable. Age of the household head, family size, education level, marital status, ethnicity, religion, occupation, and availability of unproductive labor force, number of oxen, the size of the farmland, irrigable land, perennial river water, number of livestock, market access, and use of modern agricultural input (seed, fertilizer, pesticide and herbicide) and occurrence of shocks during the last 5 years were independent variables included in the study. It can also goes in line with the variables hypothesized and indicated in the conceptual frame work developed by authors (Fig. 1).

Results and discussion

Demographic and socioeconomic characteristics of sample households

From the total sample households, female-headed households covered about 19%. Of which 22% were from Goro, 16% from Gololcha and 19.6% from Saweyna Woredas. The remaining 81% were male-headed households. Of which 78% were from Goro, 84% from Gololcha and 80.4% were from Saweyna (Table 2). With respect to educational background of respondents, 62.5% do not read and write, 30% achieved primary first cycle, 5.5% achieved primary second cycle and 0.7% achieved high school education. The ethnic background of households was Oromo. About 22% of the household head were not productive as a result of old age (67%) and disability (25%).

The mean age of the respondents was 44.04 with a standard deviation of 12.982 and the maximum and minimum value being 22 and 90 years of age, respectively. Additionally, respondents had productive family members ranging from 1 to 12 with a mean and standard deviation of 2.48 and 1.543, respectively (Table 3).

Respondents have an average family size of 7.17 (approximated to eight). The minimum value for the family size of the household was 1 and the maximum one is 21 (very large family size) (Table 4).

The agricultural practices of Bale Zone are constrained by various factors. Among these erosions, the occurrence of pests, low fertility, salinity and sandiness of the soil, frost, and water logging are some of them. However, the distributions of these agricultural constraints vary across agro-climatic zones [14]. Moreover, the ecological variation of the study area can be observed from the spatial map of the study area (Fig. 2).

Table 1 Distribution of sampled households by *Weredas* and *Kebeles*

Wereda	Agro-ecology zone	Total number of beneficiaries of PSNP	Total number of households	Number of sample households taken from the Wereda	*Name of the kebeles*	Number of samples taken from the *kebeles*
Gololcha	Dega	8993	1124	76	Dinsa	25
					Buniya	25
					Gofa	26
Goro	Weinadega	7910	847	68	Kaku	23
					Bili AKiya	23
					Goro Rayya	22
Saweyna	Kola	32,567	6636	276	Gale	92
					Biliso	92
					Arda Galma	92
Total beneficiary households	49,470	8607	420			

Total number of sampled households taken from the Zone: 420

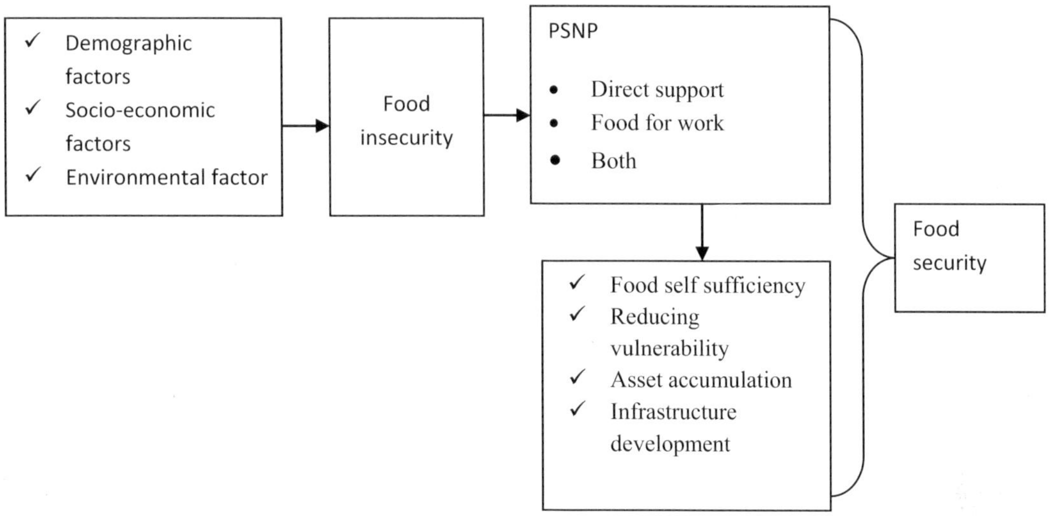

Fig. 1 Conceptual framework *Source*: developed by authors

Role of PSNP on consumption expenditure of households

PSNP helps to protect the basic level of consumption. Similarly, the program was helping households in the study areas to fulfill their consumption needs in different ways. The program increased the number of dining times and the amount of meal at each dining time. Moreover, it increased both the number of dining and the amount of meal at each dining time (Table 5).

Taking into account the overall livelihood effect of PSNP, majority (44.4%) of respondents stated that the livelihood situations of the household are a little bit better after they became a beneficiary of the program (Table 7). The program also provided a better change for 20.3% of the households. Therefore, majority of the respondents (64.7%) claim that their living standard was improved. However, 26.4% of respondents stated that their living standard has been deteriorated further while they are the beneficiaries of the program. But no evidence was found both from the primary or secondary data that realized and supported the negative impact of the PSNP (Table 6).

As depicted in Table 7 above, most of the respondents of the study confirmed that the agricultural sector in the study area has been constrained by multiple factors such as occurrence of pests, low fertility and sandiness of the soil, frost, water logging and salinity of the soil are all problems that affect agricultural productivity. However, the distributions of these agricultural constraints vary across agro-climatic zones.

Table 2 Distribution of male- and female-headed households across Weredas. *Source*: Field survey, 2016

	Sex of the HH head		Total
	Male	Female	
Wereda			
Goro			
Frequency	53	15	68
% within the Wereda	77.9%	22.1%	100.0%
% out of the total	12.6%	3.6%	16.2%
Gololcha			
Frequency	64	12	76
% within the Wereda	84.2%	15.8%	100.0%
% out of the total	15.2%	2.9%	18.1%
Saweyna			
Frequency	222	54	276
% within the Wereda	80.4%	19.6%	100.0%
% out of the total	52.9%	12.9%	65.7%
Total	339	81	420
Percentage	80.7%	19.3%	100.0%

Table 3 Descriptive statistics for age of the household head and productive member of the household. *Source*: Field survey, 2016

	N	Minimum	Maximum	Mean	SD
Age of the HH head	420	22	90	44.04	12.982
Economically productive member of the household	420	1	12	2.48	1.543

Table 4 Descriptive statistics for family size of the household. *Source*: Field survey, 2016

	N	Minimum	Maximum	Mean	SD
Family size of the HH	420	1	21	7.17	3.596

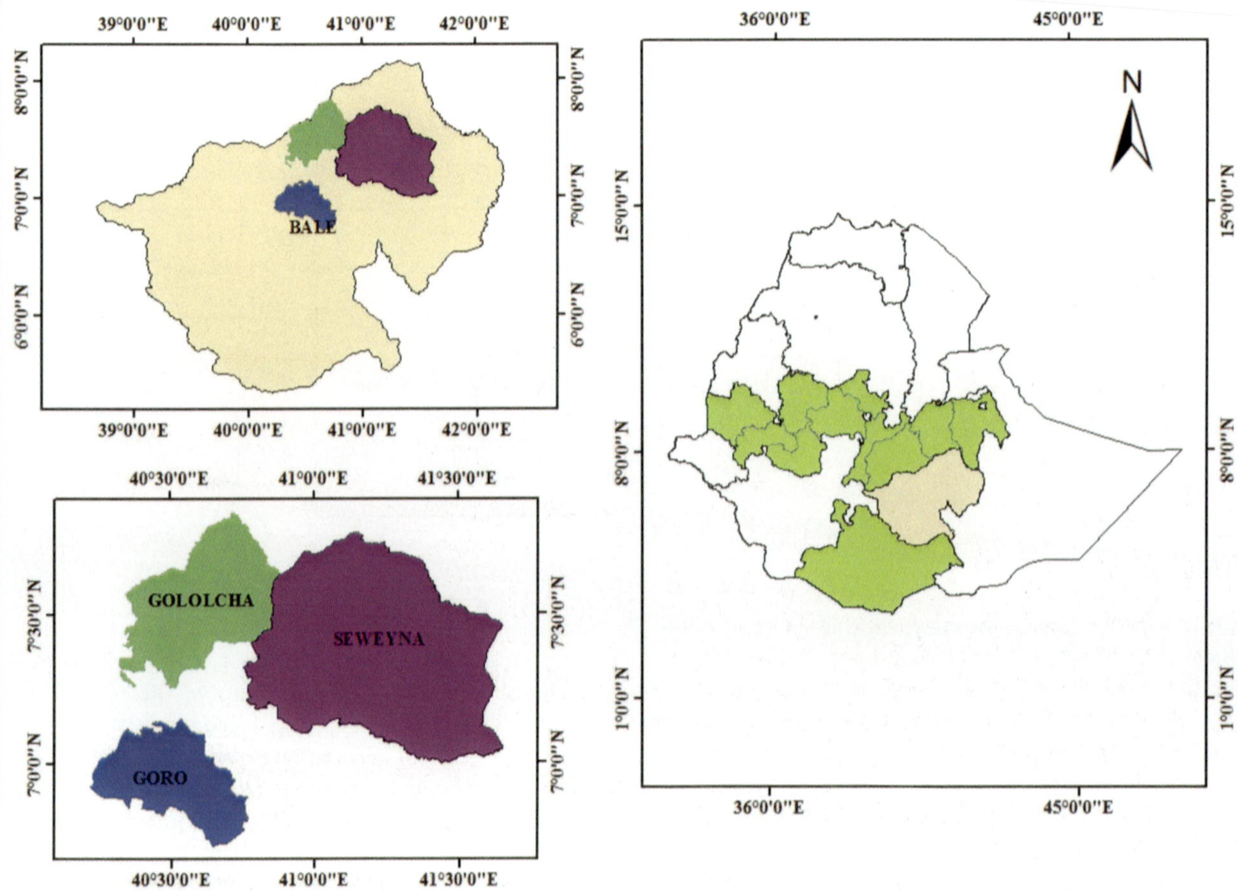

Fig. 2 Map of study area

Table 5 Role of PSNP on consumption. *Source*: Field survey, 2016

Role	Frequency	Percent	Cumulative percent
It helps the household to increase the number of dining time	108	25.7	25.7
It helps the household to increase the amount of meal at each dining time	162	38.6	64.3
It helps the household both to increase the amount of meal and dining times	150	35.7	100.0
Total	420	100.0	

Table 6 Before and after status of the beneficiary households. *Source*: Field survey, 2016

Status	Frequency	Percent	Cumulative percent
Better now	85	20.3	20.3
A little better now	186	44.4	64.7
Same	28	6.7	71.4
A little worse now	27	6.4	77.8
Much worse now	84	20.0	97.9
Don't know	10	2.1	100.0
Total	420	100.0	

PSNP provided beneficiaries the chance of using agricultural technologies such as fertilizer, improved seed, pesticide and herbicide. It is reported that the use of agricultural technologies has increased after households become beneficiaries of the program (Table 8).

Great majorities of respondents (88.6%) replied that the program has brought improvement in the expenditure of food expenses. Food security has four dimensions: sufficiency of food, access, security and time. Access refers to entitlement to produce, purchase or exchange food or receive it as a gift [7]. Therefore, PSNP provided beneficiaries the chance to buy food items through the provision of cash, even if they were unable to produce. The program also increased the food entitlement of recipient households with a direct provision of a food item (Table 9).

Table 7 Constraints of agricultural practice. ***Source*****: Field survey, 2016**

Constraints	Frequency		Percentage	
	Yes	No	Yes	No
Water logging	69	351	16.4	83.6
Soil infertility	201	219	47.8	52.2
Susceptibility to frost	75	345	18	82
Erosion	288	132	69	31
Occurrence of pests	230	190	54.8	45.2
Salinity	53	367	12.6	87.4
Sandiness	152	268	36.2	63.8

Table 8 Effect of PSNP on using agricultural inputs. ***Source*****: Field survey, 2016**

Types of agricultural inputs	Frequency		Percent	
	Yes	No	Yes	No
Fertilizers				
Before PSNP	114	306	72.8	27.2
After PSNP	262	158	62	38
Improved seed				
Before PSNP	104	352	24.8	75.2
After PSNP	273	147	65	35
Pesticide				
Before PSNP	220	200	52.4	47.6
After PSNP	325	95	77.4	22.6
Herbicide				
Before PSNP	127	293	31.2	69.8
After PSNP	250	170	59.5	40.5

Qualitative results showed that the program was playing a key role in meeting the immediate food requirements of beneficiaries through cash and food transfer. However, the long-term livelihood improvement effect of the program was less. This was similar to the finding of Daniel et al. [2], stating that beneficiaries of the PSNP did not experience faster asset growth as a result of the program. In addition, the program increased the feeling of dependency syndrome among the beneficiaries.

One among the principles of PSNP transfer was cash first principle. Cash should be the primary form of transfer when possible. This expenditure of cash in the local markets assists the stimulation and move away from food aid. Food transfers are provided at times and places when food is not available in the market, or where market prices for food are very high. This protects PSNP clients from food shortages and asset depletion [7]. Likewise, total households in Goro and 86.8% in Gololcha Woreda received cash transfer, while in Saweyna Wereda, where the availability of food in the market is limited, households have received both food and cash transfers (Table 10).

Table 9 Improvement in the expenditure of food expense for PSNP beneficiaries. ***Source*****: Field survey, 2016**

Is there any improvement in food expense of the HH?	Frequency	Percent	Cumulative percent
Yes	372	88.6	88.6
No	48	11.4	100.0
Total	420	100.0	

Asset building role of PSNP

Cattle rearing practice is one of the asset and prestige building mechanisms in the pastoral and agro-pastoral community in the Woredas under consideration. However, in the time of rain failure and occurrence of food shortage, depending on the seriousness of the problem people either sell from small ruminants to larger livestock or directly use the animals for immediate consumption. Moreover, when transfer schemes like PSNP and some other direct transfer programs are in place at the time of food gaps, the households will not be forced to sell their livestock, which in turn prevent asset depletion and enable people to retain in their locality.

According to one of the members of group discussants in Saweyna district, PSNP had shown the relief effect and absence of loss in livestock during crop failure.

The maximum value, mean and standard deviation of the estimated value of livestock before and after the program have shown a great difference. The estimated maximum value of livestock before the program was 83,400, while it has increased to 157,100 after households become the beneficiary of PSNP. The mean of two values has been increased from 6052.5065 to 15,859.2947 birr. The standard deviation has also shown increment from 7767.20564 to 16,472.58395. However, the minimum value remained constant, 0.00. Zero value of the estimated value of livestock, both before and after the program, means there are individuals that do not have livestock at all and uses the benefit that they obtain from the program solely for consumption (Table 11).

The paired t test result showed that there was a 9729.68ETB difference in the mean of the estimated value of livestock before and after the PSNP. This difference was supported by the hypothetical analysis with a p value of 0.0000; this result shows there was a significant difference in the estimated total value of livestock before and after PSNP in Bale Zone (Table 12). Some farmers participating in cash transfer from PSNP also engaged in a poultry farm and significantly improved their household income.

Table 10 Wereda and type of provision obtained (cross-tabulation). ***Source*****: Field survey, 2016**

	Types of provision			Total
	Cash	Food	Food and cash	
Wereda				
Goro				
Frequency	68	0	0	68
% within the Wereda	100.0%	0.0%	0.0%	100.0%
% within the type of provision	41.0%	0.0%	0.0%	16.2%
Gololcha				
Frequency	66	0	10	76
% within the Wereda	86.8%	0.0%	13.2%	100.0%
% within the type of provision	39.8%	0.0%	4.8%	18.1%
Saweyna				
Frequency	32	46	198	276
% within the Wereda	11.6%	16.7%	71.7%	100.0%
% within the type of provision	19.3%	100.0%	95.2%	65.7%
Total				
Frequency	166	46	208	420
% within the Wereda	39.5%	11.0%	49.5%	100.0%
% within the type of provision	100.0%	100.0%	100.0%	100.0%

Community development role of PSNP

PSNP was helping the local development in two ways. First, the program allocated a certain proportion of its budget for the construction of local infrastructures. Second, households that have able-bodied adult labor engage in public works and receive transfers for 6 months of the year. Public works focus on integrated community-based watershed development activities such as soil and water conservation measures, rangeland management (in pastoral areas) and development of community assets such as roads, water infrastructure, schools and clinics. These works contribute to improved livelihoods (through increased availability of natural resources, including water and cultivatable land, soil fertility, increased agricultural production and improved market access), strengthened disaster risk management and climate resilience, and nutrition [12].

As per the principle of PSNP, food for work beneficiary households performed different community development activities such as road construction, natural resource conservation and afforestation, ground water (locally known as *haroo*), soil conservation through soil and stone bund, and check dam. Rangeland management through area closure in Saweyna Wereda (pastoralist area) was also performed. In a similar practice, the PSNP beneficiaries of the study Woredas particularly the public work participants with the guidance of the Kebele administration and development agents highly involved in different infrastructural development activities of their identified priority activities in their respective areas. These all community-based infrastructures were helping the livelihood of the community.

According to the information obtained from key informants at Goro and Saweyna Woredas, under the public work activities/interventions like area closure, construction of dry season feeder roads, small-scale irrigation, gabion (a wire mesh structure filled with stone or earth material to protect soil erosion), soil band, stone band, soil dams, borehole, ponds, construction of schools, farmers training centers, water shade management were performed based on the priority set by the public and approval of Woreda task force.

Determinants of food security level of beneficiaries

From proposed variables as determinants of being graduated from PSNP, only three variables (age of household head, education of household head and occurrence of shock within 5 years in the household) were selected for final full-model logistic regression using a p value ≥ 0.25 (Table 13).

From the final logistic regression model, the selected variables, the age of household head, education of household head and occurrence of shock within 5 years in the household were significant determinants for graduation status with a p value of less than 0.05 (Table 13).

As the age of household heads increases by 1 year, the graduation status of the households will increase by 3% with a p value of 0.002. Households whose head were educated were 64% more likely to be graduated from the program. Those households with no experience of shock within the five consecutive years were 3.09 times more likely to be graduated (Table 13).

Challenges of PSNP implementation

According to Goro Wereda key informant experts, the management of development projects constructed

Table 11 Descriptive statistics for the total value of livestock of the HH before and after the PSNP (in Ethiopian birr). ***Source*****: Field survey, 2016**

The total value of livestock	*N*	Minimum	Maximum	Mean	SD
Before the PSNP	420	0.00	83,400.00	6052.5065	7767.20564
After the PSNP	420	0.00	157,100.00	15,859.2947	16,472.58395

Table 12 Paired *t* test result for estimated value of livestock. *Source*: Own survey, 2016

Variable	Obs	Mean	Std. Err.	SD	[95% conf. interval]	
Q19ATo ~ l[a]	420	15,729.85	834.0329	16,172.5	14,089.88	17,369.81
Q19BTo ~ l[b]	420	6000.16	401.1064	7777.742	5211.46	6788.859
diff	420	9729.686	798.3131	15,479.87	8159.955	11,299.42
mean(diff) = mean(Q19ATotal – Q19BTotal)	$t = 12.1878$					
Ho: mean(diff) = 0	Degrees of freedom = 420					
Ha: mean(diff) < 0	Ha: mean(diff) ! = 0	Ha: mean(diff) > 0				
$Pr(T < t) = 1.0000$	$Pr(\|T\| > \|t\|) = 0.0000$	$Pr(T > t) = 0.0000$				

[a] Total estimated value of livestock after PSNP

[b] Total estimated value of livestock before PSNP

Table 13 Logistic regression result for determinants of success of beneficiaries. *Source*: Own survey, 2016

Graduated	Odds ratio	Std. Err.	*z*	*P* > |*z*|	[95% conf. interval]	
Education	0.360695	0.0889668	−4.13	0.000	0.2224272	0.5849143
Age	0.9721281	0.0086928	−3.16	0.002	0.955239	0.9893159
Shock	3.098867	0.9794122	3.58	0.000	1.667925	5.75744
_cons	0.8333807	0.458823	−0.33	0.741	0.2832755	2.451759

through public works was weak because less attention is given to the members of the community for which the infrastructure is constructed on their farm plots and area closures were not protected from damages caused by animals. In addition, the communities focused on activities that generate immediate benefit. Activities performed by the beneficiaries did not have a clear-cut standard and it resulted in the development of lower-quality infrastructures. Moreover, they were obstructed by lack of monitoring and unsustainable management. Free grazing and lack of cut-and-carry systems were also the main problems that affect the conservation processes of natural resources and endangered the newly planted trees.

It is recognized that small transfers of cash or food are more likely to be consumed than investing, while the assets constructed by the public works activities will contribute to an improved enabling environment (feeder roads will stimulate trade and integrate fragmented markets, for example) rather than directly generating additional income. One of the beneficiary respondents from Goro district argued that the level of payment was perceived by the beneficiary as low:

> *...the amount of cash payment was not enough. We have received 117 birr per individual per month. This amount of money had nothing to do to change our living. The payment also lasted only for six months. Therefore, we did not have any other source of income for the remaining six months. But the program was assisting us to fulfill our daily basic consumptions for certain months.*

Key informants of Gololcha Wereda confirmed that the main and important objective of PSNP is achieving the graduations of chronically food-insecure households through participation in public work if he/she has an estimated asset of 18,000 Birr either in cash or in kind. On the other hand, during the period of graduation households who have accumulated asset level of more than the minimum threshold, 18,000 birr may continue being the beneficiary of the program. Besides, households who do not have the minimum level of an asset for graduation (estimated 18,000 birr) will be graduated for political purpose what is called 'political graduation' (KII 2016).

Similarly, the Saweyna Woreda PSNP coordinator has pointed out that there was a false inclusion and exclusion of beneficiaries, asserting '...PSNP was helping section of the community that is affected by the high level of food insecurity. Initially, the program beneficiaries were recruited by the community themselves, Kebele administrators and Woreda task force. However, better off individuals were found to be wrongly included, particularly in pastoral areas and individuals affected worst by food insecurity may wrongly be excluded. Corrupted officials, clan politics and quota allocation (for both inclusion and graduation) of program beneficiaries were the main cause for wrong inclusion and exclusion.'

Table 14 Participation of households in rural non-farm economic activities. ***Source*****: Field survey, 2016**

Participation	Frequency	Percent	Cumulative percent
Yes	19	2.6	3.1
No	401	82.4	100.0
Total	420	85.0	

The primacy of transfer was also challenged by a lack of necessary logistics for the program such as transportation and lack of coordination among the concerned bodies, natural resource degradation, and erratic rainfall, water stress, especially in pastoral Woredas, erosion, and population growth (KII 2016). Rainfall variability was also one among the many problems. The lowland areas were affected by the seasonal variability of rainfall and water shortage, while the highland areas such as Gololcha were affected by erosion and high rainfall during the harvesting time, lack of development agents in remote areas and low participation of the beneficiaries in off-farm activities (Table 14).

The participation of households in rural non-farm economic activities was very limited. Only 4.5% of respondents had household members engaged in activities other than crop production and animal rearing (Table 14). These small proportions of rural non-farm economic activity participants were engaged in activities like the petty trade. The main reason for meager income for the household from the non-agricultural activities were a lack of start-up capital, lack of the option of understanding for sectors that households can engage in beside agriculture and low skill of the household.

Conclusion

In this study, the effect of the Productive Safety Net Program (PSNP) on food security of the beneficiary households has been explored. PSNP was helping the local development in two ways. First, the program allocated a certain proportion of its budget for the construction of local infrastructures. Second, able-bodied beneficiaries contribute their labor for the construction of these infrastructures. Through the provision of cash, food or both, PSNP was helping households to fulfill the consumption needs of the households. The program increased the number of dining times and the amount of meal and food at each dining time. The program has also increased the food expenditure and level of consumption. By keeping the minimum level and smoothing consumption, PSNP has improved the food security status of the beneficiary households.

The participation of beneficiary households in public work activities such as soil and water conservation measures, rangeland management (in pastoral areas) and development of community infrastructures such as roads, water infrastructures, schools and clinics contributed to the improvement in livelihoods through the increasing availability of natural resources, increasing agricultural production and improved market access, strengthened disaster and risk management and climate resilience. These all community-based infrastructures were helping in improving the livelihood of the community.

Wrong targeting (wrong inclusion and exclusion of beneficiaries), poor conservation, monitoring and sustainable management of locally constructed infrastructures, lack of awareness of local people, low level of cooperation of concerned officials, absence of man power in remote areas are among the challenges. Furthermore, soil bunds constructed in the farm land of the farmers were not given due to attention by the farmer themselves. Free grazing and lack of cut-and-carry systems were also the main problems that affect the conservation processes of natural resources and endangered the newly planted trees.

From among the many variables as determinants of food insecurity (for beneficiary households being graduated or not), age of household head, education level of household head and occurrence of shock within 5 years were found significant with p value of less than 0.05. As the age of household heads increases by 1 year, the graduation status of the household will increase by 3% with a p value $= 0.002$. Households whose heads educated were 64% more likely to be graduated from the program. Those households with no experience of shock within the five consecutive years were 3.09 times more likely to be graduated.

The study also revealed that small proportions of beneficiaries were participating in rural non-farm economic activities. The participation of households in no-farm economic activities was hindered by lack of start-up capital and low understanding of options.

Recommendations

The following has been suggested as a means of improving the effectiveness of the PSNP in reducing the problem of food insecurity:

- Increase a culture of savings and accumulation of assets in addition to consumption.
- Engagement of beneficiary households in diversified and asset building livelihood strategies (off-farming income generating activities) are very important.
- Care should be taken in targeting and graduating PSNP beneficiaries minimizing wrong inclusion and exclusion.
- The management and monitoring practices of locally constructed community development infrastructures

should be enhanced, so that the long-term benefit of the structures will be promoted.

- Continuous capacity building training and awareness creation, raising participation, proper allocation of development agent at the grass root to facilitate proper planning, book keeping and reporting.

Abbreviations

ETB: Ethiopian birr; HCE: household consumption expenditure; HDI: Human Development Index; HH: household; PSN: Productive Safety Net; PSNP: Productive Safety Net Program.

Authors' contributions

DW designed and led the study, structured the concepts, reviewed much of the studies and analyzed much of the qualitative data. KM and KH further developed the manuscript, identified and developed important concepts, validated and helped design the arguments, conceived and helped design of the study, conducted quantitative study and edits the final research. All authors read and approved the final manuscript.

Author details

[1] Department of Rural Development and Agricultural Extension, College of Agriculture and Natural Resources , Madda Walabu University, Bale-Robe, Ethiopia. [2] Department of Public Health, College of Medicine, Madda Walabu University, Bale-Robe, Ethiopia.

Acknowledgements

We would like to thank Madda Walabu University Research and Community Service Directorate for the arrangement of such opportunity through the identification of thematic area of the university, financial support, necessary revision and constructive comments to undertake this research. Our gratitude also goes to Bale Zone and Wereda Agricultural Office for giving us necessary information on the program in the study area and for their unlimited support. We also thank all data collectors and supervisors for their unreserved effort for the successful completion of collection of the data. Finally, the authors would like to acknowledge all other concerned bodies that have shared even a piece of support for the successful development and completion of this study.

Competing interests

The authors declare that they have no competing interests.

Ethical approval and consent to participate

Ethical clearance letters were collected from Madda Walabu University research and community service directorate and Bale Zone administrative office so as to safeguard both the study participants and the researchers. All participants of the research including survey households, case studies, enumerators, the supervisors and key informants were fully informed about the objectives of the study. They all were approached friendly and in a fraternal way. Their informed consent was obtained before their involvement in the study. The researchers were developed confidentiality with all participants, enumerators and survey households. The questionnaire was designed to collect information directly related to the research questions and objectives. As a result, privacy of the participants was ensured, and no personal data were collected. The questionnaire was free from any degrading, discriminating, or any other unacceptable words that could be offensive to the participants. Finally, any phrases or paragraphs, concepts and quotations not belonging to the researcher and used in any part of the study were fully acknowledged.

References

1. CIDA's Food Security Strategy. Increasing food security. Canada; 2010.
2. Gilligan DO, Hoddinott J, Taffesse AS. An analysis of Ethiopia's Productive Safety Net Program and its linkages. International Food Policy Research Institute, 2033 K Street, NW, Washington, DC; 2008.
3. Gebremedhin B. Evaluation of Ethiopia's Food Security Program: documenting progress in the implementation of the Productive Safety Nets Programme and the Household Asset Building Programme. International Food Policy Research Institute and Institute of Development Studies; 2001.
4. FAO, IFAD and WFP. The state of food insecurity in the world. Food and Agriculture Organization; 2012.
5. Government of Ethiopia. Ethiopia's progress towards eradicating poverty: an interim report on poverty analysis study (2010/11). Ministry of Finance and Economic Development; 2012a.
6. FAO. Rome declaration on world food security and world food summit plan of action. World Food Summit. 13–17 Nov 1996. Rome; 1996.
7. Tolossa D. Rural livelihoods communities, poverty and food insecurity in Ethiopia; a case study at Erenssa and Garbi in Oromiya Zone, Amhara National Regional State. Ph.D. Thesis, Department of Geography, Faculty of Social Sciences and Technology Management, Norwegian University of Science and Technology, NTNU, Trondheim, Norway (2005).
8. FAO, IFAD and WFP. FAO statistical yearbook. World Food and Agriculture. Food and Agriculture Organization; 2013.
9. Devereux S, Sabates-Wheeler R, Tefera M, Taye H. Ethiopia's Productive Safety Net Programme (PSNP), Trends within targeted households. Sussex: Institute of Development Studies and Addis Ababa: Indak International Pvt. L. C.; 2006.
10. Devereux S. Social protection for enhanced food security in Sub-Saharan Africa. United Nations Development Programme. Working Paper 2012-010; 2000.
11. Sabates-Wheeler R, Devereux S. Cash transfers and high food prices: explaining outcomes on Ethiopia's Productive Safety Net Programme. Working Paper. UK: Institute of development studies and center for social protection, University of Sussex; 2010. http://www.future-agricultures.org. Accessed 3 Oct 2015
12. MOA. Productive Safety Net Programme phase IV programme implementation manual Addis Ababa; 2014.
13. Sarah C, et al. Summary of ESSP working paper 24; "targeting food security interventions when 'everyone is poor': the case of Ethiopia's Productive Safety Net Programme". ESSP RESEARCH NOTE 26. June 2013.
14. Bale Zone Finance and Economic Development. Physician and socio-economic profile of Bale Zone. Unpublished; 2005.

2

The intersection of food insecure populations in the Midwest U.S. and rates of chronic health conditions

S. R. Dominick[1], Nicole J. Olynk Widmar[1*], Audrey Ruple[2], Jonathan Z. Grennell Weir[1] and Lalatendu Acharya[3]

Abstract

Background: Food insecurity is the state of having insufficient access to adequate food in order to maintain a healthy lifestyle due to limited economic resources. This study expands upon the annual survey conducted by the USDA ERS, while providing evidence that additional factors, notably, medical or health-related issues play vital roles in the relationship between households and food security.

Methods: The data for this study were generated by surveying Midwestern residents. The sample of respondents was targeted to be representative of the Midwest in terms of sex, age, and income and was completed by 1265 respondents. The survey included the CPS Food Security Supplement to assess the food security of respondents. All respondents were asked the 10 household focused food security questions and respondents who indicated having children were asked the additional 8 child focused questions. Additionally, respondents were asked demographic, food security, and health status questions.

Results: Of the sample, 25% were considered food insecure. Being male, middle aged, having children, having household diabetes, having a household eating disorder, and having household depression/anxiety are significant determinants of decreased food security.

Conclusions: Establishing clear relationships between health and food can help to inform legislation. This analysis suggests the inclusion of chronic illness and health information to improve metrics and inform food security legislation.

Keywords: Food insecurity, Regional, Chronic health, Midwest

Background

During election years, media and news sources overflow with information about candidates and platforms, covering a spectrum of political, social, and economic issues. Forbes featured an article presenting data from Pew Research about the 2016 presidential campaign, which found that 84% of voters deemed the "economy" the leading ballot issue which was followed by "terrorism," "foreign policy," and "health care" [1]. One component of the legislature that impacts the economy and addresses various aspects of food-related issues is ratified between elections—the Farm Bill. In 2014, President Barack Obama signed the latest Farm Bill into effect and commented to an audience at Michigan State University, "the Farm Bill is not just about helping farmers," and explained that it covers economic initiatives such as jobs, innovation, and infrastructure [2]. Importantly, the largest portion of the Farm Bill is dedicated to nutrition assistance programs. According to the United States Department of Agriculture's (USDA) "Budget Summary and Annual Performance Plan: FY 2016," expenditures to meet mandatory and discretionary Farm Bill programs for 2016 were estimated to be $148 billion and 73% of budget outlays were for nutrition assistance [3]. President Obama explained "this country has helped Americans put food on the table when they hit a rough patch, or when they're working hard but aren't making enough money to feed their kids" [2].

*Correspondence: nwidmar@purdue.edu
[1] Department of Agricultural Economics, Purdue University, West Lafayette, IN, USA
Full list of author information is available at the end of the article

Food security and chronic health

One example of a "rough patch" President Obama included was household illness [2]. It is understood that food assistance programs, such as the Supplemental Nutrition Assistance Program (SNAP), are designed to be temporary solutions, covering short-term crisis by providing aid for a few months, of which illness is considered a temporary economic strain on households [2].

A number of studies have analyzed and discussed a relationship between food security and illness. Using the USDA Economic Research Service (ERS) 30-day measure (a derivative survey of the Current Populations Survey (CPS) Food Security Supplement), Knight et al. [4] found 17.0% of respondents with diabetes were food insecure, and food insecure individuals were more likely to be medically uninsured. Furthering health considerations, Knight et al. [4] found that 18.9% of respondents with diabetes scrimped on medication (delaying filling prescriptions, inability to afford medication, taking less medication) and scrimping was positively and strongly correlated with food insecurity.

A similar study, also using the USDA ERS 30-day food security measure, found that 22% of adults with chronic diseases were food insecure or had cost-related medication under use and 11% reported both [5]. Berkowitz et al. [5] also found that participants of Women Infant Children (WIC) programs had decreased instances of food insecurity/cost-related medication underuse and suggest investigating the program dynamics. Muldoon et al. [6] found that there were higher odds of mental illness (specifically depression and anxiety) among those who experience insufficient food and hunger compared to those without hunger. A study done by Dharamasen et al. [7] found correlations linking food insecurity to health, including positive correlations between food insecurity and adult obesity, adult obesity and poverty, and adult obesity and unemployment [7].

Food security and regional impact

While President Obama referred to the USA as a whole in 2014, most food assistance is distributed at regional levels, so it is important to understand the impacts regional characteristics may have on food security. One study by Moore and Diez Roux [8] compared *local food environments* across regions in three states and found a number of impactful differences. Comparatively, North Carolina had the largest population and highest median income of the study, while New York was the most urban [8]. In terms of food, New York had the highest number of food stores per square mile, compared to Maryland and North Carolina, and grocery stores were the most common stores in New York and Maryland, but convenience stores were most common in North Carolina [8]. Lower-income areas were found to have more grocery stores, meat and fish markets, and liquor stores than higher-income areas but fewer supermarkets, vegetable markets, bakeries, natural food stores, and specialty stores [8]. Dharamasen et al. [7] found that food insecurity was positively correlated with race, poverty, and the number of grocery stores, but negatively correlated with expenditures at full service and fast food restaurants.

Other studies suggest that neighborhood relationships have little to no impact on food security. A study of Canadian neighborhoods was unable to link food security to neighborhood characteristics such as grocery stores, but concluded that household resource constraints and social capital contributed more to food insecurity risk [9]. Similarly, in a study done in Philadelphia, PA, researchers concluded that food access (nearness to quality food) did not impact food security, but access to aid programs and household financial constraints did [10].

Given that the economy (including food security) and health care (including non-temporary chronic illness) are leading issues in the USA, it is important to understand the overlap of household food security and health status. This study sought to quantify food security and chronic healthfulness of an extended Midwestern sample. This study focused on obtaining a sample of respondents, which was targeted to be representative in terms of sex, age, income, and state of residence. No screening of any kind related to food security was employed, which makes this data collection effort distinct from many others which focus on households with income limits or utilize screening questions (related to food security) for at least some portion of potential respondents. This study also pursued an exploration of the intersection of food security and chronic healthfulness in order to understand possible relationships between the two. This study also included a county-level poverty indicator in order to explore the relationship between community and household food security.

Methods

Measuring food security

The USDA ERS measures food security for the USA. In the "Household Food Security in the United States in 2014" report, it was reported that 14.0% of US households were food insecure or did not have "access at all times to enough food for an active, healthy life for all household members" due to limited economic resources [11].

The USDA ERS uses a survey generally referred to as the Current Populations Survey (CPS) Food Security Supplement which contains eighteen total questions: ten adult/household-specific questions and eight additional questions if children are present in the household. The questions ask whether certain behaviors involving food acquisition and diet quantity and quality occurred in the household over the last 12 months [11]. The complete eighteen questions and possible response options as presented by the CPS Food Security Supplement are included in "Appendix." Using the ten adult/household-specific questions, food security was calculated at four levels: high food security (HFS), marginal food security (MFS), low food security (LFS), and very low food security (VLFS). Responses of "yes," "often," "sometimes," "almost every month," and "sometimes but not every month" were tallied, and each respondent was given a raw score ranging from one to ten [12]. Scores of zero were ranked as HFS, one or two as MFS, three to five as LFS, and six or higher as VLFS. When using all eighteen questions, scores of zero were ranked as HFS, one or two as MFS, three to seven as LFS, and eight or higher as VLFS [12].

An important deliberation in food security assesses the way it is measured. Burchi and De Muro [13] wrote "The way food security is theorised, measured and finally analyzed affects the typology of policies that will be adopted." The CPS Food Security Supplement was designed to measure three food security domains: uncertainty, quality, and quantity [14, 15]. Coates et al. [14] compared 22 food security measures across 15 countries, including the CPS Food Security Supplement and several variations, and found a number of comparisons: 18 of the 22 measures include uncertainty measures, 16 of the 22 addressed the ability to eat healthy or proper diets, 21 of the 22 contained questions about running out of food and perceptions that there was not enough food for each member to eat as much as they should or want. Despite comparability, Coates et al. [14] ultimately concluded that the CPS Food Security Supplement may not adequately cover all domains important to food insecurity but offers a good foundation.

Other criteria for measuring food security have been assessed. Burchi and De Muro [13] reviewed a number of food security measures and concluded a capability approach can improve measurements by accounting for more direct and indirect drivers. However, they concede including all factors requires an ability to access large amounts of information and can be improbable [13]. A capability approach assesses food entitlements (employment status, assets, skill sets, etc.), basic capabilities (food access/ quality, health status, decision making, etc.), and the capability to be food secure (nutrition knowledge, cultural and religious beliefs, etc.) [13]. Relatedly, Diansari and Naseki [16] found that household-head education and household-head nutritional knowledge were significant variables when predicting subjective household food security; increasing either factor increased the likelihood the household would be subjectively more food secure. Headey and Ecker [17] evaluated food measures, including leading US tools, and concluded that dietary diversity is a leading food security indicator and can be used to measure trends, shocks, seasonality, and individuals, in a way that poverty and calorie availability subjective indicators (i.e., affordability and satisfaction) cannot. The CPS Food Security Supplement has a number of accepted measurement criteria [14, 17], but could be improved with more information.

Survey instrument and data collection

A survey instrument was designed to collect data for this analysis and was hosted at Purdue University using Qualtrics. Lightspeed GMI facilitated obtaining a sample of Midwest residents through a large proprietary opt-in database in February 2016. The respondent's sex, age, annual pretax household income, and state of residence were targeted to be representative of the population of the Midwest region based on the U.S. Census Bureau [18]. The survey attracted 7277 total link clicks, 7 people did not start the survey, 29 were under the age of 18, 578 were pushed out because of the state of residence quota, 5377 were pushed out because of the income quota, and 21 gave questionable or extreme response upon evaluation of the data. The final sample of completed surveys used in this analysis was 1265.

This study defines the Midwest according to the U.S. Census Bureau region definition and represents the states of Illinois, Indiana, Iowa, Kansas, Michigan, Minnesota, Missouri, Nebraska, North Dakota, Ohio, South Dakota, and Wisconsin [19], but also includes Midwestern neighbors Kentucky and Tennessee. Looking at the 2012–2014 state averages presented by Coleman-Jensen et al. [11], the Midwest contained a wide range of food insecurity proportions, with North Dakota having the lowest Midwestern and national rate of 8.4%, and Ohio with the highest of the Midwest at 16.9%. Kentucky and Tennessee had rates of 17.5 and 16.3% food insecurity, respectively [11]. A comparison of food insecurity rates for this study's sample and those reported by Coleman-Jensen et al. [11] are provided in the results.

Demographic questions included in the survey asked about sex, age, annual pretax household income, state of residence, and household composition. For annual pretax household income, seven income categories were included and were condensed into three categories: low income: less than $35,000, mid-income: $35,000–$75,000, and high income: $75,000 or more. County of residence was collected in addition to respondents' state of residence. Each county was paired with a U.S. Census Bureau Poverty estimate [20] in order to evaluate community conditions, in relationship with household specifics. The estimate represents the percent of the county population of all ages living at or below the national poverty line. For this study, the proportions were grouped into three categories 0–10, 11–20, and 21% or greater. Like the food insecurity proportions presented in Coleman-Jensen et al.

[11], the Midwest has a wide range of county-level poverty. For this sample, North Dakota had the lowest county average poverty of 10.8% and Ohio with the highest of 16%, and Kentucky had the highest rates of the sample with a county average of 20.6% living in poverty.

The food security questions explored were based on the CPS Food Security Supplement [11]. All respondents were asked the ten adult/household-specific questions. The results were calculated using the methods outlined in the "U.S. Household Food Security Survey Module: Three-stage Design, with Screeners" report [21]. The report suggests a screening question and income-based screening options. These options were left out of this study in order to obtain a broader understanding of the food security status of the complete sample of respondents.

For comparative purposes, two subsamples, households with and without children, were generated. All respondents were asked the ten household focused food security questions, and the sample of respondents who indicated that there were no children living in the household ($n=749$) were asked no further questions. Respondents who indicated having children (persons younger than 18 years old) ($n=360$) formed a second subsample and were asked the full eighteen-question survey, and the four levels were calculated with the score adjusted to accommodate the added questions. No supplementary calculations were done for the remaining respondents who did not indicate clearly if children were present in the household ($n=150$).

Respondents were asked questions about the prevalence of health conditions in their households. To assess health conditions, respondents were asked "Please indicate if you or someone in your household have any of the following conditions" which included a list of responses: diabetes, Crohn's disease, celiac disease, eating disorder, depression/ anxiety, high blood pressure, and high cholesterol.

Data analysis

Summary statistics were generated for each question, and cross-tabulations were performed in SPSS statistical software [22]. An ordered logit model was estimated using Stata/SE 14.1 [23] in order to identify determinates that contribute to the likelihood that a respondent will fall into increasing levels of food insecurity severity.

Ordered logits have been used to assess the probability of ranked outcomes. Migliore et al. [24] used an ordered logit to evaluate the likelihood of increased organic food purchases using quality conventions and income as independent variables. Diansari and Nanseki [1] used an ordered logit to predict household subjective food security status in Indonesia using insecure, somewhat insecure, somewhat secure, secure, and highly secure as the ranked dependent variable. Peterson et al. [25] employed an ordered logit to understand consumer factors (credence attributions, time, conveniences) influencing the likelihood a consumer will choose a specific local food retailer, among US and French consumers.

For each respondent, the food security level was converted into a numeric value in order to generate a discrete dependent variable: HFS equaled zero, MFS one, LFS two, and VLFS three. The independent variables were primarily discrete binary variables where one equaled the variable descriptor. The variables were: male, age 18–24, age 25–44, age 45–64, low income less than $35,000, mid-income: $35,000–$74,999, diabetes, Crohn's disease, celiac disease, eating disorder, depression/anxiety, high blood pressure, and high cholesterol. Percent of county population in poverty was the only continuous variable with the potential to range from 0 to 100%.

Ordered logits estimate the likelihood that an outcome will fall between or beyond estimated thresholds and the thresholds were calculated using the ranked dependent variable [26]. For this study, y represents the dependent variable and can take on the values: $y=0$, 1, 2, or 3.

Rank would be established using k to represent thresholds of j,

$$\text{If } k_{j-1} < y^* < k_j \quad \text{then } y = j, \quad \text{for } j = 1, \ldots, 2, \text{ and}$$
$$\text{If } k_2 < y^* \quad \text{then } y = 3,$$

where y^* is a latent variable [26] and will be estimated by the model. The food security rank for each respondent is represented by y^*. For each survey respondent i, food security can be explained by variables X_i,

$$y^* = \beta X_i + u_i.$$

The probability of each rank, j, can then be estimated and depends on the regression outcome falling between k_j and k_{j-1} [26]

$$Pr(y_i = j) = Pr(k_{j-1} < \beta X_i + u_i < k_i).$$

Results

A sample targeted to be representative of the Midwest population was collected in February of 2016 and consisted of 1265 completed surveys. Summary statistics on demographics reported are provided in Table 1. Forty-nine percent of the sample was male, and 28% of households reported having children. Income was collected by providing seven principal categories, but for conciseness the income categories were condensed into three groups: low income: less than $35,000 (32%), mid-income: $35,000–$75,000 (34%), and high income: more than $75,000 (34%).

U.S. Census Bureau Poverty estimates for each county [20] were evaluated and condensed into three categories.

Table 1 Sample demographics ($n=1265$). *Source*: Population percentages obtained from: U.S. Census Bureau; American Community Survey, 2014 American Community Survey 1-Year Estimates, Table S1901; generated by S. R. Dominick; using American FactFinder; http://factfinder2.census.gov; 21 Sept 2015

Variable description	Survey (% of respondents)	U.S. Census Bureau, 2014 American Community Survey 1-year estimates (%)
Male	48	49
Age		
18–24	08	13
25–44	33	31
45–64	38	36
65 and older	21	20
Income		
Less than $25,000	21	24
$25,000–$34,999	11	11
$35,000–$49,999	14	14
$50,000–$74,999	20	19
$75,000–$99,999	13	12
$100,000–$149,999	13	12
$150,000 or more	09	08
Household composition		
No children in household	59	
Children in household	28	
Unstated or ambiguous	12	
Poverty level % of county population		
Low: less than 10%	22	
Mid: 11–20%	64	
High: more than 21%	13	
You or someone in your household have any of the following conditions		
Diabetes	24	
Crohn's disease	08	
Celiac disease	07	
Eating disorder	08	
Depression/anxiety	25	
High blood pressure	44	
High cholesterol	41	

Twenty-two percent of respondents lived in a county where less than 10% of the population lived at or below the poverty line (low poverty), 64% lived where 11–20% of the population lived at or below the poverty line (mid-poverty), and 13% lived where those living at or below poverty make up 21% or more of the population (high poverty). Survey participants were asked about a number of chronic health conditions. Forty-four percent of respondents indicated that they or someone in their household had high blood pressure. Personal or household member high cholesterol was selected by 41%, depression/anxiety by 25%, diabetes by 24%, Crohn's disease and eating disorders were each selected by 8% of the sample, and celiac disease was selected by 7%.

Table 2 summarizes the percent of respondents from each state who were calculated to be in food insecure households and compares that proportion with the estimations reported by the USDA [11]. For most states, the percent of respondents who were found to be in food insecure households were higher than those estimated by the USDA. Comparatively, the highest state food insecurity rate estimated by both this study and the USDA was for Kentucky and the lowest for North Dakota. The results for this sample could be higher because the respondents were not screened by income or using the optional screening question [12]. Table 2 also summarizes the percent of the population of each state living at or below the poverty line, for which the highest rate

Table 2 State-by-state food insecurity comparison with 2012–2014 USDA estimated averages

	% of sample respondents per state ($n = 11{,}265$)	USDA ERS 2012–2014 state averages (%)	% of county population living at or below poverty (sample averages)
Illinois	35	11.7	14
Indiana	24	14.6	15
North Dakota	00	8.4	11
South Dakota	21	11.9	15
Ohio	24	16.9	16
Kansas	28	15.9	13
Nebraska	16	13.9	12
Iowa	06	11.4	12
Missouri	31	16.8	14
Michigan	27	14.7	16
Minnesota	14	10.4	11
Wisconsin	14	16.3	13
Kentucky	39	17.5	21
Tennessee	27	11.4	18
Mean	22	14	

Source of Comparison USDA ERS 2012–2014 Average: Coleman-Jensen et al. [11]. Household Food Security in the United States in 2014, ERR-194, U.S. Department of Agriculture, Economic Research Service

is also in Kentucky and the lowest are North Dakota and Minnesota.

Analyzing food security status across household characteristics

A food security status was calculated for the survey sample ($n = 1265$) as well as for two subsamples (households with reported children $n = 360$ and households without children $n = 749$). A summary of the proportions can be found in Fig. 1. Sixty-five percent of the total sample of HFS, 25% were food insecure, with 10% of LFS and 15% of VLFS. For households with children, 44% were food insecure compared to 18% of households without children. Households with children had the highest proportion of respondents in each food insecure group, 15% were of LFS, and 28% were of VLFS. Households without children had 8% of LFS and 10% of VLFS.

The total sample and the two household composition subsamples food security statuses, not including MFS for concision, were cross-tabulated with four demographic categories: sex, age, income level, and percent population living at or below the poverty line for the respondent's county of residence. A full summary can be found in Table 3.[1] Z tests were performed to test the statistical difference in proportions across demographic categories. For additional understanding, each of the ten questions and subsequent responses were cross-tabulated with the same demographics and can be found in Table 4.

Females received HFS scores of zero more frequently than males. When looking at the ten adult/household-specific questions, males answered "often true," "almost every month," and "yes" more frequently when compared to females.[2] Sixty-seven percent of females had a calculated score of zero compared with 61.8% of males. Of the total sample, 18.8% of males and 12.3% of females reported being of VLFS. In terms of specific severity, males more frequently selected "yes" in response to the questions "You were hungry but didn't eat because there wasn't enough money for food," "Did you lose weight because there wasn't enough money for food," and "Did you or other adults in your household ever not eat for a whole day because there wasn't enough money for food?" The trends were similar for households with children. While there is less security overall for each sex, a greater portion of females (53.5%) were considered to be of high food security compared with males (38.4%).

For the total sample, younger respondents had higher rates of food insecurity than older respondents. Thirty-five percent of 18–24-year-olds and 40.2% of 25–44 had a calculated score of three or higher and were considered food insecure, compared with 18.1% of 45–64-year-olds and 10.4% of those 65 years old or

[1] For reference: for both the UDSA ten-question and eighteen-question surveys, respondents in the High Food Security group had a total score of zero and those considered food insecure had a score of three or higher.

[2] For reference: a selection of "yes," "often," "sometimes," "almost every month," and "sometimes but not every month" contribute to increased food insecurity score.

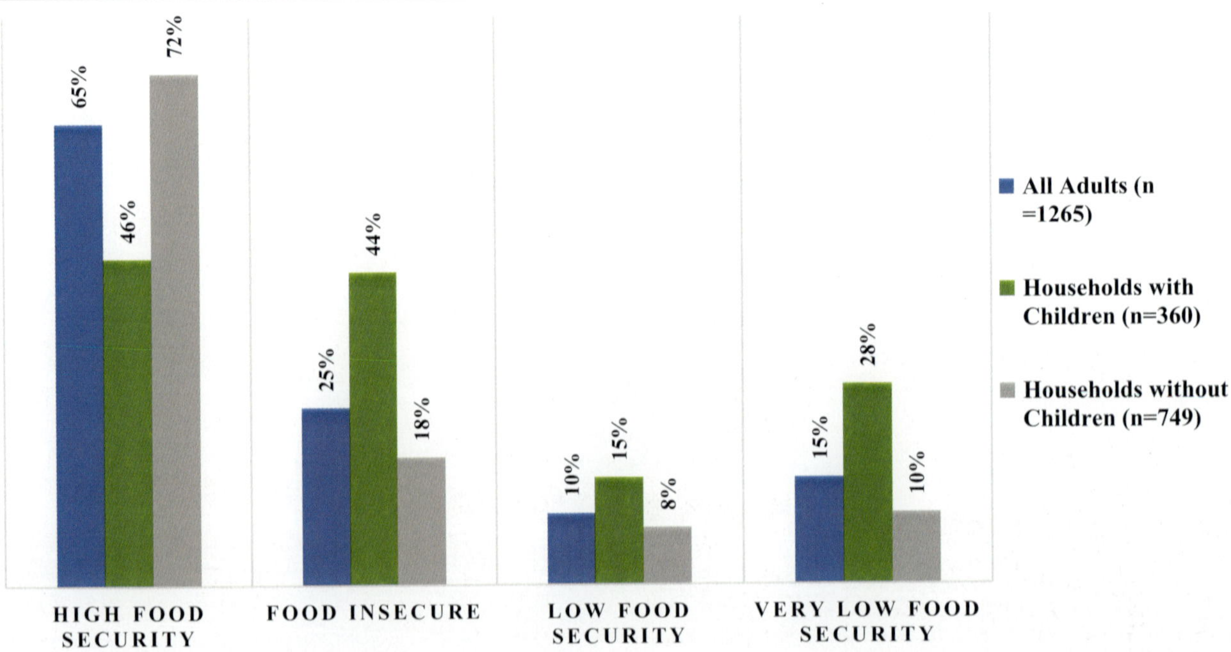

Fig. 1 USDA food security score (% of respondents $n = 1265$). High food security and food insecure do not sum to 100 (due to the marginal food security category). The difference in "households with Children" and "households without children" can be accounted for by a group of respondents for whom specific household make up could not be determined, and these are known as ambiguous households ($n = 156$)

older. For the statement "The food that (I/we) bought just didn't last, and (I/we) didn't have money to get more," 18.3% of 25–44-year-olds selected "Often true," and was over double the frequency of 18–24-year-olds (7.3%), 45–64-year-olds (3.5%), and those 65 years old or older (2.6%). The youngest age groups also had the highest proportions of VLFS, 18.8% of 18–24-year-olds and 27.2% of 25–44-year-olds, compared with 4.5% of those 65 years old or older. Distinctively, for the questions "Did you (or other adults in your household) ever cut the size of your meals or skip meals because there wasn't enough money for food," "Did you eat less than you felt you should because there wasn't enough money for food?" "You were hungry but didn't eat because there wasn't enough money for food," the younger two categories selected "yes" at higher rates compared to the older two categories.

The subsample of households with children showed similar trends with the youngest age groups more frequently in food insecure categories. Respondents considered of HFS made up 32.1% of 18–24-year-olds and 42.8% of 25–44-year-olds compared with 56.0% of 45–64-year-olds and 85.7% of those 65 years old or older. Respondents 25–44 had the largest portion of VLFS individuals with 33.2%, and this was not different from the 25% of 18–24-year-olds but was different and higher than those 45–64 years old.

Less of a gap between high food security and food insecure respondents was found across ages for the subsample of households without children. Fifty-six percent of 18–24-year-olds and 58.6% of 25–44-year-olds were considered of HFS, compared with 71.8% of 45–64-year-olds and 85.1% of those 65 years old or older. Of those who reported being 18–24 and being 25–44 years old, 30.4 and 30.7% were considered food insecure, compared with 16.7% of respondents 45–64 years old and 9.9% of 65-year-olds or older. The youngest age groups also had the highest proportions for LFS and VLFS.

Respondents in the low- and middle-income levels, largely, selected "often true," "almost every month," and "yes" with higher frequency than the remaining income groups. The low-income group more frequently selected "often true" and "sometimes true" for the statement "We worried whether our food would run out before we got money to buy more," but, the high-income group was the next most frequent selector of "often true" and the lowest selector of "sometimes true." For the statement "The food that (I/we) bought just didn't last, and (I/we) didn't have money to get more." Both the low-income group (8.8%) and the high-income group (11.3%) selected "often true" more frequently compared to the mid-income (5.2%) and were not statistically different from each other. The results for the question "Did you (or other adults in your household) ever cut the size of your meals or skip meals

Table 3 USDA food security status and demographics cross-tabulation

Food security level	Sex		Age				Income levels			Poverty level % of county pop.		
	Male (A)	Female (B)	18–24 (C)	25–44 (D)	45–64 (E)	65+(F)	Low (G)	Middle (H)	High (I)	Low (J)	Mid (K)	High (L)
All Adults												
Count	607	658	96	415	485	269	407	425	433	280	815	170
High food security	61.8^{B}	67.9^{A}	51.0^{EF}	48.7^{EF}	70.9^{CDF}	84.4^{CDE}	47.2^{HI}	68.7^{GI}	78.1^{GH}	72.9^{KL}	63.3^{J}	60.0^{J}
Food insecure	28.5^{B}	21.9^{A}	35.4^{EF}	40.2^{EF}	18.1^{CDF}	10.4^{CDE}	34.4^{HI}	22.1^{G}	19.2^{G}	17.5^{KL}	26.0^{J}	31.7^{J}
Low food security	9.7	9.6	16.7^{EF}	13.0^{EF}	7.4^{CD}	5.9^{CD}	14.5^{I}	10.4^{I}	4.4^{GH}	6.8	10.4	10.6
Very low food security	18.8^{B}	12.3^{A}	18.8^{EF}	27.2^{EF}	10.7^{CDF}	4.5^{CDE}	19.9^{HI}	11.8^{G}	14.8^{G}	10.7^{KL}	15.6^{JL}	22.4^{JK}
Households with children												
Count	190	170	28	250	75	7	52	132	176	77	239	54
High food security	38.4^{B}	53.5^{A}	32.1^{EF}	42.8^{EF}	56.0^{CD}	85.7^{CD}	17.3^{HI}	44.7^{G}	54.5^{G}	53.2	43.7	42.6
Food insecure	48.9^{B}	37.6^{A}	57.1^{EF}	46.4^{E}	32.0^{CD}	14.3^{C}	59.6^{HI}	40.2^{G}	41.5^{G}	35.1	45.4	48.1
Low food security	12.6	18.2	32.1^{D}	13.2^{C}	16.0	14.3	34.6^{HI}	18.2^{GI}	7.4^{GH}	19.5	14.0	14.8
Very low food security	36.3^{B}	19.4^{A}	25.0	33.2^{E}	16.0^{D}	0.0	25.0	22.0^{I}	34.1^{H}	15.6^{KL}	31.4^{J}	33.3^{J}
Households without children												
Count	343	406	46	140	341	222	297	243	209	166	486	97
High food security	72.3	72.4	56.5^{EF}	58.6^{EF}	71.8^{CDF}	85.1^{CDE}	51.2^{HI}	80.2^{GI}	93.3^{GH}	79.5^{KL}	70.8^{J}	68.0^{J}
Food insecure	19.0	17.5	30.4^{EF}	30.7^{EF}	16.7^{CDF}	9.9^{CDE}	31.0^{HI}	14.4^{GI}	4.3^{GH}	10.2^{KL}	18.9^{JL}	27.8^{JK}
Low food security	7.6	8.4	15.2^{EF}	15.0^{EF}	5.9^{CD}	5.4^{CD}	13.5^{HI}	7.0^{GI}	1.4^{GH}	2.4^{KL}	9.3^{J}	11.3^{J}
Very low food security	11.4	9.1	15.2^{F}	15.7^{F}	10.9^{F}	4.5^{CDE}	17.5^{HI}	7.4^{GI}	2.9^{GH}	7.8^{L}	9.7^{L}	16.5^{JK}

For each demographic category (i.e., sex, age), each value of one column marked with a superscript of another column denotes significant difference at the .05 level. No superscript denotes no difference. No calculations were done across demographic categories. High food security and food insecure do not sum to 100. The difference is for a category called marginal food security. The difference in "households with Children" and "households without children" can be accounted for by a group of respondents for whom specific household make up could not be determined; these are known as ambiguous households ($n = 156$)

Table 4 USDA 10 demographic, income, and poverty comparisons

	% of resp.	Sex		Age				Income levels			Poverty level % of county pop		
		Male (A)	Female (B)	18–24 (C)	25–44 (D)	45–64 (E)	65+ (F)	Low (G)	Middle (H)	High (I)	Low (J)	Mid (K)	High (L)
Count		607	658	96	415	485	269	407	425	433	280	815	170
We worried whether our food would run out before we got money to buy more													
Often true	08	9.6	7.3	9.4^{F}	16.1^{EF}	4.7^{D}	2.6^{CD}	13.0^{HI}	4.5^{GI}	7.9^{GH}	3.9^{KL}	8.7^{JL}	14.1^{JK}
Sometimes true	17	20.8^{B}	14.4^{A}	22.9^{DF}	24.8^{EF}	15.7^{DF}	7.4^{CDE}	24.8^{HI}	16.9^{GI}	11.1^{GH}	14.3	18.3	18.8
Never true	72	66.9^{B}	76.3^{A}	62.5^{EF}	55.9^{EF}	77.9^{CDF}	88.5^{CDE}	57.5^{HI}	76.9^{G}	80.1^{G}	80.0^{KL}	70.4^{J}	64.7^{J}
The food that (I/we) bought just didn't last, and (I/we) didn't have money to get more.													
Often true	08	10.9^{B}	6.2^{A}	7.3^{DF}	18.3^{CEF}	3.5^{D}	2.6^{CD}	8.8^{H}	5.2^{GI}	11.3^{H}	4.3^{KL}	8.5^{JL}	15.3^{JK}
Sometimes true	17	18.5	15.5	31.3^{EF}	22.2^{EF}	15.1^{CDF}	7.1^{CDE}	27.3^{HI}	15.8^{GI}	8.3^{GH}	12.9	17.8	19.4
Never true	72	67.5^{B}	76.9^{A}	58.3^{EF}	55.7^{EF}	80.2^{CDF}	89.2^{CDE}	60.4^{HI}	76.9^{G}	79.2^{G}	81.4^{KL}	71.0^{J}	64.1^{J}
I/we couldn't afford to eat balanced meals.													
Often true	08	8.6	7.9	9.4^{F}	13.5^{EF}	7.0^{FD}	1.9^{CDE}	13.5^{HI}	4.9^{G}	6.5^{G}	4.6^{K}	9.4^{J}	8.2
Sometimes true	19	21.1	17.8	26.0^{EF}	28.9^{EF}	15.3^{CDF}	9.7^{CDE}	28.3^{HI}	18.8^{GI}	11.5^{GH}	15.4^{L}	19.4	25.9^{J}
Never true	70	67.5^{B}	72.8^{A}	59.4^{EF}	54.2^{EF}	76.5^{CDF}	87.7^{CDE}	55.0^{HI}	74.1^{GI}	80.8^{GH}	78.2^{KL}	68.7^{J}	64.7^{J}
Did you (or other adults in your household) ever cut the size of your meals or skip meals because there wasn't enough money for food?													
Yes	18	20.3	16.1	20.8^{F}	29.9^{EF}	13.4^{DF}	7.4^{CDE}	26.0^{HI}	14.1^{G}	14.5^{G}	12.9^{KL}	18.7^{J}	24.1^{J}
No	79	76.4^{B}	81.3^{A}	71.9^{EF}	65.8^{EF}	84.9^{CDF}	91.1^{CDE}	68.6^{HI}	83.5^{G}	84.3^{G}	85.0^{KL}	77.8^{J}	74.7^{J}
Of those who said yes: How often did this happen?													
Almost every month	47	60.2^{B}	32.1^{A}	40.0	58.9^{E}	27.7^{D}	45.0	34.0^{I}	40.0^{I}	76.2^{GH}	47.2	44.7	56.1
Some months	38	30.1^{B}	48.1^{A}	45.0	29.8^{E}	52.3^{D}	40.0	45.3^{I}	45.0^{I}	20.6^{GH}	33.3	40.8	34.1
Only 1 or 2 months	13	7.3^{B}	18.9^{A}	5.0	11.3	16.9	15.0	17.9^{I}	13.3^{I}	3.2^{GH}	11.1	13.8	9.8
Did you eat less than you felt you should because there wasn't enough money for food?													
Yes	18	19.6	17.2	26.0^{EF}	25.8^{EF}	15.1^{CD}	10.0^{CD}	27.8^{HI}	15.8^{G}	12.0^{G}	14.3^{L}	18.9	22.4^{J}
No	79	77.3	80.1	67.7^{EF}	69.4^{EF}	82.9^{CDF}	89.6^{CDE}	67.3^{HI}	81.4^{GI}	86.8^{GH}	83.2^{L}	77.9	75.3^{J}
You were hungry but didn't eat because there wasn't enough money for food?													
Yes	14	16.3^{B}	11.2^{A}	18.8^{EF}	22.7^{EF}	10.3^{CDF}	4.1^{CDE}	18.9^{HI}	12.0^{G}	10.4^{G}	9.6^{KL}	14.2^{J}	17.6^{J}
No	83	79.2^{B}	85.7^{A}	75.0^{EF}	71.3^{EF}	87.4^{CDF}	94.1^{CDE}	74.9^{HI}	84.5^{G}	88.0^{G}	87.1^{KL}	81.3^{J}	81.2^{J}
Did you lose weight because there wasn't enough money for food?													
Yes	10	14.0^{B}	6.8^{A}	10.4^{F}	16.9^{EF}	8.0^{DF}	4.1^{CDE}	12.0	8.5	10.4	7.9^{L}	10.2	14.7^{J}
No	86	81.4^{B}	89.8^{A}	79.2^{EF}	77.1^{EF}	89.5^{CDF}	94.8^{CDE}	80.8^{HI}	88.0^{G}	88.2^{G}	88.6^{L}	85.8	81.2^{J}
Did you or other adults in your household ever not eat for a whole day because there wasn't enough money for food?													
Yes	09	11.9^{B}	5.8^{A}	8.3^{DF}	$17.8C^{EF}$	$4.7D^{F}$	1.9^{CDE}	8.4	7.8	9.9	4.6^{KL}	9.2^{J}	12.9^{J}
No	89	84.8^{B}	92.1^{A}	87.5^{DEF}	77.6^{CEF}	93.6^{CDF}	97.0^{CDE}	87.2	89.9	88.7	93.2^{KL}	88.0^{J}	84.1^{J}

Table 4 (continued)

	% of resp.	Sex		Age				Income levels			Poverty level % of county pop		
		Male (A)	Female (B)	18–24 (C)	25–44 (D)	45–64 (E)	65+ (F)	Low (G)	Middle (H)	High (I)	Low (J)	Mid (K)	High (L)
Of those who said yes: How often did this happen?													
Almost every month	43	43.1	42.1	12.5^{D}	50.0^{CE}	26.1^{D}	60.0	38.2	39.4	48.8	38.5	42.7	45.5
Some months	47	44.4	52.6	87.5^{D}	43.2^{C}	47.8	40.0	47.1	48.5	46.5	46.2	48.0	45.5
Only 1 or 2 months	05	8.3	0.0	0.0	2.7^{E}	17.4^{D}	0.0	8.8	6.1	2.3	0.0	6.7	4.5

For each demographic category (i.e., sex, age), each value of one column marked with a superscript of another column denotes significant difference at the .05 level. Responses for the category "don't know/ prefer not to answer can be found by subtracting the sum of responses shows per question from 100%

because there wasn't enough money for food?" and the sub question of "how often did this happen?" are particularly interesting. While the low-income group more frequently selected "yes," the high-income group more frequently selected "almost every month."

Looking at the total sample, the low-income group had the lowest proportion of HFS respondents and the highest portions of LFS and VLFS respondents. Thirty-four percent of the respondents in the low-income group were considered food insecure; specifically, 14.5% were of LFS and 19.9% of VLFS. For the mid-income group, 22.1% were food insecure, which was not different from the 19.2% of high-income respondents who were food insecure. Also 10.4% of mid-income respondents were of LFS and 11.8% were of VLFS. The high-income group had the largest proportion of respondents in the HFS group with 78.1%.

The lowest-income group still reported the largest proportions of food insecurity, but the proportions for mid- and high incomes were also larger when looking specifically at the subsample of households with children. Fifty-nine percent of the low-income group, 40.2% of the mid-income group, and 41.5% of the high-income group were considered food insecure. Interestingly, one of the higher proportions of VLFS for households with children was reported in the high-income group. Thirty-four percent of the high-income group had a calculated VLFS status, and was not different from the 25.0% reported in the low-income group, but is different from the proportion of mid-income respondents with 22.0%. When looking at households without children, the low-income group reported the highest food insecure proportion. HFS respondents made up 51.2% of the low-income group and 31.0% were food insecure, compared with 93.3 and 4.3% of the high-income group, respectively. The low-income group also reported the highest LFS and VLFS proportions with 13.5 and 17.5%.

Respondents living in a county with a higher percent of people living at or below poverty have higher rates of food insecurity. Seventy-two percent of the total sample living in areas of low poverty were considered of HFS, compared to 63.3% of respondents living at mid-poverty and 60.0% of respondents living at high poverty. Respondents living in a high-poverty county were more frequently of VLFS (22.4%) compared with those living in a low-poverty county (10.7%) and a mid-poverty county (15.6%). Those living in a high-poverty county also selected "yes" more frequently to the questions "Did you eat less than you felt you should because there wasn't enough money for food?" "You were hungry but didn't eat because there wasn't enough money for food?" "Did you lose weight because there wasn't enough money for food?"

For the subsample of households with children, 31.4% of respondents living in a mid-poverty county and 33.3% of respondents living in a high-poverty county were of VLFS, almost double the rate of respondents living in a low-poverty county (15.6%). Households without children living in a low-poverty county were more frequently of HFS, 79.5% compared with 70.8% of respondents living in a mid-poverty county, and 68.0% living in a high-poverty county. Respondents living in a high-poverty county were also more frequently food insecure with 27.8% of the sample and were more frequently of VLFS (16.5%).

Food security and chronic health indicators

If food aid programs are intended to support temporary conditions, it is important to consider the impact of chronic health conditions on food security status. Respondents were asked to indicate whether they or someone in their household had any of seven common conditions, which may impact an afflicted person's relationship with food. The conditions were: diabetes, Crohn's disease, celiac disease, eating disorder, depression/ anxiety, high blood pressure, and high cholesterol.

Figure 2 displays a preliminary analysis of the hypothesis that a relationship exists between chronic health and food security, even among the high-income groups. Each income level was proportioned into two groups: those for whom someone in the household had at least one of the health conditions and those for whom no one is afflicted. Each of those six groups was then proportioned into three levels of food security: HFS, MFS, and food insecure.

For the low-income group, 132 respondents did not report any instances of the seven chronic health conditions and 275 reported at least one. Of those who did not report any instances, 54% were of HFS, 11% were of MFS, and 36% were food insecure. Of the 275 respondents that reported a chronic health condition, 44% were of HFS, 22% were of MFS, and 34% were food insecure. In the mid-income group, 163 respondents reported no chronic health conditions and 262 reported at least one. Of those who did report a health condition, 68% were considered of HFS, 10% were of MFS, and 23% were considered food insecure. Of those who did not report a condition, 71% were of HFS, 9% were of MFS, and 21% were considered food insecure. Among the high-income group, 139 respondents did not report a health condition and 294 indicated at least one. Of those who did not report a condition, 91% were considered of HFS, 0% were of MFS, and 9% were considered food insecure. Of those who did report a condition, 72% were of HFS, 4% were of MFS, and 24% were considered food insecure. Since this population was of particular interest, a comparative Fishers exact test was performed and revealed that the

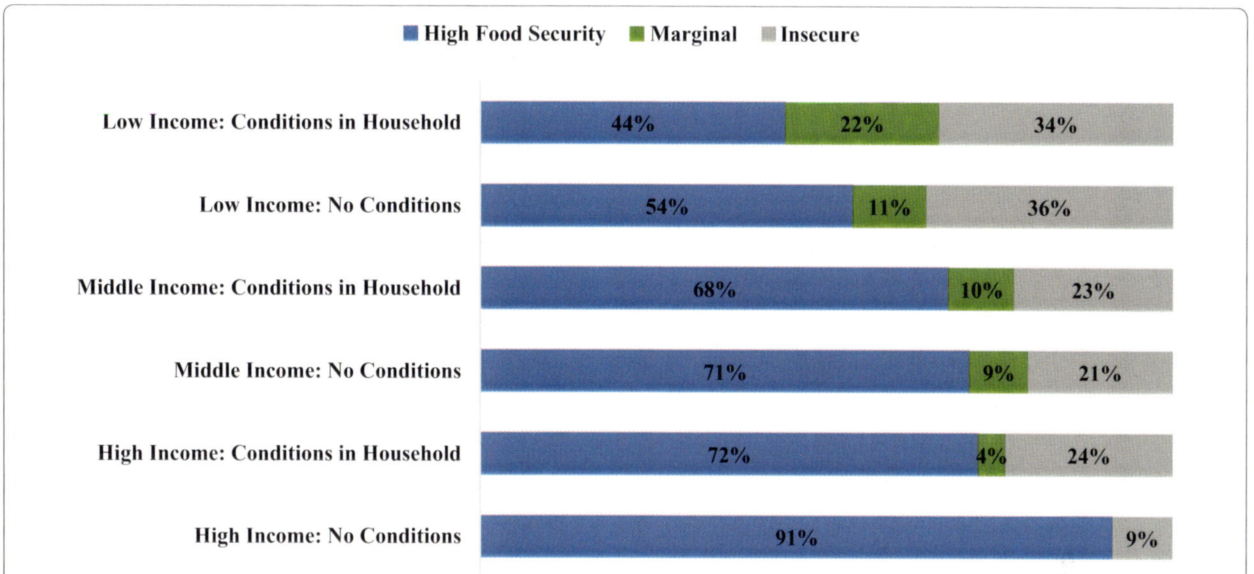

Fig. 2 Proportionality of food security among respondents with and without chronic health conditions (% of respondents $n = 1265$)

proportions of respondents who were high income, had a household health condition, and were food insecure were statistically different and higher compared to those who were of high income, had no household health conditions, and were food insecure. In short, the intersection between food insecurity and health conditions consistently emerges as important and necessary to consider if aiming to improve household's well-being, and this consideration exists even among high-income groups.

To understand the relationship between food security and health, cross-tabulations and *Z* test were performed to determine statistical difference between the proportions of respondents (Table 5). Generally, higher rates of food insecurity existed for respondents who indicated that at least one health condition was present in the household.

Respondents who indicated having diabetes or that someone in the household had diabetes had larger proportions of food insecurity overall and larger proportions of VLFS. These relationships were stronger for the subsample of households with children. Of households with children, 66.7% of respondents who reported household diabetes were food insecure, compared with 33.3% of respondents without household diabetes. Over half the respondents with household diabetes were of VLFS, 52.3% compared with only 17.7% of respondents with no instance of diabetes. The subsample of households without children has a similar relationship. Twenty-three percent of respondents with household diabetes were food insecure, compared with 16.7% of respondents without an instance of diabetes. Also 14.6% of respondents with household diabetes were of VLFS, compared with 9.0% without household diabetes.

Respondents with household Crohn's disease were more frequently food insecure (62.2%) than those without household Crohn's disease (21.9%). Over half the respondents with Crohn's disease were of VLFS (51.0% compared with 12.4%). This relationship is also true for the subsample of respondents with children. Eighty percent of respondents with children and with household Crohn's disease were also food insecure, and 68.7% were of VLFS. Households with children but without instances of Crohn's disease were HFS, at 52.2%. For households with celiac disease, 65.5% were food insecure and 54.8% were of VLFS. Comparatively, 67.8% of respondents without instances of celiac disease were HFS. For the subsample of households with children, 82.5% were food insecure and 73% were of VLFS when celiac disease was also present in the household.

Of respondents who indicated they or someone in their household experienced an eating disorder, 26.0% were HFS and 64.0% were food insecure, in contrast to respondents who did not report household eating disorders (68.3% were HFS). Over half the respondents who indicated household eating disorders were also of VLFS (51.0%). Similarly, of households with children, 52.0% of respondents who did not report an eating disorder were HFS and 83.3% of households which reported an eating disorder were food insecure, with 73.3% of VLFS. In households without children, the trend is similar, though less extreme. Only 38.7% of households without children who reported household

Table 5 USDA food security status and chronic condition cross-tabulation

Food security level	Diabetes		Crohn's disease		Celiac disease		Eating disorder		Depression/ anxiety		High blood pressure		High cholesterol	
	No (A)	Yes (B)	No (A)	Yes (B)	No (A)	Yes (B)	No (A)	Yes (B)	No (A)	Yes (B)	No (A)	Yes (B)	No (A)	Yes (B)
All adults														
Count	958	307	1167	98	1181	84	1165	100	949	316	709	556	752	513
High food security	69.8^{B}	49.8^{A}	68.1^{B}	27.6^{A}	67.8^{B}	25.0^{A}	68.3^{B}	26.0^{A}	72.4^{B}	42.7^{A}	65.6	64.2	65.6	64.1
Food insecure	20.9^{B}	38.1^{A}	21.9^{B}	62.2^{A}	22.2^{B}	65.5^{A}	21.7^{B}	64.0^{A}	19.6^{B}	41.5^{A}	25.2	24.8	25.5	24.4
Low food security	9.4	10.4	9.5	11.2	9.6	10.7	9.4	13.0	8.6^{B}	12.7^{A}	11.4^{B}	7.4^{A}	11.8^{B}	6.4^{A}
Very low food security	11.5^{B}	27.7^{A}	12.4^{B}	51.0^{A}	12.6^{B}	54.8^{A}	12.4^{B}	51.0^{A}	11.0^{B}	28.8^{A}	13.8	17.4	13.7^{B}	17.9^{A}
Households with Children														
Count	249	111	293	67	297	63	300	60	244	116	236	124	246	114
High food security	55.8^{B}	22.5^{A}	52.2^{B}	16.4^{A}	52.2^{B}	14.3^{A}	52.0^{B}	13.3^{A}	54.1^{B}	27.6^{A}	53.0^{B}	31.5^{A}	52.0^{B}	31.6^{A}
Food insecure	33.3^{B}	66.7^{A}	35.2^{B}	80.6^{A}	35.4^{B}	82.5^{A}	35.7^{B}	83.3^{A}	35.2B	61.2^{A}	36.0^{B}	58.1^{A}	38.2^{B}	55.3^{A}
Low food security	15.7	14.4	16.0	11.9	16.5	9.5	16.3	10.0	14.8	16.4	16.5	12.9	17.1	11.4
Very low food security	17.7^{B}	52.3^{A}	19.1^{B}	68.7^{A}	18.9^{B}	73.0^{A}	19.3^{B}	73.3^{A}	20.5^{B}	44.8^{A}	19.5^{B}	45.2^{A}	21.1^{B}	43.9^{A}
Households without children														
Count	592	157	726	23	735	14	594	19	593	156	396	353	426	323
High food security	73.8	66.9	72.9	56.5	72.4	71.4	73.4^{B}	48.4^{A}	77.6^{B}	52.6^{A}	70.2	74.8	71.8	73.1
Food insecure	16.7^{B}	23.6^{A}	17.8	30.4	18.2	14.3	17.3^{B}	38.7^{A}	14.3^{B}	32.7^{A}	20.5	15.6	19.7	16.1
Low food security	7.8	8.9	8.0	8.7	8.0	7.1	7.7	16.1	6.6^{B}	13.5^{A}	9.3	6.5	10.6^{B}	4.6^{A}
Very low food security	9.0^{B}	14.6^{A}	9.8	21.7	10.2	7.1	9.6^{B}	22.6^{A}	7.8^{B}	19.2^{A}	11.1	9.1	9.2	11.5

For each demographic category (i.e., sex, age), each value of one column marked with a superscript of another column denotes significant difference at the .05 level. No superscript denotes no difference. No calculations were done across demographic categories. High food security and food insecure do not sum to 100. The difference is for a category called marginal food security. The difference in "households with Children" and "households without children" can be accounted for by a group of respondents for whom specific household make up could not be determined, and these are known as ambiguous households ($n = 156$)

eating disorders were food insecure, and 22.6% were of VLFS. Comparatively, 17.3% of respondents who did not report an eating disorder were food insecure and 9.6% were of VLFS.

Of respondents who reported household depression/anxiety, 42.7% were considered HFS, while 72.4% of respondents who did not report household depression/anxiety were considered HFS. Respondents who reported household depression/anxiety were also more frequently of LFS (12.7%) and of VLFS (28.8%) compared with respondents who did not report household depression/anxiety (8.6 and 11.0%, separately). Similarly, of the subsample of households with children, 44.8% of those who reported household depression/anxiety were of VLFS and 61.2% were food insecure in total. More frequently, households without reported depression/anxiety were food secure (54.1% compared with 27.6%). Households without children who reported household depression/anxiety were more frequently food insecure. Thirty-two percent of households without children who reported household depression/anxiety were food insecure compared with 14.3% of households without children and without depression/anxiety. Also, more respondents without children were frequently of VLFS, 19.2% compared with 7.8%.

For the total sample, food security for respondents with household high blood pressure was not statistically different from respondents without household instance of high blood pressure except in the category LFS. Eleven percent of respondents without household high blood pressure were more frequently of LFS, compared with 7.4% of respondents with household high blood pressure. This result is counterintuitive to the results presented for the previous conditions. Households with children showed results similar to all other reported conditions. Of households with children, 53.0% of respondents who did not indicate instance of high blood pressure were HFS, compared with 31.5% of respondents who did indicate. Fifty-eight percent of respondents with household high blood pressure were food insecure and 45.2% were of VLFS.

Respondents without instances of high cholesterol were more frequently considered of LFS (11.8%) compared with respondents who reported household high cholesterol (6.4%). Conversely, respondents with household high cholesterol were more frequently considered of VLFS. Households with children were more frequently food insecure and of VLFS if they also reported that they or someone in their household had high cholesterol. Respondents without instance of household high

Table 6 Estimated level of food insecurity ordered logit ($n = 1265$)

Variables	Coefficient (SE)	Marginal effects			
		High food security	Marginal food security	Low food security	Very low food security
		dy/dx (SE)	dy/dx (SE)	dy/dx (SE)	dy/dx (SE)
Male	.2498* (.1315)	– .0522 (.0275)	.0149 (.0079)	.0159 (.0084)	.0214 (.0113)
Age 18–24	1.1435*** (.2971)	– .2393 (.0616).	.0684 (.0190)	.0728 (.0196)	.0980 (.0256)
Age 25–44	1.7554*** (.2419)	– .3673 (.0494)	.1050 (.0177)	.1117 (.0175)	.1504 (.0213)
Age 45–64	.9182*** (.2150)	– .1921 (.0444)	.0549 (.0139)	.0584 (.0143)	.0787 (.0184)
Low income less than $35K	2.1023*** (.1903)	– .4399 (.0377)	.1258 (.0168)	.1338 (.0156)	.1802 (.0175)
Middle income: $35K–$74,999	.6972*** (.1809)	– .1459 (.0373)	.0417 (.0116)	.0444 (.0118)	.0597 (.0155)
Household with children	.9298*** (.1614)	– .1945 (.0335)	.0556 (.0110)	.0592 (.0112)	.0797 (.0142)
Poverty level % of county population	.0225* (.0124)	– .0047 (.0026)	.0013 (.0007)	.0014 (.0008)	.0019 (.0010)
Diabetes	.5086*** (.1736)	– .1064 (.0362)	.0304 (.0108)	.0323 (.0113)	.0436 (.0150)
Crohn's disease	.3183 (.3896)	– .0666 (.0815)	.0190 (.0233)	.0202 (.0248)	.0272 (.0334)
Celiac disease	.1444 (.4642)	– .0302 (.0971)	.0086 (.0277)	.0091 (.0295)	.0123 (.0398)
Eating disorder	.9906*** (.3497)	– .2072 (.0732)	.0592 (.0217)	.0630 (.0229)	.0849 (.0302)
Depression/anxiety	.5097*** (.1543)	– .1066 (.0323)	.0305 (.0097)	.0324 (.0101)	.0436 (.0134)
High blood pressure	– .0188 (.1654)	.0039 (.0346)	– .0011 (.0099)	– .0012 (.0105)	– .0016 (.0141)
High cholesterol	.1336 (.1715)	– .0279 (.0358)	.0079 (.0102)	.0085 (.0109)	.0114 (.0147)
Cut one	3.9113 (.3257)				
Cut two	4.5523 (.3329)				
Cut three	5.3128 (.3413)				
Pseudo-R^2	.1588				
Prob > χ^2	.0000				
Log likelihood	–1089.1399				

For STATA, the ordered logit thresholds are reported as "cuts." Margins predicted at sample means

p values: * $p < .10$, ** $p < .05$; *** $p < .01$

cholesterol were more frequently HFS. Respondents in households without children and with high cholesterol in the household were statistically different from, and less frequently of LFS, than households without children and without household instance of high cholesterol.

Identifying determinants of food insecurity status with ordered logit model

An ordered logit was run in order to understand the significance of household chronic health status as a contributing factor to short-term (12-month) food security status, and the results can be found in Table 6. The model run was found to be significant overall, and the likelihood ratio χ^2 results are presented.

All demographics included were significant determinates. Being male increased the likelihood the household would have increased food insecurity. At the margin, being male decreased the likelihood of a household receiving a score of zero (or being high food security) and increased the likelihood of receiving a score greater than zero. Being in either of the included age categories increased the likelihood that a respondent would be food insecure, and being 25–44 had the greatest marginal contributions. The lower-income categories were also significant variables for predicting food insecurity and being in the low-income group had the higher marginal contributions. The presence of a child in the household increased the likelihood that a respondent would be food insecure.

The poverty-level score was the only continuous variable, and the logit suggests that as the poverty level of the county increases, a household living in that county was more likely to be food insecure. Of all seven chronic illnesses, diabetes, eating disorders, and depression/anxiety were statistically significant. Having either of these conditions present in the household increases the likelihood that a respondent will be food insecure. At the margin, having an eating disorder had the highest contribution. All of the logit results supported the findings of the cross-tabulations.

Discussion

The results of the cross-tabulations for income were consistent with findings from the USDA in that lower incomes generally experience reduced food security. However, there were areas where closer examination of each question suggested reduced security could be experienced across broader income ranges. Recall, there were comparable results and statistical sameness between the high- and low-income groups for the statements "We worried whether our food would run out before we got money to buy more," "The food that (I/we) bought just didn't last, and (I/we) didn't have money to get more" and for the question "Did you (or other adults in your household) ever cut the size of your meals or skip meals because there wasn't enough money for food?" and the sub question of "how often did this happen?"

In order to consider the relationship between high income and food security further, the household composition of high-income respondents was investigated. For high-income respondents whose household compositions were known ($n = 385$), respondents were divided into two groups, those in a household with four or more people ($n = 140$) and those with a household composition of less than four ($n = 245$). A Fisher's exact test was performed, and high-income respondents with a household composition of four or more had statistically different proportions of respondents in each food security category, compared to high-income respondents with a household composition of less than four people. Of the 140 respondents who were high income and had four or more people in their household, 39% were considered food insecure, compared to only 11% of the 245 respondents who were high income but had less than four household members. The relationship between food insecurity and household composition is important to consider for households at all income levels, even high-income groups. According to the USDA Food and Nutrition Service (FNS), household size and income are taken into consideration when dispensing benefits. According to the FNS, households of four people must earn at most $2633 gross income a month to qualify, which is roughly $31,596 annually [27]. As shown in this analysis, even households of $75,000 or more can experience food insecurity, especially with four or more household members, but they may not qualify for benefits.

The impact of children on food security was reflective of the household composition findings. In all demographic comparisons and chronic health comparisons, households with children had higher instances of food insecurity and higher rates of more severe food insecurity than households without children. Children were also a significant variable in the logit estimations. This finding was supported by Coleman-Jensen et al. [11] who found that households with children had higher rates of food insecurity.

Comparable to the literature, chronic illness has an impactful relationship with food security. The unique finding in this study was that the results suggest that the nature of the illness may be important; however, it may be unrelated to food inherently. Diabetes was a significant indicator of food insecurity and was similar to the results found by Knight et al. [4]. The results of the cross-tabulations and logit estimations were reflective of Muldoon et al. [6]; generally, respondents with household depression/anxiety had higher rates of food insecurity. Diabetes, eating disorders, Crohn's disease, celiac disease, high blood pressure, and high cholesterol all have food management considerations, but only diabetes and eating disorders were significant in the logit predictions. Depression/anxiety may be perceived to be the furthest removed from food and eating of the chronic illnesses investigated, but was identified as a significant variable. This suggests that the nature of food security and chronic illness may go beyond the intuitive relationship with food and that the specific nature of the illness may be important. As mentioned, aid programs were intended to support temporary conditions, but the results here suggest it is important to consider the impact of chronic conditions.

Another interesting result was the higher instance of food insecurity among males in the sample. This may seem counterintuitive to Coleman-Jensen et al. [11] who reported higher rates of food insecurity and very low food insecurity among households headed by single females. Marital status of the sample was not studied, although further analysis could shed light on this comparison. The results of this study were also similar to those found in Knight et al. [4] with higher rates of food insecurity among 18–44-year-olds.

This study also found that an individual was more likely to be food insecure as the poverty level of their county of residence increased. These results suggest that communities play a significant role in food security. These findings were not contradictory of the findings of Mayer et al. [10] and Kirkpatrick and Tarasuk [9], who suggest neighborhood-level impacts might not play a role in food security because both focused on grocery access. However, this was reflective of Dharamasen et al. [7] and Moore and Diez Roux [8] who considered regional income and poverty impacts. Overall food insecurity maybe related to limited food access due to resource constraints, both in the region and in the household, and not the specific availability of food markets.

Conclusion

It was found that 25% of the total sample were considered food insecure. Food insecurity was higher among males, middle-aged individuals, households with children, and low incomes. In the analysis of regional impacts, it was found that respondents living in counties with greater portions of the population living at or below the poverty line were more likely to be food insecure. When looking at chronic health, diabetes, eating disorders, and depression/anxiety were found to be significant variables predicting health score.

The researchers of this study acknowledge a number of health issues can be alleviated by improved food quality and access, so addressing chronic illness as a contributor of food security can improve the metrics of food security measurement and can improve policy designed to ease food access constraints on households. This study focused on the Midwest, but similar studies could be applied to measures around the world and inform global programs. As pointed out by Coates et al. [14], the CPS Food Security Supplement used by the USDA is comparable to a majority of measures used around the world. The Midwest represented a wide range of US food security, as measured by the CPS Food Security Supplement. The results reported here suggest that the CPS Food Security Supplement may underestimate the nature of food security by missing key factors such as income constraints (even in higher incomes), local community factors, and the impact of chronic illness. If the CPS Food Security Supplement underestimates US food security, measures similar to it around the world may also be underestimating.

While voter's ballots and election polls suggest food security and health care are separate issues, it is clear from this study and a number of the studies reviewed, the two issues are related in significant ways. The current Farm Bill, the leading food and nutrition legislation in the USA, will be in effect until at least 2018 [28]. Establishing clear relationships between health and food can inform large pieces of legislation like the Farm Bill. The results presented here suggest the inclusion of chronic illness and health to improve metrics and inform food security legislation because of the impactful relationship between chronic illness and food security, specifically with respect to diabetes, eating disorders, and depression/anxiety.

Special consideration should also be given to high-income populations who may be ineligible under current policy for benefits but who could still be hungry due to household composition and household health conditions. Uncovering the impacts, especially impacts that were previously unknown or seem counter to convention (i.e., food insecurity challenges in high-income households), that current food policies have on Midwestern households is imperative in informing impactful future policies.

Policies evaluating food security should also include evaluations of the community. While this study focused on poverty indicators, reviewed studies have found other community variables that relate to food security. For effectiveness and efficiency, policy should consider all factors, case in point, policy addressing local food access by increasing food supply may not alleviate the issue if the target region is most impacted by income constraints related to limited health care. This study included children and household size, but other household dynamics such as marital status, the specific ages of the children, and number of generations in the household should be studied in future research.

Future studies of both health and food security could and should include an indicator for the other factor. Future research should also directly ask whether respondents actively chose between medication and food, and under what circumstances. It is also important to note that other illnesses require treatment that is not limited to medication and can lead to financial hardship, such as therapy. Future research should broaden the scope of illness and treatment and explore the impact on food security status. Additionally, future studies of health and food security should also include regional components in order to understand the impact of location on each. Policy makers should consider how food and health overlap when measuring food security and when generating programs to alleviate it.

Authors' contributions
SD participated in data collection, data analysis, and manuscript preparation. NJOW participated in data collection, data analysis, and manuscript preparation. AR participated in manuscript preparation. JZGW aided in data analysis. LA participated in data collection and manuscript preparation. All authors read and approved the final manuscript.

Author details
[1] Department of Agricultural Economics, Purdue University, West Lafayette, IN, USA. [2] Department of Comparative Pathobiology, Purdue University, 725 Harrison Street, West Lafayette, IN 47907, USA. [3] Department of Consumer Sciences, Purdue University, West Lafayette, IN, USA.

Acknowledgements
None.

Competing interests
The authors declare that they have no competing interests.

Funding
This activity was funded, by Purdue, as part of AgSEED Crossroads funding to support Indiana's Agriculture and Rural Development.

References
1. McCarthy N. The top issues for voters in the 2016 presidential election. Forbes. 2016. Accessed 2 Oct 2016 by S. R. Dominick from http://www.forbes.com/sites/niallmccarthy/2016/07/11/the-top-issues-for-voters-in-the-2016-presidential-election-infographic/#23cb2b508535.
2. Office of the Press Secretary (OPS): The White Wouse. Remarks by the president at signing of the Farm Bill—MI. 2014. Michigan State University, East Landing Michigan. Accessed 2 Oct 2016 by S. R. Dominick from https

://www.whitehouse.gov/the-press-office/2014/02/07/remarks-president-signing-farm-bill-mi.
3. U.S. Department of Agriculture "Budget Summary and Annual Performance Plan: FY 2016." 2015. S. R. Dominick. http://www.obpa.usda.gov/budsum/budget_summary.html. Accessed 10 Oct 2016.
4. Knight CK, Probst JC, Liese AD, Sercy E, Jones SJ. Household food insecurity and medication "scrimping" among US adults with diabetes. Prev Med. 2016;83:41–5.
5. Berkowitz SA, Seligman HK, Choudhry NK. Treat or eat: food insecurity, cost-related medication underuse, and unmet needs. Am J Med. 2014;127:303–10.
6. Muldoon KA, Duff PK, Fielden S, Anema A. Food insufficiency is associated with psychiatric morbidity in a nationally representative study of mental illness among food insecure Canadians. Soc Psychiatry Psychiatr Epidemiol. 2013;48:795–803.
7. Dharamasen S, Bessler DA, Capps O Jr. Food environment in the United States as a complex economic system. Food Policy. 2016;61:163–75.
8. Moore LV, Diez Roux AV. Associations of neighborhood characteristics with the location and type of food stores. Am J Public Health. 2006;96(2):325–31.
9. Kirkpatrick SI, Tarasuk V. Assessing the relevance of neighbourhood characteristics to the household food security of low-income Toronto families. Public Health Nutr. 2010;13(7):1139–48.
10. Mayer VL, Hillier A, Bachhuber MA, Long JA. Food insecurity, neighborhood food access, and food assistance in Philadelphia. J Urban Health Bull N Y Acad Med. 2014;91(6):1087–97.
11. Coleman-Jensen A, Rabbitt MP, Gregory C, Singh A. Household Food Security in the United States in 2014, ERR-194. U.S. Department of Agriculture, Economic Research Service; 2015.
12. USDA Economic Research Service. U.S. Household Food Security Survey Module: Three-stage Design, with Screeners. 2012. Accessed 1 June 2016 by S. R. Dominick http://www.ers.usda.gov/datafiles/Food_Security_in_the_United_States/Food_Security_Survey_Modules/hh2012.pdf.
13. Burchi F, De Muro P. From food availability to nutritional capabilities: advancing food security analysis. Food Policy. 2016;60:10–9.
14. Coates J, Frongillo EA, Rogers BL, Webb P, Wilde PE, Houser R. Commonalities in the experience of household food insecurity across cultures: what are measures missing? J Nutr. 2006;136:1438S–48S.
15. Hamilton WL, Cook JT, Thompson WW, Buron LF, Frongillo EA, Jr, Olson CM, Wehler CA. Household food security in the United States in 1995 technical report of the food security measurement project September. 1997. 1997 USDA Food and Consumer Service Office of Analysis and Evaluation 1–c1.
16. Diansari P, Nanseki T. Perceived food security status—a case study of households in North Luwu, Indonesia. Nutr Food Sci. 2015;45(1):83–96.
17. Headey D, Ecker O. Rethinking the measurement of food security: from first principles to best practice. Food Secur. 2013;5:327–43.
18. U.S. Census Bureau; American Community Survey, 2014 American Community Survey 1-Year Estimates, Table S1901; generated by S. R. Dominick; using American FactFinder; http://factfinder2.census.gov. 21 Sept 2015.
19. U.S. Census Bureau (USCB). Geography "Geographic Terms and Concepts—Census Divisions and Census Regions". 2015. Accessed 8 Aug 2015 by S. R. Dominick from https://www.census.gov/geo/reference/gtc/gtc_census_divreg.html.
20. U.S. Census Bureau (USCB 2). State and County Estimates for 2015. 2015. Accessed 8 Aug 2015 by S. R. Dominick from http://www.census.gov/did/www/saipe/data/statecounty/data/2015.html.
21. USDA Economic Research Service. U.S. household food security survey module: three-stage design, with screeners. 2012. Accessed 1 June 2016 by S. R. Dominick http://www.ers.usda.gov/datafiles/Food_Security_in_the_United_States/Food_Security_Survey_Modules/hh2012.pdf.
22. IBM Corp. Released 2015. IBM SPSS Statistics for Windows, Version 23.0. Armonk: IBM Corp.
23. StataCorp. Stata statistical software: release 14. College Station: StataCorp LP; 2015.
24. Migliore G, Schifani G, Cembalo L. Opening the black box of food quality in the short supply chain: effects of conventions of quality on consumer choice. Food Qual Prefer. 2015;39:141–6.
25. Peterson HH, Taylor MR, Baudouin Q. Preferences of locavores favoring community supported agriculture in the United States and France. Ecol Econ. 2015;119:64–73.
26. Baum CF. An introduction to modern econometrics using Stata. College Station: Stata Press; 2006.
27. USDA FSN Food and Nutrition Service. "Supplemental Nutrition Assistance Program (SNAP): Eligibility. 2016. Accessed 12 Dec 2016 by S. R. Dominick from http://www.fns.usda.gov/snap/eligibility#Income.
28. USDA Economic Research Service. Agricultural Act of 2014: highlights and implications. 2016. Accessed 1 Nov 2016 by S. R. Dominick https://www.ers.usda.gov/agricultural-act-of-2014-highlights-and-implications/.

Assessment of postharvest loss along potato value chain: the case of Sheka Zone, southwest Ethiopia

Benyam Tadesse[1*], Fayera Bakala[2] and Lamirot W. Mariam[3]

Abstract

Background: Ethiopia has possibly the highest potential for potato production than any country in Africa. Postharvest loss (20–25%) is one of the major problems in the potato production. Therefore, this study was conducted with the objective of assessing postharvest losses along potato value chain actors and identifying its determinants in the study area.

Results: The descriptive result indicated that the quantity of postharvest losses at producer, local trader, wholesaler and retailer level was 21.724, 1.838, 3.406 and 4.07 kg/qt, respectively. The average gross margin with loss of producers, local traders, wholesalers and retailers was 6464.70, 282,169.89, 219,644.61 and 345,826.36 Birr, respectively, which is less than the average gross margin without loss (10,146.12, 284,015.83, 221,274.69 and 352,986.62 Birr, respectively). Distance to the nearest market, area allocated for potato and total output determine postharvest loss positively, and sex, experience, family size of working age, selling price and access to credit determine postharvest loss negatively.

Conclusion: In the study area, lack of storage facilities for potato was raised by farmers and other actors as a priority problem. Intervention of government from input supply until the end consumers is paramount and preparing storage mechanism is a must.

Keywords: Gross margin, Multiple linear regressions, Postharvest loss, Potato

Background

Achieving food security continues to be a challenge as it is affected by a complexity of factors [1–3]. Increasing the food availability is therefore not only increasing the productivity in agriculture, but also a need to lower the losses [4–6]. Food losses after harvest until the food reach the consumer are significant [7]. A large amount of food and products are not reaching the consumer particularly due to postharvest losses [4, 7] during harvesting, handling, transporting, storage, processing, packaging and distribution. Postharvest losses reduce the availability of food crops and income that could be generated by selling these products; thus in terms of quantity they are linked to food security [7–9]. According to FAO [10], postharvest losses in developing countries can range from 15 to 50%.

Potato (*Solanum tuberosum* L.) is the fourth most important food crop in the world on the basis of production after maize, rice and wheat with annual production accounts of nearly 300 million tonnes [11]. Out of these, over half of production occurs in developing countries [12]. During the production year of 2015/2016, it ranks first in area coverage and third in both total production and productivity among the root crops grown in Ethiopia [13]. It produces considerably more energy and protein than cereals [14]. It is also the fastest growing staple food crop and source of cash income for smallholder farmers in Ethiopia [15, 16]. It has a short cropping cycle and a large production per unit area in a given time. It provides more nutritious food per land unit in less time and often under more adverse condition than other food crops due to its efficient water use. It is one of the most efficient

*Correspondence: tadessebenyam7@gmail.com
[1] College of Agriculture and Natural Resources, Department of Agro Economics, Mizan-Tepi University, Mizan-Aman, Ethiopia
Full list of author information is available at the end of the article

crops in converting natural resources, labor and capital into a high-quality food. Potato provides more food per unit area than any other major staple crop. They are the perfect food and one of the few that can actually sustain life on its own. Thus, it has significant impact on providing nutrition to families, increasing household income and providing surplus to the wider market [17].

Ethiopia has possibly the highest potential for potato production than any country in Africa with 70% of the 13.5 million ha of arable land suitable for potato cultivation. Over one million highland farmers could grow potatoes in Ethiopia. Two of the three known agroecologies Woyina Dega (1500–2300 masl) and Dega (above 2300 masl) exhibit the best out-grower potato production [18]. However, the potato is widely regarded as a secondary non-cereal crop in part because it has never reached the potential that it has in supporting food security. It is estimated that 296,577.59 ha is now planted and 36,576,382.69 qt production annually [13]. Hence, it contributes toward efforts of ensuring food and nutrition security. In Ethiopia, potato is becoming a prominent source of income since the crop is the most important cash crop for smallholder farmers in the mid-altitude and highland areas of the country [19, 20]. Ethiopia has a much higher potential to increase agricultural production of the crop through use of improved seeds and undertaking technological innovation that facilitate the management and reduce postharvest losses [17, 21].

Postharvest loss (20–25%) is one of the major problems in the potato production. This entails that significant loss is incurred to the small holders that could have helped in nutrition, food security and income generation [22]. Potato yield productivity has increased far more than 24 tons per hectare due to adoption of new varieties [17]. However, postharvest loss reduction efforts have not been tailored well. Harvesting loss reduction helps increasing income, achieve food security and subsequent storage lose reduction [23]. Thus, to reduce postharvest losses, appropriate technologies should be developed and promoted. With the reduction in postharvest losses by 50%, food availability would be increased by 20% without cultivating an additional hectare of land for increasing crop yield [24]. Until recently, knowledge of postharvest handling of fruits and vegetables such as improved storage, packaging, transport and handling techniques in developing regions like southwest Ethiopia was virtually nonexistent for perishable crops in most areas, thus allowing for considerable losses of produce. Postharvest losses have been highlighted as one of the determinants of the food problem [25]. According to Oyekanmi [26], postharvest loss prevention technology techniques becomes paramount as more produce is transported to non-producing areas to supply the growing population as well as storing for longer period to obtain a year-round supply. Despite the remarkable progress made in increasing world food production at the global level, approximately half of the population in the third world does not have access to adequate food supplies. There are many reasons for this; one of which is food loss occurring in the postharvest and marketing systems. Evidence suggests that these losses tend to be highest in countries where the need for food is greatest [25]. Food supply can be improved either by increase in production or more importantly, reduction in loss. Since many researches show that great effort is being made in the area of food production, especially in the developing countries, the decline in food production therefore can be traced to postharvest losses. Reduction in postharvest losses therefore will increase food availability, hence alleviation of food shortage problems. Managing the effect of postharvest losses has the potential tendency to reduce the effect of the efforts put into production and increase marketing efficiency [27]. Therefore, this study seeks to examine postharvest losses along potato value chain, underlying factors that contribute to the massive postharvest losses, taking into consideration the postharvest handling practices and how it affects the income of potato farmers in southwest Ethiopia.

Methods

Description of study area

This study was conducted in southwest Ethiopia, Masha District of Sheka Zone. Masha is one of the three districts in Sheka Zone of Southern Nation Nationalities and Peoples Region (SNNPR), which is located at 677 km to southwest of Addis Ababa. Sheka Zone is known by its dense forest coverage, and the agroecology is 70% mid-altitude, 20% high altitude and 10% low altitudes. The zone composes three districts, viz., Yeki, Andracha and Masha. Its altitude ranges from 950 to 3300 m above sea level. Mean rainfall level is more than 2000 mm. Regarding rainfall distribution, high rainfall occurs at June, July, August, September, medium at October, November, March, low at April and May and little/no at December and January. Major crops grown in the area are enset, sorghum, maize, coffee, potato, field pea, fava bean, wheat, barley, haricot bean and teff. Agricultural farming system is enset-production-based farming system. Total land coverage of the area is 217,527.15 hectare and from this 26% is for cultivation, 41.3% plantations, 2.24% pasture, 8.96% cultivable and 3.5% uncultivable land. The area was selected for this study due to exceptionally high potential for potato production in the southwestern part of Ethiopia. The district is composed of 19 rural kebeles; among these the study was conducted in 3 kebeles, viz., Gatimo, Atiso and Shibo, as indicated in Fig. 1.

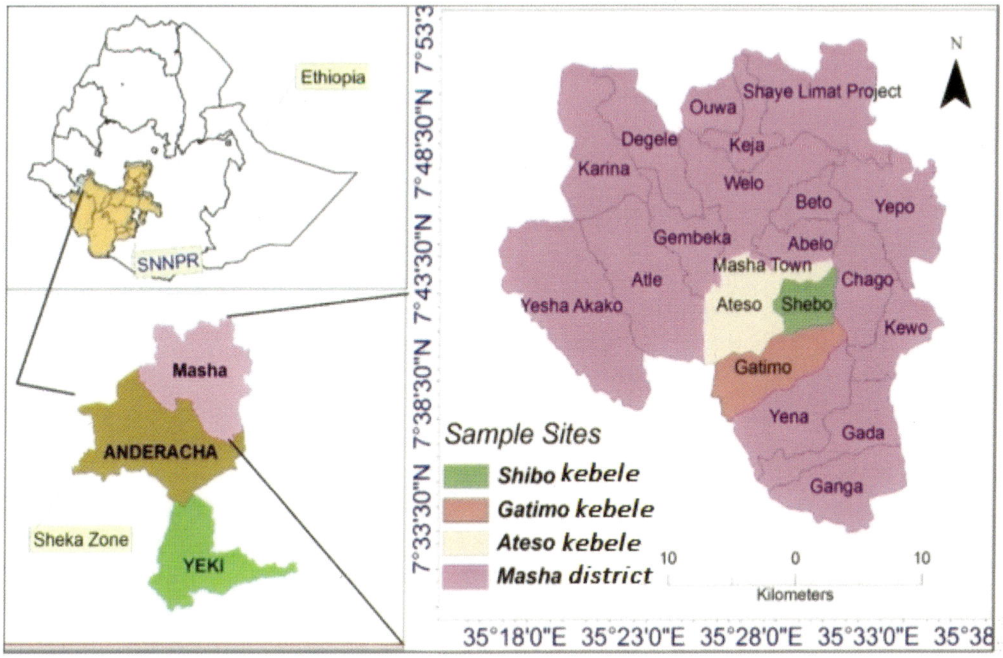

Fig. 1 Map of study area

Data source and instruments

In this study, both primary and secondary data sources were used. Different approaches of primary data collection methods were used. This includes survey, focus group discussions, key informant interviews, field observations and market assessments. Individuals from agricultural development and cooperative offices as well as the local people who have knowledge and experience about potato production and marketing were selected as key informants and interviewed on the issues related to potato production and marketing.

Sampling procedure

In this study, three-stage sampling technique was used. In the first stage, high-quantity potato-producing district was purposively selected. At the second stage, among the 19 rural kebeles of district, 3 kebeles were again purposively selected based on the intensity of production and marketing of potato. In the third stage, household heads producing potato were selected randomly from the total potato producers from three kebeles. The sample size was determined by rule of thumb suggested by Greene [28] $N \geq 50 + 8$ m, where N is sample size and "m" is the number of explanatory variables. The sample size was determined by proportional to total population within three kebeles. Thus, using the total potato producers household list 193 producers were selected proportionally to total population of three kebeles (Tables 1, 2).

Table 1 Sample size of producers. *Source*: own survey, 2017

Kebeles	Total number of potato producers (population)	Sample size	Percent	Cum.
Gatimo	116	60	31.09	31.09
Shibo	108	56	29.02	60.1
Atiso	149	77	39.9	100
Total	373	193	100	

Table 2 Sample size of traders. *Source*: own survey, 2017

Market place	Commission agents	Traders	Wholesalers	Retailers
Masha town	5	7	7	5

Regarding other value chain actors sample size, 7 local traders, 5 wholesalers and 8 retailers were selected by snowball techniques.

Data analysis

Data were analyzed by descriptive statistics, gross margin and ordinary least squares regression analysis. Descriptive statistics such as frequency distribution, percentages and mean were used in analyzing socioeconomic characteristics of respondents and quantity of potato lost at each value chain actor's level, while gross margin analysis was used to estimate the profit made by potato value chain actors in the study area. The gross profit of a business is estimated as the difference between the total sales and variable cost incurred.

$$GM = TR - TVC,$$

where *GM* is gross margin and *TR* is total revenue [value of output (amount realized from the sale of potato)]. It was obtained by multiplying the quantity of potato sold by the unit selling price. *TVC* is total variable cost which includes cost of all inputs (preharvest and postharvest labor wage, transportation costs and other input costs).

Econometric model

Econometrically, multiple linear regression model was used to examine the relationship between postharvest loss of potato and explanatory variables. The general form of multiple linear regressions is:

$$Y = f(X_1, X_2, X_3, X_4 \ldots X_k)$$
$$Y = \beta_o + \beta_1 X_1 + \beta_2 X_2 + \beta_3 X_3 + \cdots + \beta_4 X_k + \varepsilon i$$

Y represents postharvest loss and X_1, X_2, X_3, $X_4 \ldots X_k$ represents independent or explanatory variables, and εi is the disturbance factor. It is possible to write the function as:

Quantity of potato lost = f (Sex, Age, farming experience, Education_level, Distance to nearest market, FMSZ_working_age, Area allocated for potato production, total_output, selling price, local_seed, Improved seed, CREDITS and training). Based on this quantity lost function of potato, the econometric model for quantity of potato lost was written as:

$$\begin{aligned} &\text{Quantity of potato lost} \\ &= \beta_o + \beta_1 \text{Sex} + \beta_2 \text{Age} + \beta_3 \text{experience} + \beta_4 \text{Education} \\ &+ \beta_5 \text{Distance} + \beta_6 \text{FMSZ} + \beta_7 \text{AOLP} + \beta_8 \text{total\_output} \\ &+ \beta_9 \text{sellingprice} + \beta_{10} \text{local\_seed} + \beta_{11} \text{Improved seed} \\ &+ \beta_{12} \text{CREDITS} + \beta_{13} \text{DIUGETTR} + \varepsilon i \end{aligned}$$

where β_o = intercept term.

Results

Socio-demographic characteristics of respondents

The average age of interviewed respondents was 37.6632 years, while the average family size of the respondents was 6 in numbers. The education levels of respondents in the study areas were generally low (mean of 5.3990 years of schooling). On average, producer households had 7.0415 years of experiences in potato production and marketing. The mean distance from the nearest market center of producers in the study area was 5.5026 km. From the total land owned by the respondent, the land allocated for potato production is 0.55 ha/household, that means 27.94% of total.

From the total respondents, only 2% were Females and the rest were males. About 96.4% of studied respondents were married, while the remaining were not married (Table 3). Only half of producers had access to market information. From the total respondents, 2.6% of respondents had access to credit, while the rest 97.4% of respondents had no access of credit. Concerning other services, about 60% of producers had access to extension (Table 4).

Table 3 Demographic characteristics of respondents for continuous variables. *Source*: Own survey, 2017

Variables	Mean	SD
Age	37.6632	10.06259
Education level	5.399	3.17402
Distance from market	5.5026	2.72742
Experience	7.0415	4.34554
Total family size	6.8622	2.05808
Family size in working age	3.4218	1.48497

Table 4 Socioeconomic characteristics of producers (categorical variables). *Source*: own survey, 2017

Variables	Description	Frequency	Percent
Access to extension service	Yes	116	60
	No	77	40
Access to market information	Yes	96	50
	No	97	50
Sex	Male	190	98
	Female	3	2
Access to credit	Yes	5	3
	No	188	97

Socioeconomic characteristics of traders

In this study, trader refers to local traders, retailers and wholesalers. As indicated in Table 3, the average age of local traders, retailers and wholesalers was 31.43, 29.13 and 35 years, respectively. Local traders, retailers and wholesalers averagely have 4 family members. Averagely, wholesalers were more experienced (4.2 years) as compared to local traders (2.71 years) and retailers (2 years). But in education level retailers were more educated than wholesalers. The initial working capital of local traders, retailers and wholesalers was 6500.14,[1] 359.38 and 21,928 Birr, respectively, but the respective current working capital was 59,000, 3410.34 and 65,034 Birr. Table 6 shows all traders and wholesalers were male, while 63% of retailers were female and the rest were males. All traders, retailers and wholesalers had market information; however, none of traders had access to credit sources (Table 5).

According to the survey result in Table 6, all traders responded that they have access to market information. All local traders and wholesalers were males only, but 38% of retailers were females. All local traders, wholesalers and retailers responded that they have no access for credit service in the study area.

Assessment of postharvest loss along potato value chain

The descriptive result indicated that the mean value of the amount of potato postharvest loss at producer level was 9.31 qt per year per household which means 21.72%. When we estimate it in ETB, one household lose 3683.11 Br[1] per year due to potato postharvest loss (Fig. 2).

Next to producer, postharvest loss of potato was higher at retailer level. The postharvest loss assessment of local traders revealed that the quantity of potato lost in quintal per household was 3.34 qt which accounts 0.59%. It indicated that local traders lose 1324.64 ETB per household per year due to potato postharvest loss. Wholesalers lose 2.5 qt per household per year due to postharvest loss which estimated 0.65%. It indicated that one wholesaler lose 1630.1 ETB per year. Retail level losses were about 1.92 percent of the total produce of potato in the study area. The causes of loss were physical injury during harvest, rotting and disease infection, lack of storage area and poor handling area. The discarded potato fetched no economic value to the retailers. The aggregate postharvest loss from production (farm get level) to consumption level is 24.88% (Table 7).

Table 5 Socio-demographic characteristics of traders (continuous variables). ***Source*****: own survey, 2017**

Variables	Local traders (N = 7)		Wholesalers (N = 5)		Retailers (N = 8)	
	Mean	SD	Mean	SD	Mean	SD
Age	31.43	2.82	29.13	3.682	35	6
Family size	4.14	1.773	4.5	1.773	4	1.96
Experience	2.71	1.704	2	1.165	4.2	1.67
Education	8	1.915	2.63	1.408	8.4	1.48
Initial working capital [Birr]	6500.14	2491.86	359.38	166.636	21,928	4657.781
Current working capital [Birr]	59,000	62,804	3410.3	3279.721	65,034	14,761.95

Table 6 Socio-demographic characteristics of traders (categorical variables). ***Source*****: own survey, 2017**

Variables	Local traders (N = 7)			Retailers (N = 8)			Wholesalers (N = 5)		
	Description	Frequency	%	Description	Frequency	%	Description	Frequency	%
Gender	Male	7	100	Male	5	63	Male	5	100
	Female	0	0	Female	3	38	Female	0	0
Access to credit	Yes	0	0	Yes	0	0	Yes	0	0
	No	7	100	No	8	100	No	5	100

[1] The basic unit of money in Ethiopia; equal to 100 cents.

Fig. 2 Harvesting of potato by plowing the cultivated area

Table 7 Assessment of postharvest loss along potato value chain. ***Source*****: own survey, 2017**

	Producer	Local trader	Wholesaler	Retailer
Postharvest loss in quintal per year	9.31	3.34	2.5	4.07
Postharvest loss in % per year	21.72	0.59	0.65	1.92
Postharvest loss in Birr per year	3683.11	1324.64	1630.1	2856.03

Causes of postharvest losses at producer level

Harvesting

In Masha District, potato is harvested by different mechanisms. The first round potato is harvested by hand and suffers mechanical injury during harvest period because of digging tools, wooden sticks, hoes or forks. At the second round, producer collected remaining potato from the first round harvest by plowing potato cultivated area and/or digging via hoe. This kind of harvesting mechanisms leads the crop to damage due to injuries, cutting through or scraping away the outer skin of produce which provide entry points for molds and bacteria causing decay, increase water loss from the damaged area and cause an increase in respiration rate, and thus heat production finally hastens its senescence.

Mechanical injury

The high moisture content and soft texture of root crops like potato make them susceptible to mechanical injury, which occurred from production to retail marketing because of poor harvesting practices in the study area (Figs. 2, 3).

Curing

The producers in Masha Zone usually stayed their harvested potato without any shade at farm gate for 2–4 days due to low local market demand, and since the area is recognized by its heavy rainfall, the harvested potato was easily spoiled again. They did not cure their harvested product in the sided shade to get a quality product. One of the simplest and most effective ways to reduce water loss and decay during postharvest storage of root, tuber and bulb crops is curing after harvest. The type of wound also affects periderm formation: Abrasions result in the formation of deep, irregular periderm; cuts result in a thin periderm; and compressions and impacts may entirely prevent periderm formation. And also crops are most likely to be injured at harvest by the digging tools, which may be wooden sticks, machetes, hoes or forks. Researchers observed huge amount of potato thrown away or discarded at the farm gate; these all are neither consumed nor marketed in any form (Fig. 4).

Fig. 3 Harvesting, collecting and storage habit of potato in the study area

Fig. 4 Discarded potato at the farm gate

Selection and grading

All potatoes showing greening decay or severe damage owing to harvesting or pest attack should be discarded at harvest. Immature tubers and those showing minor damage or wetted by rain should be put aside for immediate consumption. Potatoes to be stored for food, or seed should be fully mature and free from any visible damage or decay. But in the study area the reality is different (Fig. 5).

Storage

The term "storage" as now applied to fresh produce is the holding of produces under controlled conditions. Usually in the study area, producers took the harvested potato to home if they had not get buyers and wait for sometimes until they get buyers and there is a chance to store potatoes for seed for subsequent production period. They place the produce directly on to the soil, especially wet soil, use dirty harvesting or field containers contaminated with soil, crop residues or decaying produce and dirty containers. There was paramount potato postharvest loss occurred due to improper and lack of storing area, insects and worms as well poor handling techniques. There are also situations where commission agents and traders collect potato together in open area around accessible load trucks from producers after bargained the selling price, but they promise to pay the price after it loaded. In this case, if traders face some hindering problems to load the collected potato, there is a probability to keep for 2–3 days while beaten by rainfall until it loaded. Significant amount of spoiled potato was observed due to these circumstances, and even sometimes traders refuse to take collected potato and free from losses. Commission agents are working for traders and/or wholesalers, and they never incur loss in potato marketing system (Fig. 6).

Fig. 5 Selection and grading of potato

Transportation

The primary modes of transportation of potatoes in the study area to the markets were by loading it on back of animals like horses and donkeys. Due to over-packing or under-packing of field or marketing containers, careless handling, such as dropping or throwing or walking on produce and sitting on packed containers during the process of grading, transport or marketing, and the perish ability nature of potato, there is high postharvest loss (Fig. 7).

Diseases

Late blight caused by *Phytophthora infestans* was a major fungus disease in the study area causing huge yield losses. The climate is favorable for the development of late blight, particularly during the two rainy seasons. High costs of chemicals to control the disease limit the use of fungicides, particularly by small-scale farmers. Producers had weak habit of preventive applications of fungicides rather they often start spraying when the disease is already present in the field making control very difficult. Infested fields are then the source of inoculums (spores) for other fields, and in this way late blight can spread rapidly throughout the production areas. The situation is further aggravated by poor

Fig. 6 Storing system in the study area

Fig. 7 Potato loading and unloading system in the study area

spraying techniques and the use of cheaper and less effective chemicals. Introducing varieties with some degree of resistance was important, but they still need fungicide applications to control the disease. It has been observed that late maturing varieties tend to have more resistance to late blight than early varieties, but farmers in the study area prefer early maturing varieties. It can be concluded that methods to control late blight are available, but they are not adequately applied by producers with limited resources for inputs and lack of knowledge (Fig. 8).

Comparison between gross margin with loss and without loss in potato value chain

The average gross margin with loss (6464.70 Birr) was less than the average gross margin without loss (10,146.12 Birr). This shows that postharvest losses reduce the income of farmers in the study area. The percentage loss of income incurred by the farmers was 36.3%. The average gross margin with loss of local traders, wholesalers and retailers was 282,169.89, 219,644.61 and 345,826.36 Birr which was less than the average gross margin without loss which is indicated in Table 8.

Fig. 8 Sample of potato lost due to disease

Table 8 Gross margin with loss and without loss. ***Source*****: authors computation, 2017**

	Producers	Local traders	Wholesalers	Retailers
Total variable cost	1,312,689.89	210,979.52	264,965.05	163,190.01
Total revenue without loss	3,270,888.61	2,199,090.307	1,813,887.84	2,987,082.96
Total revenue with loss	2,560,376.64	2,186,168.733	1,802,477.29	2,929,800.82
Gross margin without loss	1,958,198.72	1,988,110.79	1,548,922.8	2,823,892.95
Gross margin with loss	1,247,686.75	1,975,189.213	1,537,512.24	2,766,610.81
Average margin without loss	10,146.12	284,015.83	221,274.69	352,986.62
Average margin with loss	6464.70	282,169.89	219,644.61	345,826.36

Determinants of postharvest loss

Multiple linear regression model was used to assess factors that determine postharvest losses of potato at producer level in the study area. To exclude heteroscedasticity problem, we regress the variables by using robust but before running the regression we checked multicollinearity problem, and the result showed that there was no serious multicollinearity problem. Model is generally significant as given by the F coefficient and its probability. The *R*-squared is high showing that the explanatory variables explain about 81.9% of the variation in the dependent variable.

Out of 13 hypothesized variables, eight variables were found to determine postharvest loss of potato in the study area. These are sex (SEX), distance to the nearest market (DISFM), Experience (EXPR), and family size of working age (FMSZ), area allocated for potato (AOLP), total output (TOTP), selling price (PRICE) and access to credit (ACTC). Access to extension service (ACCEXT), improved seed (IMPSD), local seed (LOSD), age (AGE) and education level (EDUCL) were not significantly associated with dependent variable (Table 9).

The regression result of sex indicated that the variable had negative relationship with postharvest loss and significant at less than 10% significance level. This implies that female-headed households are likely to experience postharvest losses for these crops as compared to the male-headed households. The coefficient of the variable indicated that potato producer being a male should reduce potato postharvest loss by 1.914101 quintal. As similar to the hypothesis, Experience had negative effect on potato postharvest loss and was significant at < 10% probability level. The negative relationship between quantity lost and farm experience indicated that the increase in farm experience resulted in a decrease in quantity lost. The coefficient of the variable indicated that the increase in farm experience by 1 year resulted in a decrease in quantity lost by 0.0957481 quintal. The better postharvest handling practices also reduce postharvest loss of potato. Family size of working age was statistically significant at < 1% significance level. As expected, the variable has a negative effect on quantity postharvest lost of potato. The negative and significant relationship indicates that as potato production was labor-intensive activity, larger number of working age family size provides higher labor force to undertake potato production and postharvest handling activities. The coefficient of the variable indicated that the increase in one number of working age family size reduced quantity postharvest by 2.249992 quintal or the decrease in one number of working age family size increased quantity postharvest loss by 2.249992 quintal. Distance to nearest market had positive effect on potato postharvest loss and found to be statistically significant at < 1% significance level. The positive relationship indicates that the farther is a household from the market, farmers

Table 9 Determinants of losses at producer level. ***Source*****: own survey, 2017**

Number of obs. = 193
F**(13, 179) = 122.33**
Prob > *F* = 0.0000
R**-squared = 0.8191**
Root MSE = 3.8426

Qlost		Robust			
		Coeff.	SE	*T*	*P* > *t*
AGE	0.031,197	0.031197	0.0316204	0.99	0.325
SEX	− 1.914101	− 1.914101	1.039497	− 1.84	0.067
EDUCL	− 0.0597035	− 0.0597035	0.0878469	− 0.68	0.498
DISFM	0.6984369	0.6984369	0.1966846	3.55	0.000
EXPR	− 0.0957481	− 0.0957481	0.0539875	− 1.77	0.078
FMSZ	− 2.249992	− 2.249992	0.4184121	− 5.38	0.000
AOLP	1.118513	1.118513	0.3593258	3.11	0.002
TOTP	0.1334658	0.1334658	0.0194476	6.86	0.000
PRICE	− 0.0130786	− 0.0130786	0.0066856	− 1.96	0.052
LOSD	0.2912362	0.2912362	0.746762	0.39	0.697
IMPSD	− 0.7174505	− 0.7174505	1.215861	− 0.59	0.556
ACTC	− 2.845566	− 2.845566	1.454018	− 1.96	0.052
ACCEXT	0.6302327	0.6302327	0.8539101	0.74	0.461
_cons		20.605	4.538065	4.54	0.000

forced to transport or store their product and it leads to potato postharvest loss. The coefficient of the variable indicated that the increase in distance of market by 1 km resulted in a decrease in potato postharvest loss by 0.6984369 quintal. Area allocated for potato production was significant at < 1% significance level, and as expected it had a positive relationship with quantity of potato postharvest loss. The coefficient of the variable illustrated that the increase in 1 ha of area allocated for potato production resulted in 1.118513 quintal of quantity of potato postharvest loss. Total output was statistically significant at < 1% significance level and had positive effect on quantity of potato postharvest loss. The coefficient of the variable showed that the increase in the total output of potato by 1 quintal increased quantity postharvest of potato by 0.1334658 quintal. Selling price of potato was significant at 5% significance level. As hypothesized, it had negative effect on quantity postharvest loss of potato. The coefficient of the variable indicated that the increase in selling price of potato resulted in the decrease in quantity of potato postharvest lost by 0.0130786 Birr. Access to credit was significant at 5% significance level and had negative effect on quantity of postharvest lose. The coefficient of the variable indicated that access for credit resulted in 2.845566 quintal of potato postharvest loses.

Discussion

The average age of interviewed respondents was 37.6632 years. The result indicates a good supply of agile workforce in potato production in the study area. This result is in line with finding of Ayandiji et al. [24]. Babalola et al. [29] argued that age is a very important demographic characteristic because it determines the size and quality of the labor force. Older farmers are expected to use their farming experience to decide on appropriate postharvest handling practices and hence an overall reduction in postharvest losses [30].

The descriptive result indicated that the mean value of the amount of potato postharvest loss at producer level was 9.31 qt per year per household which means 21.72%. Postharvest losses in fruits and vegetables are about 25–40% [31]. 75% losses were occurred at field level [32]. 30% of fruits, 10% of vegetables, 50% of root crops and 60% of cash crops were lost [33]. The results are in line with reports of Admassu [34], Abebe and Bekele [35] and Humble and Reneby [36]. According to Hodges et al. [9], postharvest losses in developing countries can range from 15 to 50%. Horticultural crops are perishable products, and they are more prone to greater losses than for non-perishable crops [4]. Postharvest losses at producer level occur due to harvesting injuries; for example, for the first round potato is harvested by hand and some simple digging material and at this stage loss could occur. At the second round, producers collect potato output remained from the first round harvest; hence, they plow potato cultivated area and/or dig via hoe. These potato harvesting mechanisms trigger potato loss at farm gate level. In addition to this, after harvested usually potato stayed at farm gate for 2–4 days due to lack of potato market demand in the area without any shade, and since the area is recognized by its heavy rainfall, the harvested potatoes were easily spoiled again. Researchers observed huge amount of potatoes thrown away or discarded fruits at the farm gate; these all are neither consumed nor marketed in any form. Estimated losses at farm, market and consumption level were reported as 38.6, 35.9 and 25.5% of the total losses, and total postharvest losses were 31% of the total production [31]. Martey et al. [37] found that about 75% of the total postharvest losses occurred at the farm level and about 25% at the market level.

Next to producer, postharvest loss of potato was higher at retailer level in the study area. Losses at retailer level are obvious because he is unaware about the daily sales and he buys potato according to his experience and bears losses in shape of unsold quantity. In the study area, retailers lack shops for performing their activities and they sell potato in the market place which is exposed to sunlight. Bari [31], FFTC [38], Liu [39] also indicate that unsold quantity was lost daily and also retailers don't have enough resources to store their unsold commodity. These results are also similar to FFTC [38] and Gajanana et al. [40]. Postharvest losses reduce the income of farmers in Masha District of southwest Ethiopia. This result is in line with the finding of Ayandiji et al. [24]. The average gross margin with loss was less than the average gross margin without loss for all potato value chain actors. The factors that influence the postharvest losses significantly at the farm level were identified in the study area.

The regression result of sex indicated that the variable had negative relationship with postharvest loss and significant at < 10% significance level. Married household heads are thought to have an advantage with regards to labor availability for their production and postharvest handling activities, which in turn could minimize postharvest losses [41]. Takane [42] contented that usually single-headed households are female-headed households, and because of the absence of husbands, female-headed households have fewer economically active household members and are in a disadvantageous position relative to their male-headed counterparts in deploying family labor for farm activities. Smallholder farmers find themselves at a major disadvantage because many do not understand the market well, how it works and why prices fluctuate, they have little or no information about market conditions and prices, they are not organized collectively, and they have no experience of market negotiation [43]. There is therefore a need for these farmers to be made aware of various products they can grow in relation to their climatic conditions and market demand [44]. Male-headed households are likely to experience postharvest losses for these crops as compared to the female-headed households. This result contradicts the finding of Liu [39] who concluded that female farmers were found to be more prone to high levels of losses than their male counterparts. A study conducted by Ortmann and King [41] revealed a positive and statistically significant relationship between sex of household head and both cabbage and spinach postharvest losses. Considering the fact that more time and careful handling are required to minimize mechanical damages in leafy and fruit vegetables, females are likely to encounter minimal losses since they are careful handlers and are considered to be more patient as compared to their male counterparts.

The education levels of respondents in the study areas were generally low (mean of 5.3990 years of schooling). This finding is consistent with the finding of Ortmann and King [41]. Education is the key in understanding consumer quality expectations in niche markets which is necessary also because consumers may have different expectations and acceptance of the same food product produced using different technologies [45]. The more educated the farmer, the less will be potato postharvest loss. It has had positive effect on the adoption of appropriate agricultural technologies and skills to the farming population over the years. This agrees with the findings of Oduekun [45] that the level of education influences participation in agricultural productive activities, adoption, transfer and application of innovations. This could be a contributory factor to the high postharvest losses in potato production in the study areas because only farmers with good education often appreciate and use most postharvest technologies available. This result is consistent with the findings of Fawole and Fasina [46], Basavaraja et al. [47], Adesina and Baidu-Forson [48]. Education level of farmers also had significant impact on postharvest losses [37].

As potato production was labor-intensive activity, larger number of working age family size provides higher labor force to undertake potato production and postharvest handling activities. This is because farmers who had larger household sizes tended to have lower levels of postharvest losses because they have relatively high amount of family labor. Distance from the farm to the market was positively and significantly related to tomato postharvest losses. Similar results were reported by Liu [30].

On average, producer households had 7.0415 years of experiences in potato production and marketing. As similar to the hypothesis, Experience had negative effect on potato postharvest loss and was significant at < 10% probability level. More experienced farmers are expected to have minimal postharvest losses as compared to the inexperienced ones since they can utilize the gained experience to make important handling decisions as well as having market contacts to ensure that harvested produce is sold quickly [41]. The research result implied that as the farmer become more experienced, postharvest losses diminish. An experienced farmer population implies good knowledge and adoption of postharvest handling technology among the farmers [30]. As revealed by [41], older household heads have farming experience and adopted new technologies than young farmers; therefore, farming experience was expected to have a negative relationship with postharvest losses. Farming experience is thought to positively influence technology adoption [37, 49].

The mean distance from the nearest market center of producers in the study area was 5.5026 km. Distance to nearest market had positive effect on potato postharvest loss and found to be statistically significant at < 1% significance level. PHLs also varied depending upon the distance to the market. This pushes to higher quantity of potato postharvest loss. When they are marketed to medium-distance markets, PHLs were 5.15%, whereas for long-distance markets they were 8.17% [50]. According to Ortmann and King [41], the distance between the farm and the market had a positive and statistically significant relationship with cabbage and tomato postharvest losses. The farther the market is from the farm, the longer it takes for the produce to reach the market and hence an increase in postharvest losses due to heat build-up and in transit mechanical injuries [4]. According to Ayandiji et al. [24], increase in the distance from the farm to the market will increase the quantity of fruit loss; this is because the longer the distance of the farm to the market, the longer the time it will take for the produce to get to the market, and so the losses will increase because of the congestion and packaging of the tomato together for a long time. The longer the distance the more the time it will take for the produce to get to the market and so, the losses will increase because of congestion of the product and build up of heat [30].

From the total land owned by the respondent, the land allocated for potato production is 0.55 ha/household (i.e., 27.94% of total farmland). Area allocated for potato production was significant at < 1% significance level, and as expected it had a positive relationship with quantity of potato postharvest loss. When the area for production increases, total output will increase and also it will increase potato postharvest loss. Higher land holdings serve as an incentive to produce surplus for markets [42].

With respect to postharvest losses, large land holdings imply large volumes being produced, and the higher the production volumes, the higher the losses since farmers face the constraints of poor handling practices and limited storage facilities [30, 43]. The larger the area put into cultivation leads to higher production and greater chance of losses due to poor handling and lack of proper storage [41]. The larger the farm size, the higher the likelihood for postharvest losses. Similar results were reported by Gajanana [40] and Ortmann and King [41]. They reported that as production scale increases farmers will have to contend with the problem of storage and transportation; and where these facilities are not adequate, losses are imminent. The larger the area put into cultivation, the higher the quantity harvested and chances of losses due to poor handling and lack of proper storage [30]. Takane [42], however, argued that farm size may have indirect positive impacts on market participation by enabling farmers to generate production surpluses and overcome credit market, thus reducing postharvest losses. Overall, an increase in farm size tends to have a significant increase in tomato postharvest losses [41]. According to the same source instead of increasing farm size or land allocated to tomatoes, reducing the farm size or land allocated to tomatoes has an overall effect of tomato postharvest loss reduction. The larger the area put into cultivation the higher the quantity harvested and chances of higher losses due to poor handling and lack of proper storage. Good hygiene practices such as hand washing and postharvest handling equipment washing minimize chances of produce contamination, hence a reduction in postharvest losses [4]. This was in line with study of Begum et al. [51] which was studied with the title of Economic Analysis of Postharvest Losses in Food Grains for Strengthening Food Security in Northern Regions of Bangladesh. Increase in the number of baskets harvested of fruits also results in increase in the losses because there is no effective method of storage; hence, the more the quantity of harvested produce, the more the spoilage [24].

Conclusion

The descriptive result of assessment of postharvest losses along potato value chain actors indicated that 21.72, 0.59, 0.655 and 1.92% loss was estimated at producer, local trader, wholesaler and retailer levels, respectively. The average gross margin with loss of producers, local traders, wholesalers and retailers was 6464.70, 282,169.89, 219,644.61 and 345,826.36 Birr which was less than the average gross margin without loss (10,146.12, 284,015.83, 221,274.69 and 352,986.62 Birr). Out of suggeted variables sex, distance to the nearest market, experience, and family size of working age, area allocated for potato, total output, selling price and access to credit. Farmers in the study area lack skills of pre- and postharvest management and disease prevention of potato. Therefore, training on cultivation, disease prevention and pre- and postharvest management are important factors for enhancement of potato productivity and reducing postharvest losses in the study area. In the study areas, lack of storage facilities for potato was raised by farmers and other actors as a priority problem. There is the need for provision of good storage facilities to store the produce that are harvested before they are taken to the market. This will help to reduce the losses that occur at the farm level. Therefore, it is recommended to expand DLS in high potato-producing areas as per standard DLS design and construction. Through technical support to the farmers, cost effective mechanism of expanding DLS should be considered. Postharvest technology should be introduced to reduce the losses. In the potato market survey, it was observed that potato is transported over long distance either spread on floor of the truck, back of horse or put in congested sacks. During loading and unloading, there is mishandling of the products which lead to quick spoilage and high loss. Therefore, it is important to establish potato transportation standards and enforce it. There should be ready market for the produce. The markets must be well organized and also the road network must be improved in order to aid easy transportation of their produce. Roads linking farms to market should be improved to reduce transit losses. Establishment of farmers market and cooperative marketing should be encouraged to reduce losses related to marketing functions.

Abbreviations

PHLs: postharvest losses; DLS: diluted light storage; ETB: Ethiopian Birr; SNNPR: Southern Nation Nationalities and Peoples Region.

Authors' contributions

The corresponding author carried out the responsibilities of proposal writing, data collection and preparation of the manuscript. The co-authors FB and LWM were participated in activities like communicating with the donors, commenting on issues to be raised on the proposal and questioner preparation and participating in the write-up of the manuscript. All authors read and approved the final manuscript.

Author details

[1] College of Agriculture and Natural Resources, Department of Agro Economics, Mizan-Tepi University, Mizan-Aman, Ethiopia. [2] College of Agriculture and Natural Resources, Department of Natural Resource Economics, Mizan-Tepi University, Mizan-Aman, Ethiopia. [3] College of Agriculture and Natural Resources, Department of Horticulture, Mizan-Tepi University, Mizan-Aman, Ethiopia.

Acknowledgements

This study received financial support from Mizan-Tepi University. DAs and government office workers of Masha District are acknowledged for their cooperation during the data collection.

Competing interests

The authors declare that they have no competing interests.

Funding
This work was funded by Mizan-Tepi University from its research and community service budgets.

References

1. Nicholas H, Nisar M, James D. A review of emergency food security assessment in Ethiopia. Humanitarian policy group report, a study commissioned by and prepared for the World Food Programme, Rome. 2006.
2. FAO. How to feed the world in 2050. Rome: FAO; 2009.
3. Dercon S, Krishnan P. Vulnerability, seasonality and poverty in Ethiopia. J Dev Stud. 2015;36(6):25–53.
4. Kader AA. Increasing food availability by reducing postharvest losses of fresh produce. In: V international postharvest symposium, vol. 682. 2005. p. 2169–76.
5. Parfitt J, Barthel M, Macnaughton S. Review—food waste within food supply chains: quantification and potential for change to 2050. Philos Trans R Soc B Biol Sci. 2010;365(1554):3065–81.
6. Victor K. Post-harvest losses and strategies to reduce them. Technical paper on post harvest losses. Action Contre la Faim (ACF), member of ACF International. 2014. p. 2–25.
7. FAO/World Bank Work. Food and Agricultural Organization of the United Nations. Reducing post-harvest losses in grain supply chains in Africa: lessons learned and practical guidelines. FAO/World Bank Work. FAO Headquarters 18–19 Mar 2010, Rome, Italy. 2010.
8. Jayne TS, Mather D, Mghenyi E. Principal challenges confronting smallholder agriculture in sub-Saharan Africa. World Dev. 2010;38(10):1384–98.
9. Hodges RJ, Buzby JC, Bennett B. Foresight project on global food and farming futures: postharvest losses and waste in developed and less developed countries: opportunities to improve resource use. J Agric Sci. 2011;149:37–45.
10. FAO. IFAD (2012) The State of Food Insecurity in the World 2012: economic growth is necessary but not sufficient to accelerate reduction of hunger and malnutrition. Rome: FAO; 2014.
11. Ayalew T, Paul CS, Hirpa A. Characterization of seed potato (*Solanum tuberosum* L.) Storage, pre-planting treatment and marketing systems. The case of West Arsi. Afr J Agric Res. 2014;9(15):1218–26.
12. Devaux A, Kromann P, Ortiz O. Potato for sustainable global food security. Potato Res. 2014;57:185–99.
13. CSA (Central Statistical Authority). Agricultural sample survey 2015/16. Volume V. Report on area and production of major crops for private peasant holdings, meher season. Statistical Bulletin 578, Central Statistical Agency, Addis Ababa, Ethiopia. 2016. p. 10–19.
14. Haverkort AJ, Koesveld MJ, van Schepers HTAM, Wijnands JHM, Wustman R, Zhang XY. Potato prospects for Ethiopia: on the road to value addition. Lelystad: PPO-AGV (PPO publication 528). The Netherlands. 2012. p. 1–66.
15. Beliyu L, Tederose T. Knowledge gap in potato technologies adoption: the case of Central Highlands of Ethiopia. J Agric Ext Rural Dev. 2014;6(8):259–66.
16. Berhanu M, Getachew W. The role of local innovations to promote improved technologies: the case of potato farmers research groups (FRGs) in Jimma and Illubabor Zones, South Western Ethiopia. E3 J Agric Res Dev. 2014;4(3):39–49.
17. Vita and IPF (Irish Potato Federation). Potatoes in development: a model of collaboration for farmers in Africa. Pdf doc. 2014.
18. Tesfaye A. Potato production manual. Amharic Version printed in 1999 Ethiopian Calendar. 2008.
19. Mulatu E, Ibrahim OE, Bekele E. Policy changes to improve vegetable production and seed supply in Hararghe, Eastern Ethiopia. J Veg Sci. 2005;11(2):81–106.
20. Gildemacher P, Kagnongo W, Ortiz O, Tesfaye A, Woldegiorgis G, Wagoire W, Kakuhenzira R, Kinyae P, Nyongesa M, Struik PC. Improving potato production in Kenya, Uganda and Ethiopia: a system diagnosis. Potato Res. 2009;52:173–205.
21. Tewari VK, Ashok Kumar A, Kumar SP, Nare B. Farm mechanization status of West Bengal in India. J Agric Sci Rev. 2012;1(6):139–46.
22. BoFED (Bureau of Finance and Economic Development). Annual statistics for Amhara National Regional State. Bahir Dar, Ethiopia. 2007.
23. Hakan K. Design and management of postharvest potato (*Solanum tuberosum* L.) storage structures. Ordu Univ J Sci Technol. 2012;2(1):23–48.
24. Ayandiji A, Adeniyi OR, Omidiji D. Determinant Post harvest losses among tomato farmers in Imeko-Afon Local Government Area of Ogun State, Nigeria. 2011.
25. Babalola DA, Agbola PO. Impact of malaria on poverty level: evidence from rural farming households in Ogun State, Nigeria. Babcock J Econ Finance. 2008;1(1):108–18.
26. Oyekanmi MO. Determinants of postharvest losses in tomato production: a case study of Imeko-Afon Local Government Area of Ogun State. Unpublished B.Sc. thesis, Department of Agriculture, Babcock University. 2007.
27. Addo JK, Osei MK, Mochiah MB, Bonsu KO, Choi HS, Kim JG. Assessment of farmer level postharvest losses along the tomato value chain in three agro ecological zones of Ghana. 2015. p. 338–344.
28. Greene WH. Econometric analysis. 5th ed. London: Prentice Hall Inc; 2003.
29. Babalola DA, Makinde YO, Omonona BT, Oyekanmi MO. Determinants of post-harvest losses in tomato production: a case study of Imeko-Afon local government area of Ogun State. J Life Phys Sci. 2010;3(2):14–8.
30. Maremera G. Assessment of vegetable postharvest losses among smallholder farmers in Umbumbulu area of Kwazulu-Natal Province, South Africa. MSc Thesis, 2014.
31. Bari A. Economic assessment of post-harvest losses of mango in Multan and Rahim Yar Khan Districts. A postgraduate research thesis in department of Agricultural Economics, University of Agriculture Faisalabad. 2004.
32. Kisaka-Lwayo M. Risk preferences and consumption decisions in organic production: the case of KwaZulu-Natal and Eastern Cape Provinces of South Africa, PhD Dissertation, 2012.
33. Mezgebe AG, Terefe ZK, Bosha T, Muchie TD, Teklegiorgis Y. Post-harvest losses and handling practices of durable and perishable crops produced in relation with food security of households in Ethiopia: secondary data analysis. J Stored Prod Postharvest Res. 2016;7(5):45–52.
34. Admassu S. Post-harvest sector challenges and opportunities in Ethiopia. Addis Ababa: Ethiopian Agricultural Research Organization; 2005.
35. Abebe HG, Bekele H. Farmers' post-harvest grain management choices under liquidity constraints and impending risks: implications for achieving food security objectives in Ethiopia. Poster paper presentation, International Association of Agricultural Economists Conference, Gold Coast, Australia. 2006.
36. Humble S, Reneby A. Post-harvest losses in fruit supply chains—a case study of mango and avocado in Ethiopia. Karin Hakelius, Swedish University of Agricultural Sciences, Department of Economics. 2014.
37. Martey E, Al-Hassan RM, Kuwornu JKM. Commercialization of smallholder agriculture in Ghana: a Tobit regression analysis. Afr J Agric Res. 2012;7(14):2131–41.
38. FFTC. Post-harvest losses of fruits and vegetables in Asia, an international information centre for farmers in the Asia Pacific Region. 1993. http://www.fftc.agnet.org/library.php?func=view%26id=20110630151214. Accessed 12 May 2017.
39. Liu FW. Post-harvest handling of horticultural crops in Asia. Department of Horticulture, National Taiwan University. 1990.
40. Gajanana TM, Sudha M, Saxena AK, Dakshinamoorthy V. Post-Harvest handling, marketing and assessment of losses in papaya. In: ISHS Acta Horticulturae, 851: II international symposium on papaya, 2008. 2008.
41. Ortmann GF, King RP. Research on agri-food supply chains in Southern Africa involving small-scale farmers: current status and future possibilities. Agrekon. 2010;49(4):397–417.
42. Takane T. Labour use in smallholder agriculture in Malawi: six village case studies. Afr Study Monogr. 2008;29(4):183–200.
43. Magingxa LL, Alemu ZG, Van Schalkwyk HD. Factors influencing access to produce markets for smallholder irrigators in South Africa. Dev South Afr. 2009;26(1):47–58.
44. Thamaga-Chitja J, Hendriks SL. Emerging issues in smallholder organic production and marketing in South Africa. Dev South Afr. 2008;25(3):317–26.
45. Oduekun FK. Factors affecting the adoption of improved rice production package in Obafemi/Owode and Ifo LAG's of Ogun State. M.Sc. thesis, Department of Agricultural Extension Services, University of Ibadan. 1991.
46. Fawole OP, Fasina O. Factors predisposing farmers to organic fertilizer uses in Oyo state, Nigeria. J Rural Econ Dev. 2005;14(2):81–90.
47. Basavaraja H, Mahajanshetti SB, Udagatti NC. Economic analysis of post-harvest losses in food grains in India: a case study of Karnataka. Agric

Econ Res Rev. 2007;20:117–26.
48. Adesina AA, Baidu-Forson J. Farmers' perception and adoption of new agricultural technology: evidence from analysis in Burkina Faso and Guinea, West Africa. Am J Agr Econ. 1995;13:1–9.
49. Gangwar LS, Singh D, Singh DB. Estimation of post-harvest losses in Kinnow Mandarin in Punjab using a modified formula. Agric Econ Res Rev. 2007;20(2):315–31.
50. Kereth GA, Lyimo M, Mbwana HA, Mongi RJ, Ruhembe CC. Assessment of post-harvest handling practices: knowledge and losses of fruits in Bagamoyo district of Tanzania. Food Sci Qual Manag. 2013, p. 8–15.
51. Begum EA, Hossain MI, Papanagiotou E. Economic analysis of post-harvest losses in food grains for strengthening food security in Northern Regions of Bangladesh. Int J Appl Res Bus Adm Econ. 2012;01(03):1839–8456.

4

Concentration and formation behavior of naturally occurring formaldehyde in foods

Farrhin Nowshad, Md. Nazibul Islam and Mohidus Samad Khan*

Abstract

Background: In recent years, formaldehyde is reported to be widely used as a food preservative to increase the shelf life of fruits and fishes in tropical countries. Formaldehyde is detrimental to human health. Hence, use of formaldehyde as a food preservative is legally prohibited in most of the countries. To regulate formaldehyde application in foods, the regulatory bodies often conduct on-the-spot analytical tests to detect artificially added formaldehyde in food items. However, formaldehyde is ubiquitous in the environment and is present in many animal and plant species as a product of their normal metabolism. This naturally occurring formaldehyde may interfere in the detection of artificially added formaldehyde in foods. It is, therefore, important to study the concentration and formation mechanism of naturally occurring formaldehyde in food items.

Results: In this study, the formaldehyde contents of food samples were determined using spectrophotometric technique. The naturally occurring formaldehyde contents of a wide range of fruit, vegetable, milk, poultry, mutton and meat samples were determined. In addition, formaldehyde contents of processed food items, such as: cooked beef and poultry, beverages, and commercially available UHT milk and powdered milk samples, were also assessed and analyzed. The naturally occurring formaldehyde contents of fruit, vegetable, milk, poultry, mutton and meat samples were found up to 58.3, 40.6, 5.2, 8.2, 15.2 and 8.5 ppm, respectively. Formaldehyde contents of commercially available UHT milk, powdered milk, beverages, cooked beef and poultry were found up to 187.7, 194.1, 21.7, 4.3 and 4.0 ppm, respectively. This study also analyzed the time dynamic behavior of the formation of endogenous formaldehyde content of banana (AAB genome of *Musa* spp.), mandarin and beef.

Conclusions: The experimental results provide a baseline data of natural occurring formaldehyde content of the analyzed food items. The formation behavior of formaldehyde may vary according to food types, storage temperature, storing time, and aging pattern of the food items. The findings of this study will be useful for the consumers, researchers, legal authorities and other stakeholders working on food safety and preservation.

Keywords: Formaldehyde, Methylated compounds, Natural formation, Health risk, Formation kinetics

Background

Formaldehyde is a flammable, highly reactive and readily polymerizing colorless gas at normal temperature and pressure. It has a pungent, distinct odor and may cause a burning sensation to eyes, nose, and lungs at high concentrations [1–3]. Formalin, an aqueous solution of formaldehyde (37–40 wt%), is a colorless liquid which is used as a biological preservative [4]. Recently, it has been reported that formalin is widely used in different tropical countries as an artificial preservative for fruits, vegetables and fishes [4–13]. There are direct and indirect health hazards associated with formaldehyde and formalin consumption. Consumption of formalin on a regular basis can be injurious to the nervous system, kidney and liver, and may cause asthma, pulmonary damage and cancer [6, 14–16]. The use of formaldehyde as a food preservative is prohibited in most of the countries [6, 13, 17–21]. To restrict the use of formaldehyde as a food preservative, the regulatory bodies often collect food samples from local markets to perform on-the-spot analysis, or to send food samples to the nearby analytical laboratory for the qualitative and

*Correspondence: mohid@che.buet.ac.bd
Department of Chemical Engineering, Bangladesh University of Engineering and Technology, Dhaka 1000, Bangladesh

quantitative analysis of formaldehyde in food items [7, 19, 22]. However, formaldehyde is naturally produced in a wide variety of food items, such as: fruits and vegetables, meats, fish, crustacean and dried mushroom as a common metabolic by-product [23]. In biological systems, formaldehyde is generated from different methylated compounds by demethylases, and from interconversion of glycine and serine that is catalyzed by pyridoxal phosphate [24]. Naturally occurring formaldehyde content also varies according to food types and food conditions.

The presence of naturally occurring formaldehyde may interfere in detecting artificially added formaldehyde in foods. Thus, it is important to quantify the naturally occurring formaldehyde content in foods to estimate external formaldehyde dosage. Limited scientific information is available regarding the levels of naturally occurring formaldehyde in foods [25]. This study aims to identify and quantify naturally occurring formaldehyde content in a wide range of food items such as: fruits, vegetables, milk and meats. In addition, formaldehyde contents of processed food items, such as: cooked beef and poultry, beverages, and commercially available UHT and powdered milk samples, were also assessed and analyzed. The formaldehyde contents of food samples were determined using spectrophotometric technique [8, 9, 26]. The time dynamic behavior of naturally occurring formaldehyde formation in the food samples was also analyzed in this experimental study. This study will help consumers, nutritionists, scientists, legal authorities and other stakeholders by providing a baseline data of naturally occurring formaldehyde contents in foods, and also to help them understanding the dynamic behavior of formaldehyde formation in foods.

Possible health hazards of formaldehyde consumption

There is no set standard for the daily intake of formaldehyde from food; however, the World Health Organization (WHO) estimated it to be in the range of 1.5–14 mg/d (mean 7.75 mg/d) for an average adult [25], and according to the European Food Safety Authority (EFSA), the daily oral exposure to formaldehyde from the total diet should not exceed 100 mg formaldehyde per day [7, 27]. If consumed at a higher concentration, formaldehyde may cause damage to the GI tract, kidney, liver and lungs, and may lead to cancer [2, 4, 6, 16, 28]. Formaldehyde, when ingested, exerts an irritant action upon mucous membranes and may cause inflammatory changes in the liver and kidneys [29]. In addition, there is evidence linking formaldehyde with nasopharyngeal cancer [6, 30]. The international Agency for Research on Cancer (IARC) has classified formaldehyde (as well as formalin) as a Group 1 carcinogen [4, 31]. Table 1 describes different health hazards caused by formaldehyde consumption.

Table 1 Hazardous effects of formaldehyde [2, 4, 6, 16, 25]

Interactions	Possible health hazards
Ingestion	Excessive ingestion can cause Severe pain with inflammation, ulceration and necrosis of the mucous membranes lining almost every internal organ Nausea Vomiting blood Diarrhea with bloody stool Blood from urine Gastrointestinal lesions Acidosis Vertigo and circulation failure Systemic effects include Metabolic acidosis CNS depression and coma Respiratory distress Renal failure Liver failure Cancer and tumor development and Irreversible neurotoxicity
Inhalation	The Department of Health and Human Services (DHHS) and the International Agency for Research on Cancer (IARC) have characterized formaldehyde as a human carcinogen based on studies of inhalation exposure in humans and laboratory animals. Formaldehyde has been linked to Nasopharyngeal cancer Gastrointestinal cancer and Possible links to brain cancer and leukemia

Methods

Sample collection

Fresh fruit, vegetable and meat samples were collected from local markets (Dhaka). Pure milk sample was collected from the local dairy firm (Mymensingh). Different UTH (cow) milk samples (AARONG, MILK-VITA, IGLOO and PRAN), powdered (cow) milk samples (DANO,MARKS and DIPLOMA) and beverage samples (Instant and brewed coffee, COCA COLA, CLEMON and VITA MALT) were collected from local grocery shops (Dhaka).

Chemicals and reagents

Reagent grade 37% formaldehyde solution (Merck KGaA, Germany), ammonium acetate (Merck, Germany), acetic acid (Merck, Germany), potassium hydroxide (Merck, Germany), nitric acid (Merck, Germany), acetyl acetone (Loba-chemie, India) and trichloroacetic acid (EMPLURA grade, Merck, Germany) were used in the experimental study. Ultra-high purity de-ionized water (18.2 MΩ.cm, Purite, UK) was used for dilution and solution preparation. Whatman 42 filter paper was used to filter the sample solutions. In this study, freshly prepared 10% (wt%) trichloroacetic acid (TCA) was used for extraction of formaldehyde from meat samples. Freshly prepared Nash reagent was used as an indicator to detect the absorbance (415 nm) of formaldehyde in sample solutions [8]. Nash reagent is light sensitive and was kept in an air tight dark-glass reagent bottle at room temperature [8, 11]. 0.1 N potassium hydroxide and 0.1 N nitric acid were used to adjust the pH (6.0–6.5) of the distillate [8]. A pH meter (HANNA Instruments, USA, HI2211) was used to check the pH, and a Shimadzu UV–VIS 2600 spectrophotometer was used to measure the absorbance.

Sample preparation

Fruit and vegetable samples were peeled off, cut into small pieces, and blended with water in 1:10 ratio; the juice was separated from the residual solids using a clean cloth as sieve and then filtered using Whatman 42 filter paper. Fresh milk, UTH milk and beverage samples (except instant and brewed coffee) were used without filtration. Powdered milk samples and coffee (instant and brewed) samples were prepared by diluting the solid with water in 1:2 ratio; followed by filtration using Whatman 42 filter paper. The filtrates of fruit, vegetable, milk and coffee samples were diluted to 100 times. The pH of the diluted samples was kept within 6–6.5 [8, 11]. For the preparation of meat and lever samples, 10 g of each sample was cut into small pieces and blended with equal weight of water. After fine blending, the weight of the sample was taken and equal amount of 10% TCA was added. The sample was kept for homogenization [32]. After homogenization was done, it was filtered through Whatman 42 filter paper.

Time dynamic study

The time dynamic behavior of naturally occurring formaldehyde in foods was investigated in this study. To understand the time dynamic behavior of natural formation of formaldehyde, the formaldehyde contents of banana (AAB genome of *Musa* spp.) and mandarin samples were measured for 3 days; for beef sample, formaldehyde contents were measured for 8 weeks. During this study, beef sample was kept in frozen storage for 8 weeks (at a temperature of − 5 °C) while banana and mandarin samples were kept in a normal refrigerator for 3 days at a temperature of 4 °C. Sample preparation of the frozen items and the measurement of formaldehyde contents were carried out at room temperature following the same process mentioned in the previous section.

Detection method

The pH of freshly prepared diluted fruit, vegetable, milk and beverage samples was adjusted between 6 and 6.5 (with potassium hydroxide or nitric acid) [8, 11]. For the meat samples, 5 ml sample of the TCA extract was added to 5 ml of water and then adjusted to pH between 6 and 6.5. The solution volume was made up to 25 ml with water. Then, 5 ml of ready samples was added to 5 ml of Nash Reagent followed by heating in a water bath for 10–15 min at 60 °C, and cooling under running tap water. The formaldehyde contents of the above samples were measured using spectrophotometer (Shimadzu UVVIS 2600).

For the spectrophotometric detection of formaldehyde content, a calibration curve was generated by plotting absorbance of known formaldehyde concentration (0–10 ppm) prepared from a stock solution of 37% formaldehyde. Formaldehyde solutions of known concentrations (0–10 ppm) were added to Nash reagent to get the respective absorbance reading at 415 nm. The curve obtained by plotting formaldehyde concentrations in aqueous solutions against absorbance is shown in Fig. 1.

Results and discussions

Naturally occurring formaldehyde content in fruits and vegetable samples

Experimental results of naturally occurring formaldehyde contents of different fruit and vegetable samples are presented in Table 2 and Fig. 2. The experimental results of banana, grape, apple, pear, plum, beetroot, cabbage, cauliflower, potato, onion, kohlrabi, carrot, radish, cucumber and tomato were found compatible with results reported by Centre for Food Safety, Hong Kong [30]. No reported data of naturally occurring formaldehyde were found for pomegranate, pomelo fruit, pineapple, ripe papaya, sapodilla, guava, olive, amla, bangi fruit, green papaya, plantain and lemon; therefore, the experimental results provide the baseline data for the above food items.

Formaldehyde content of milk samples

Table 3 and Fig. 3 represent the experimentally obtained results of formaldehyde contents in pure (cow) milk, commercial UTH and powdered (cow) milk samples. The experimental results for pure cow milk (5.2 ± 3.5 ppm) were compatible with reported value (3.3 ppm) [30, 33]. The experimental results show that formaldehyde

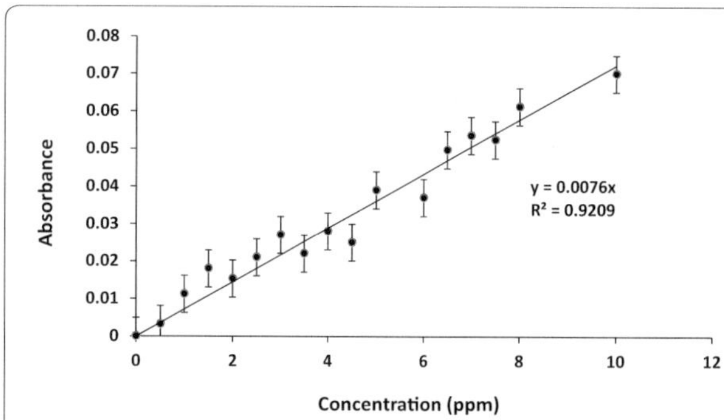

Fig. 1 Calibration curve showing the relation between absorbance and concentration of formaldehyde in aqueous solutions at 415 nm (error bars indicate SD for $n = 5$ samples)

Table 2 Naturally occurring formaldehyde contents in different fruits and vegetables (SD for $n = 5$ samples)

Serial no.	Fruit items		Vegetable items	
	Samples	**Formaldehyde content (ppm; avg ± SD)**	**Samples**	**Formaldehyde content (ppm; avg ± SD)**
1	Banana (AAA genome of *Musa* spp.)	20.7 ± 3.1	Carrot (*Daucus carota*)	10.8 ± 2.1
2	Banana (AAB genome of *Musa* spp.)	14.8 ± 1.1	Radish (*Raphanus sativas*)	6.4 ± 2.2
3	Grape (black; *Vitis vinifera*)	15.7 ± 6.2	Tomato (*Solanum lycopersicum*)	14.7 ± 6.2
4	Pomegranate (*Punica granatum*)	6.7 ± 1.1	Cucumber (*Cucumis sativus*)	6.4 ± 1.6
5	Pomelo fruit (*Citrus maxima*)	16.3 ± 2.2	Green papaya (*Carica papaya*)	40.6 ± 5.5
6	Litchi (*Litchie chinensis*)	6.7 ± 1.3	Lemon (*Citrus limon*)	Below detection limit
7	Pineapple (*Ananas comosus*)	20.8 ± 3.0	Beetroot (*Beta vulgaris*)	37.5 ± 2.1
8	Green apple (*Malus domesticus*)	13.4 ± 4.0	Cabbage (*Brassica oleracia var.capitata*)	8.1 ± 6.3
9	Red apple (*Malus domesticus*)	17.2 ± 2.6	Cauliflower (*Brassica oleracia var.botrytis*)	30.8 ± 4.5
10	Orange (*Citrus sinensis*)	56.9 ± 5.7	Potato (*Solanum tuberosum*)	16.5 ± 3.8
11	Mandarin orange (*Citrus reticulata*)	58.3 ± 3.9	Plantain (*Musa paradisiacal*)	38.9 ± 3.3
12	Pear (*Pyrus communis*)	57.7 ± 5.5	Onion (*Allium cepa*)	10.5 ± 5.5
13	Mango (Langra; *Mangifera indica*)	10.8 ± 3.3	Kohlrabi (*Brassica oleracia gongylodes*)	36.2 ± 5.4
14	Mango (Himsagor; *Mangifera indica*)	22.4 ± 5.6		
15	Ripe papaya (*Carica papaya*)	55.7 ± 3.0		
16	Guava (*Psidium guajava*)	33.3 ± 3.2		
17	Sapodilla (*Manilkara zapota*)	11.5 ± 3.2		
18	Olive (*Olea europaea*)	56.6 ± 4.2		
19	Amla (*Phyllanthus emblica*)	8.7 ± 1.2		
20	Plum (*Prunus domestica*)	10.3 ± 1.0		
21	Bangi fruit (*Cucumis melo*)	15.1 ± 2.8		

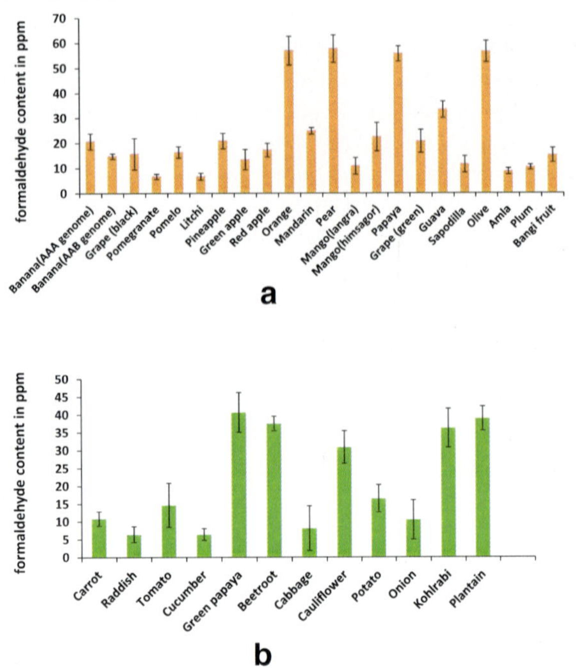

Fig. 2 Naturally occurring formaldehyde contents of different fruits and vegetable items; **a** fruits, and **b** vegetables ($n = 5$ samples, error bar indicates SD)

Table 3 Formaldehyde concentrations in different milk samples (SD for $n = 5$ samples)

Serial no.	Samples	Formaldehyde content (ppm; avg ± SD)
1	Pure milk	5.2 ± 3.5
2	DANO powder milk	129.3 ± 3.3
3	DIPLOMA powder milk	194.1 ± 6.6
4	MARKS powder milk	90.6 ± 3.5
5	PRAN UTH milk	69.9 ± 7.1
6	IGLOO UTH milk	187.7 ± 3.1
7	AARONG UTH milk	64.6 ± 5.1
8	MILK VITA UTH milk	58.7 ± 6.6

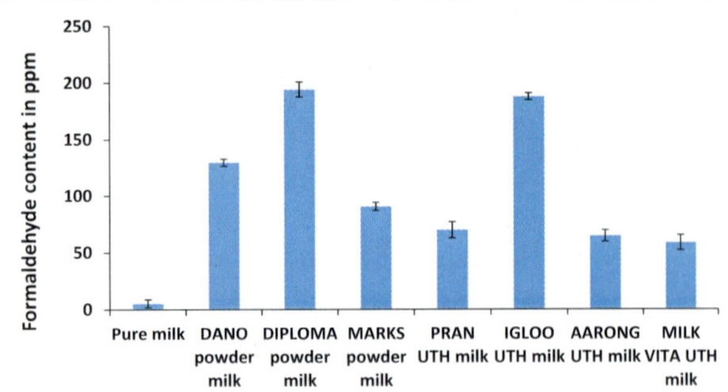

Fig. 3 Formaldehyde content of milk samples ($n = 5$ samples, error bar indicates SD)

content in UTH milk and powdered milk samples were higher (58.7–187.7 ppm) than that of pure milk sample. Possible explanations for higher formaldehyde content in commercial milk samples are dosing of formaldehyde during milk processing, preservation and/or packaging to improve the shelf life, or conversion of milk ingredient to primary aldehyde during milk processing [34–36].

Formaldehyde content of meat samples

Figure 4 and Table 4 represent experimentally obtained results for naturally occurring formaldehyde in meat and lever samples. Values obtained for poultry (8.2 ± 1.0) and beef (8.5 ± 0.6) are slightly higher than the reported values, which are 2.5–5.7 and 4.6, respectively [30]. Generally formaldehyde is introduced in ruminant feeds either as a preservative agent or as a reagent used to dietary components from ruminal degradation [33]. It is reported that significantly higher concentration of formaldehyde was obtained from the fresh muscle tissue of calves consuming 0.10% formalin-treated whey, whereas the muscle tissue of controlled calves or those consuming whey containing 0.05% formalin exhibited lower formaldehyde concentration [37]. Therefore, formalin-treated diet could be a probable reason of obtaining comparatively higher concentration of formaldehyde than the reported values. There was no reported value found for formaldehyde content in mutton. Hence, the experimental result serves as baseline data for mutton.

In cases of cooked poultry and beef, a significant drop in formaldehyde concentration was observed (Table 4, Fig. 4). Poultry and beef samples were minced and cooked in water for 6 min at 70 °C. Reduction in formaldehyde content in cooked samples could be due to the volatile characteristics of formaldehyde at high temperature (50 °C or above) [38, 39].

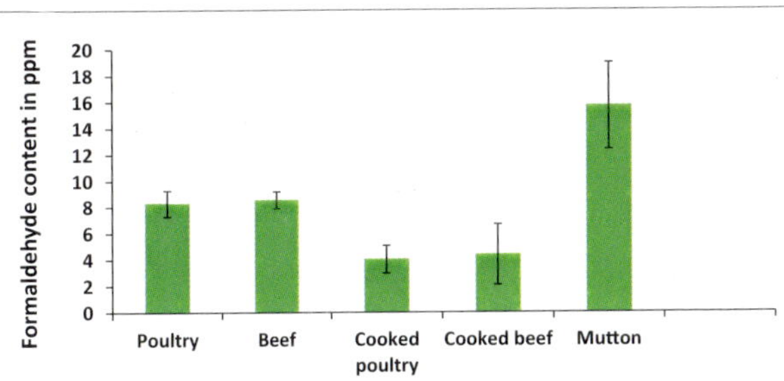

Fig. 4 Formaldehyde content of meat samples ($n = 5$ samples, error bar indicates SD)

Table 4 Formaldehyde concentrations in meat samples (SD for $n = 5$ samples)

Serial no.	Samples	Formaldehyde content (ppm; avg ± SD)
1	Poultry	8.2 ± 1.0
2	Beef	8.5 ± 0.6
3	Cooked poultry	4.0 ± 1.1
4	Cooked beef	4.3 ± 2.3
5	Mutton	15.2 ± 3.2

Formaldehyde content of beverage samples

Figure 5 and Table 5 represent the formaldehyde levels found in different beverages. The results obtained were compatible with the reported data [30]. Formaldehyde was found at slightly higher in instant coffee than in brewed coffee. This slightly greater value suggests that formaldehyde might escape from coffee during brewing [40]. There was no literature value found for formaldehyde content in malt beverages.

Time dynamic behavior of formaldehyde formation in foods

During the investigation of time dynamic behavior of naturally occurring formaldehyde in foods, banana (AAB genome of *Musa* spp.), mandarin and beef samples were analyzed. Beef sample was kept in frozen storage for eight weeks (at a temperature of − 5 °C) while banana and mandarin samples were kept in a normal refrigerator for 3 days at a temperature of 4 °C. The formaldehyde contents were measured at room temperature. Figure 6 represents the time dynamic behavior of endogenous formaldehyde contents in banana (AAB genome of *Musa* spp.), mandarin and beef samples.

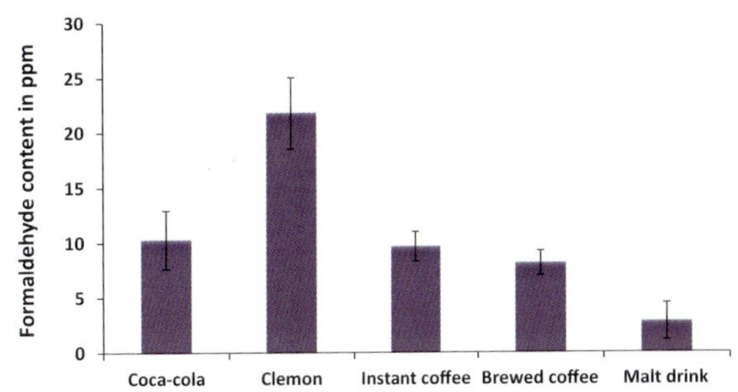

Fig. 5 Formaldehyde content of beverage samples ($n = 5$ samples, error bar indicates SD)

Table 5 Formaldehyde concentrations in beverage samples (SD for $n = 5$ samples)

Serial no.	Samples	Formaldehyde content (ppm; avg ± SD)
1	Coca-cola	10.2 ± 2.6
2	Clemon (transparent)	21.7 ± 3.2
3	Instant coffee	9.6 ± 1.3
4	Brewed coffee	8.1 ± 1.1
5	Malt drink (VITA MALT)	2.7 ± 1.6

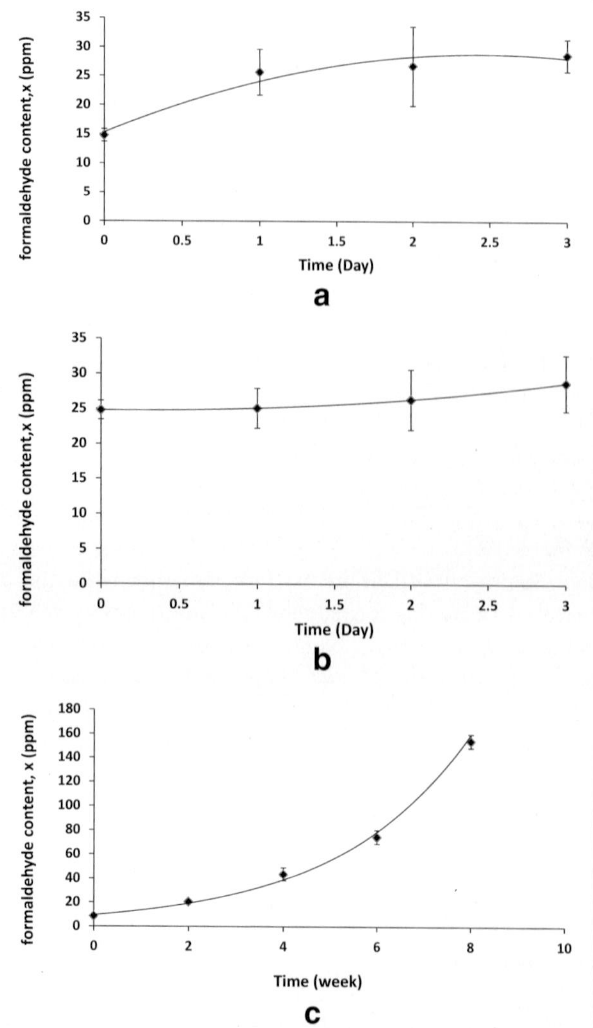

Fig. 6 Time dynamic behavior of naturally occurring formaldehyde content in **a** banana (AAB genome of *Musa* spp.), **b** mandarin and **c** beef ($n = 5$ samples, error bar indicates SD)

It was found that the formaldehyde content in banana sample gradually increased with time (Fig. 6a). The possible reason of the gradual increase in formaldehyde content is the formation of *S*-adenosyl-L-methionine (SAM) during banana ripening, which is associated with endogenous formaldehyde production [24, 41]. Ethylene is produced during the ripening process of banana [42]. SAM, a major methyl donor in cells, is associated with the biosynthesis of ethylene [43, 44]. It has been reported that, during ripening process, the SAM level increases in climacteric fruits [43]. In addition, the pH value of banana sample changes from 4.5 (at $t = 0$) to 5.2 (at $t = 3$ days) during the ripening process, which indicates a decrease in acid content [45].

Mandarin sample also exhibited gradual increase in formaldehyde content with time (Fig. 6b). The possible explanation could be the continuous formation of formaldehyde in acidic condition [46]. Mandarin is a non-climacteric and strongly acidic fruit. So in this acidic condition, acid hydrolysis of N-, O- and S-methoxy compounds takes place and increases the formaldehyde content gradually [47]. Other potential precursors of formaldehyde formation are various sulfur compounds present in fruits, for example, 1,2,4-trithiolane, 1,2,4,5-tetrathiane and dimethyl disulfide which can also undergo degradation to form formaldehyde [48].

A slow increase in formaldehyde content was observed in frozen beef sample (Fig. 6c). Formaldehyde accumulation during the frozen storage of meat could be a possible reason [30, 49]. Formaldehyde might be formed during the aging and deterioration of flesh [50]. It was reported that proteins of muscle undergo chemical and physical changes during frozen storage which may result in, loss of quality, change in flavor, odor and color; most of which changes are caused by the production of formaldehyde in the muscle [51]. It was also reported that the accumulation of formaldehyde and the resulting deterioration of different meat products during frozen storage are primarily caused by the enzymatic activity of trimethylamine oxide aldolase (TMAOase) [49]. The amount of formaldehyde formed depends mainly on the temperature of frozen storage and storing time [52].

Formaldehyde formation from methylated compounds can be represented by the following reaction (Eq. 1) [53]:

$$\text{Methylated compounds} \xrightarrow{\text{Demethylase}} \text{Formaldehyde} \quad (1)$$

$$\begin{array}{lll} \text{At } t = 0 & a & 0 \\ \text{At } t = t & a-x & x \end{array}$$

The order of the kinetics of the above reaction (Eq. 1) can be described as [53]:

$$\text{First order}: \ 2.303 \times \log (a - x) = -Kt + c \quad (2)$$

$$\text{Second order}: \ 1/(a - x) = Kt + c \quad (3)$$

The experimental results of Fig. 6 were plotted following the kinetic equations (Eqs. 2 and 3); it was found that the endogenous formaldehyde formation in banana, mandarin and beef followed second-order kinetics (Eq. 3) with a rate constant (K) 0.2332, 0.083 $ppm^{-1}day^{-1}$ and 0.010 $ppm^{-1}week^{-1}$ (0.0014 $ppm^{-1}day^{-1}$), respectively (Fig. 7a–c). Table 6 represents the rate constants obtained from graphs in tabulated form.

Conclusion

Formaldehyde is naturally present in different food items. It is important to know the concentration of naturally occurring formaldehyde in foods to determine any

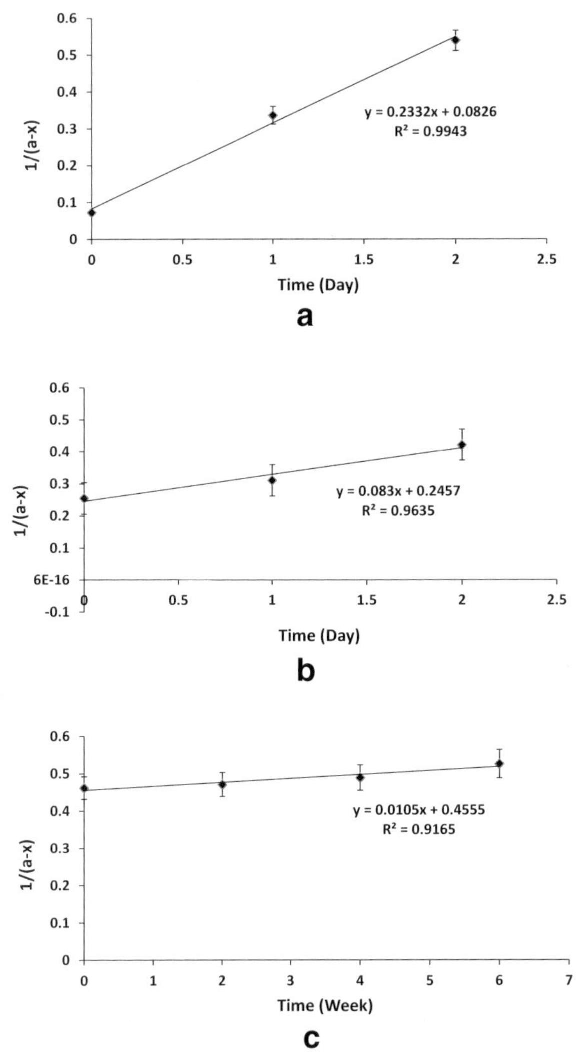

Fig. 7 Kinetics of naturally occurring formaldehyde formation in **a** Banana (AAB genome of *Musa* spp.), **b** mandarin and **c** beef ($n = 5$ samples, error bar indicates SD)

Table 6 Rate constants for banana, mandarin and beef samples

Serial no.	Samples	Rate constant (K; $ppm^{-1}day^{-1}$)
1	Banana (AAB genome of *Musa* spp.)	0.2332
2	Mandarin	0.083
3	Beef	0.0014

external formaldehyde dosage. This study offers baseline data of formaldehyde content naturally found in a wide range of food items: fruits, vegetables, milk and meats. Formaldehyde contents of processed food items, such as: commercially available UHT and powdered milk samples, beverages, and cooked poultry and beef, were also assessed and analyzed. The formaldehyde concentrations of cooked meat samples were found lower than those of fresh meats. The formaldehyde contents of the commercially available milk samples (cow) were found higher than that of pure milk sample. Addition or formation of formaldehyde during milk processing and preservation could be the possible reasons to have high formaldehyde concentrations in commercially available milk samples. Higher formaldehyde concentrations in commercial milk samples are alarming since young population is the major consumer of them. Further study is required to identify the sources of high formaldehyde contents in the commercially available milk items and associated health effects. In addition, the time dynamic behavior of the formation of endogenous formaldehyde in banana, mandarin and beef samples were analyzed. This study demonstrated that the endogenous formaldehyde formation process in banana, mandarin and beef samples followed second-order reaction kinetics. However, the formation behavior of formaldehyde may vary according to food types, storage temperature, storing time and aging pattern of the food items. The above understanding will be useful for the consumers, researchers, legal authorities and other stakeholders working on food safety and preservation.

Authors' contributions

FN carried out a major part of the literature review, experimental research and drafted the manuscript. MNI helped experimental sections and to revise the manuscript. MSK conceived the study, supervised the research project, coauthored and supervised manuscript preparation, and helped to finalize the manuscript. All authors read and approved the final manuscript.

Acknowledgements

This research was supported by BCEF Academic Research Fund, and BUET CASR Research Grant. The authors would also like to thank S. Samira, N. J. Munia, S. Kamal and M. Rahman for their technical assistance and discussion. The research and manuscript are free of conflict of interest.

Competing interests

The authors declare that they have no competing interests.

Funding

BCEF Academic Research Fund, and CASR Research Fund, BUET.

References

1. IARC Working Group on the Evaluation of Carcinogenic Risks to Humans. IARC monographs on the evaluation of carcinogenic risks to humans. Ingested nitrate and nitrite, and cyanobacterial peptide toxins. IARC Monographs on the Evaluation of Carcinogenic Risks to Humans 94 (2010): v.
2. Wilbur S, Harris MO, McClure PR, Spoo W. Toxicological Profile for Formaldehyde. Atlanta, GA: US Department of Health and Human Services. Public Health Service, and the Agency for Toxic Substances and Disease Registry. 1999. http://www.atsdr.cdc.gov/ToxProfiles/tp.asp.

3. Smoke T, Smoking I. IARC monographs on the evaluation of carcinogenic risks to humans. IARC: Lyon; 2004. p. 1–1452.
4. Alam AKMN. Post-harvest fishery losses and mitigation measures. BAU Department of Fisheries Technology; 2014.
5. Bearnot E, Khan MS. The truth about formalin. In: DhakaTribune, July 23, 2014. Dhaka 2014.
6. Mamun MAA, et al. Toxicological effect of formalin as food preservative on kidney and liver tissues in mice model. IOSR J Environ Sci Toxicol Food Technol. 2014;8(9):47–51.
7. Wahed P, et al. Determination of formaldehyde in food and feed by an in-house validated HPLC method. Food Chem. 2016;202:476–83.
8. Jaman Niloy, et al. Determination of formaldehyde content by spectrophotometric method in some fresh water and marine fishes of Bangladesh. Int J Fish Aquat Stud. 2015;2(6):94–8.
9. Noordiana N, Fatimah AB, Farhana YCB. Formaldehyde content and quality characteristics of selected fish and seafood from wet markets. Int Food Res J. 2011;18:125–36.
10. Ma J. Formaldehyde in Noodlefish in Food Safety Focus C.f.f. Safety, Editor. Centre for Food Safety, The Government of the Hong Kong Special Administrative Region: Hong Kong; 2010.
11. Uddin MM, et al. Analyzing time dynamic concentration of formaldehyde in fresh and formalin treated fish 'Labeo rohita'. In: International conference on chemical engineering 2014. Dhaka; 2014.
12. Khan MS, Rahman MS. Introduction. In: Khan MS, Rahman MS, editors. Pesticide residue in foods: sources, management, and control. Cham: Springer; 2017. p. 1–6.
13. Islam MN, Bint-E-Naser SF, Khan MS. Pesticide food laws and regulations. In: Rahman MS, editor. Pesticide residue in foods: sources, management, and control MS Khan. Cham: Springer; 2017. p. 37–51.
14. Formaldehyde and cancer risk; 2011. http://www.cancer.gov/about-cancer/causes-prevention/risk/substances/formaldehyde/formaldehyde-fact-sheet.
15. Songur A, Ozen OA, Sarsilmaz M. The toxic effects of formaldehyde on the nervous system. Rev Environ Contam Toxicol. 2010;203:105–18.
16. Abdu Hussein, Kinfu Yamrot, Agalu A. Toxic effects of formaldehyde on the nervous system. Int J Anat Physiol. 2014;3(3):50–9.
17. Formalin Banning in Europe in 2016. The Molecular Pathology WG: European Union of Medical Specialists; 2016.
18. Islam MN, Mursalat M, Khan MS. A review on the legislative aspect of artificial fruit ripening. Agric Food Secur. 2016;5(1):8.
19. Amit SK, et al. A review on mechanisms and commercial aspects of food preservation and processing. Agric Food Secur. 2017;6:51.
20. Guyomard H, et al. Eating patterns and food systems: critical knowledge requirements for policy design and implementation. Agric Food Secur. 2012;1(1):13.
21. Dutilleul FC. The law pertaining to food issues and natural resources exploitation and trade. Agric Food Secur. 2012;1(1):6.
22. Islam MN, et al. A legislative aspect of artificial fruit ripening in a developing country like Bangladesh. Chem Eng Res Bull. 2015;18(1):30–7.
23. Liteplo RG, Beauchamp R, Meek ME, Chenier R. Concise international chemical assessment document 40: Formaldehyde. World Health Organization: Geneva, Switzerland; 2002.
24. TreÂzl L, et al. Endogenous formaldehyde level of foods and its biological significance. Z Lebensm Unters Forsch A. 1997;205:300–4.
25. World Health Organization (WHO). Organic pollutants: formaldehyde. In: Theakston F (editor) Air quality guidelines for Europe, Chapter 5. Copenhagen, Denmark: WHO Regional Office for Europe; 2001, p.1-25.
26. Castell CH, Smith Barbara. Measurement of formaldehyde in fish muscles using TCA extraction and Nash reagent. J Fish Board Can. 1972;30:91–8.
27. European Food Safety Authority. Endogenous formaldehyde turnover in humans compared with exogenous contribution from food sources. EFSA J. 2014;12(2):1–11.
28. Services DoHaH. Public health statement-formaldehyde. Agency for Toxic Substances and Disease Registry; 2008.
29. Food Preservatives Committee. Formaldehyde in food. British Med J. 1924;2(3320):289–90.
30. Tang Xiaojiang, et al. Formaldehyde in China: Production, consumption, exposure levels, and health effects. Environ Int. 2009;35:1210–24.
31. Hossain MM. Consumption in rural Bangladesh: households, lifestyles, and identities. Helsinki: Department of Economics and Management, University of Helsinki; 2012.
32. Cheng Yung-Sung, et al. Chemical composition of aerosols from Kerosene heaters burning jet fuels. Aerosol Sci Technol. 2001;35(6):949–57.
33. Barry JL, Tomé D. Formaldehyde content of milk in goats fed formaldehyde-treated soybean oil-meal. Food Addit Contam. 1991;8:633–40.
34. Ahmed KMF, et al. Detection of some chemical hazards in milk and some dairy products. Afr J Food Sci. 2015;9(4):187–93.
35. Awan Adeela, et al. A study on chemical composition and detection of chemical adulteration in tetra pack milk samples commercially available in Multan. Pak J Pharm Sci. 2014;27(1):183–6.
36. Consumer Report. Formalin in Nestle Pakistan milk. In: The Network for Consumer Protection. December 13, 2016, 017/06. Islamabad. 2006.
37. Buckley Katherine E, Fisher Lorne J, MacKay VG. Levels of formaldehyde in milk, blood, and tissues of dairy cows and calves consuming formalin treated whey. J Agric Food Chem. 1988;36(6):1146–50.
38. Wiglusz R, Grazyna Nikel ES, Jarnuszkiewicz I, Igielska B. The effect of temperature on the emission of formaldehyde and volatile organic compounds from laminate flooring-case study. Build Environ. 2000;37(1):41–4.
39. Mason DJ, et al. Determination of naturally-occurring formaldehyde in raw and cooked Shiitake mushrooms by spectrophotometry and liquid chromatography-mass spectrometry. Taylor Francis Food Addit Contam. 2004;21:1071–82.
40. Shibamoto T. Formaldehyde in Coffee. In: Linskens H-F, Jackson JF (editors) Analysis of Nonalcoholic Beverages. Berlin, Heidelberg: Springer; 1988. p. 173-183.
41. Huszti Z, Tyihik E. Formation of formaldehyde from S-adenosyl-L-[methyl-3-H]methionine during enzymic transmethylation of histamine. Fed Eur Biochem Soc. 1986;209(2):362–6.
42. Bouzayen M, et al. Mechanism of fruit ripening. In: Pua EC, Davey MR, editors. Plant developmental biology—biotechnological perspectives. Berlin: Springer; 2010.
43. Van de Poel B, et al. S-adenosyl-L-methionine usage during climacteric ripening of tomato in relation to ethylene and polyamine biosynthesis and transmethylation capacity. Physiol Plant. 2013;148(2):176–88.
44. Van de Poel B, et al. Determination of S-adenosyl-L-methionine in fruits by capillary electrophoresis. Phytochem Anal. 2010;21(6):602–8.
45. Prasanna V, Prabha TN, Tharanathan RN. Fruit ripening phenomena—an overview. Crit Rev Food Sci Nutr. 2007;47(1):1–19.
46. Yamazaki H, Ogasawara Y, Sakai C, Yoshiki M, Makino K, Kishi T, Kakiuchi Y. Studies on formaldehyde in Lentinus edodes. J Food Hyg Soc Jpn. 1980;21:165–70.
47. Tyihak E, Blunden G, Yang M-H, Crabb TA, Sardi E. Formaldehyde, as its dimedone adduct, from Ascophyllum nodosum. J Appl Phycol. 1996;8:211–5.
48. Yasumoto K, Iwami K, Mitsuda H, et al. A new sulfur-containing peptide from Lentinus edodes acting as a precursor to lenthionine. Agric Biol Chem. 1971;35:2059–69.
49. Norliana S, Abdulamir AS, Abu Bakar F, Salleh AB. The health risk of formaldehyde to human beings. Am J Pharmacol Toxicol. 2009;4(3):98–106.
50. Jianrong L, Junli Z, Lifang Y. Determination of formaldehyde in squid by high-performance liquid chromatography. Asia Pac J Clin Nutr. 2007;16:127–30.
51. Trezl L, Pipek J. Formation of excited formaldehyde in model reactions simulating real biological systems. J Mol Struct (Theochem). 1988;170:213–23.
52. Sotelo CG, Pineiro C, Perez-Martin RT. Denaturation of fish protein during frozen storage: role of formaldehyde. Z Lebensm Unters Forsch. 1995;200:14–23.
53. Bahl A, Bahl BS, Tuli GD. Essentials of physical chemistry. New Delhi: S. Chand and Company; 2009.

Agricultural mechanization, a key to food security in developing countries: strategy formulating for Iran

Mohammad Emami[1], Morteza Almassi[1*], Hossein Bakhoda[1] and Issa kalantari[2]

Abstract

Background: Mechanization is a multi-dimensional concept and widely used in agriculture. There is, however, a major difference between the application of mechanization in developed and developing countries The developing countries tend to design their own strategies in food security given the challenges they face in all aspects of their economy including feeding a growing population, reducing poverty, protecting the environment, managing the effects of climate change and fighting malnutrition all which may further contribute to a reduction in economic growth and political instability. The goal of the strategies, with the help of appropriate technologies, is to lead to a sustainable agricultural development and, ultimately, food security. The policy making in each country should be based on its own conditions. This article uses Iran as an example of a developing country and considers the country's specific climate as well as political and economic conditions to present development-oriented policies for achieving sustainable food security based on agricultural mechanization that may be adaptable to other developing countries. The main objective of this paper is to identify and provide guidelines to the current and future challenges of Iran's food security, and it argues that for any strategy to succeed in producing a sustainable agricultural production, it will need a proper analysis and a formulation of an appropriate mechanization plan.

Methods: To achieve the objective of a self-sustaining agricultural mechanization strategy, a SWOT analysis technique was used to identify the strengths, weaknesses, opportunities and threats and help us provide a framework by which policies can be defined. The framework includes internal and external factors that affect the development of agricultural mechanization and seek to provide ideas for agricultural development with the help of mechanization. These factors were then prioritized using the Hierarchical Analysis Method, and based on the obtained results, the final strategies were extracted and prioritized by the Technique for Order of Preference by Similarity to Ideal Solution (TOPSIS), a multi-criteria decision analysis method.

Results and conclusion: According to the results, weaknesses and threats were the most important factors. Environmental threats, especially water shortages, economic problems as well as availability of the mechanization fleet and compatibility of the equipment within the country's agricultural system were identified as the most important factors affecting the agricultural development. In order to achieve sustainable food security, with regards to the identified factors the necessary recommendations and Governmental-support policies in the agricultural sector were presented: 1. reforming the country's planting pattern according to the climatic conditions considering the relative advantage of agricultural production, 2. modernizing the mechanization fleet; 3. investing in research and development of agriculture and modern knowledge; the production and import of agricultural machinery and modern technologies according to their suitability with the country's conditions and future needs of the country; strengthening of the supply chain and maintenance services.

*Correspondence: morteza.almassi@gmail.com
[1] Department of Agricultural Mechanization, Science and Research Branch, Islamic Azad University, Tehran, Iran
Full list of author information is available at the end of the article

Keywords: Food security, Agricultural mechanization, Policy making, Strategy formulating, Developing countries, Iran

Background

Agricultural mechanization today has a very broad meaning. This broad meaning includes production, distribution and utilization of a variety of tools, machinery and equipment for the development of agricultural land, planting, harvesting and primary processing [3, 15, 19, 25]. Today, the debate on development of agricultural mechanization turns into the debate on improving of agricultural techniques as well as helping them improve the sustainability of the entire agricultural system [19]. Evidence suggests that mechanization has a major impact on demand and supply of farm labor, agricultural profitability, and a change in rural landscape [24] and can be defined as an economic application of engineering technology to increase the labor efficiency and productivity. The United Nations Food and Agriculture Organization (FAO) and the United Nations Industrial Development Organization (UNIDO) concluded that the goal of agricultural mechanization is to reduce labor. Increasing productivity by updating executive operations in order to gain more power, increasing the level of cultivated land, moving toward industrialization and strengthening the market for rural economic growth and ultimately improving the livelihoods of farmers are the goals of mechanization [11, 12, 15].

In the pre-industrial stage in Western countries, one of the strategies to increase agricultural production was mechanization. At this stage, the agricultural sector used high-capacity machinery for crop operations that were suitable for large land and replacement of labor. History shows that agricultural mechanization has led to rapid industrialization in the western hemisphere. More recently in the twenty-first century, many Asian countries have embraced this western thinking and implemented mechanization policies in accordance with their own particular circumstances [5]. Mechanization technology changes with industrial growth in the country and economic and social progress of the farmers. While the loss of interest in agriculture by land owners and the lack of access to agricultural labor force for farm operations are among the most important social and economic issues in highly industrialized countries, increasing the area of cultivation and increasing labor productivity are the requirements of mechanization in developing countries. Therefore, mechanization technology requires dynamic and regional conditions [27]. For example, mechanization in countries such as the USA and Canada has dramatically changed from the perspective of cultivating based on the time of initial deployment, but in many developing countries, agriculture still has a strong dependence on labor [24]. To this end, developing countries, on their path to achieving food security, need to design their own strategies for agriculture.

Achieving food security in an environmentally sustainable way is one of our greatest challenges [4]. To this end, policies must be selected that, with the help of appropriate technology, will lead to sustainable development of agricultural production in developing countries and ultimately will lead to sustainable food security.

World food security and developing countries

In the first half of the twenty-first century, the world faces numerous challenges to feed the growing population, reduce poverty, protect the environment and face climate change. These challenges can sustain hunger and malnutrition, reduce economic growth—lead to political instability and irreversible damage to the environment and human survival [19, 28]. The latest estimate of hunger is 795 million people [14], and as the world's population grows to over 9 billion people by 2050, it is anticipated that the need to meet the ever-increasing demand for more food is urgent and necessary [26]. As some sources predict, global food demand for that year will be twice as much as it is now [16]. As productivity growth is not enough in many developing countries to respond to rising demand, imported food in these countries are expected to increase dramatically [23]. Today, despite considerable investment in development, food insecurity is widespread throughout East and West Africa [22]. In Sub-Saharan Africa, we have witnessed an increase in imports over the last forties, where food products have not been responsive to growing populations [13]. Also, the Middle East, as one of the developing regions of the world, must cope with many of the unintended and unforeseen consequences of modernization. Increasing birth rate and rapidly growing population in the area, paying more attention to urbanization among policymakers and neglecting rural development, because of increasing rural-to-city migration, unbridled expansion of governments, etc., are among these consequences. The development of agriculture in this area has not been sufficient to provide a satisfactory level of national supply for achieving food security and the availability of domestic food resources [1]. On the other hand, food security in many countries of the region, such as Iraq, is heavily influenced by the oil-based economy, over three decades of war and its policies [31]. From Iran in the east to Morocco in the west, the Middle East has witnessed a deepening dependence on food imports, and this trend continues at a high rate. However, the Arab Spring showed that availability of food at affordable prices is still a prominent feature of many economic, social and political perspectives in the Middle East [1]. On the other hand, the World Food Crisis (2006–2006) has led many countries to reassess their dependence on imports for a significant portion of their food consumption [9]. These countries imposed trade restrictions and moved toward self-sufficiency. Due to the high prices and fluctuations of the market during this period, it was obvious that these countries would be concerned about the excessive dependence on the world food markets and take on such policies.

But on the contrary, some policymakers believe that food security will be achieved by considering comparative advantage in agricultural production, reducing trade restrictions and strengthening and development of international food trade [7]. This article, based on evidence from Iran, provides policies for achieving sustainable food security in developing countries. These development policies are based on agricultural mechanization.

Iran is the second largest country in the Middle East, both in terms of size and population. A quarter of Iran's population lives in rural areas (Fig. 1). And 21.2% of the total working population of the country are employed in the agricultural sector [10]. Despite having oil revenues, agriculture plays an important role in the economy of Iran. Despite the declining share of agricultural sector in gross domestic product (GDP), this figure is estimated at about 12% in 2017 (Fig. 2).

Agricultural development has long been at the heart of Iran's food and agricultural policies. But statistics show the deep dependence of food security on imports. For example, the dependence ratio for grain imports in Iran is 28.7% in 2014, and the country's food imports are still high (Fig. 3). The exploitation of water resources in Iran, with more than 70% above the global average, is inefficient [17]. Meanwhile, 74% of Iran's land, about 120 million hectares, is inappropriate for agriculture [18]. It is expected that in the years to come, there will not be enough water even for such a small amount of agricultural land. These conditions along with its political and international issues and its foreign policies, make the future of agriculture and food security in Iran more difficult and challenging. Under these circumstances, Iran must identify current challenges and focus on the future of food and agriculture.

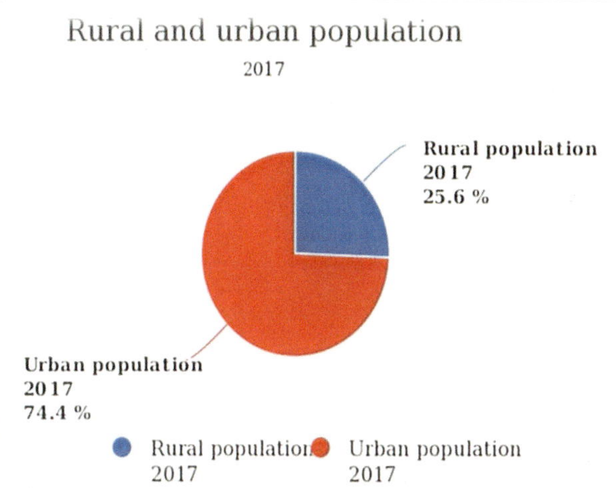

Fig. 1 Rural and urban population of Iran

Agricultural mechanization strategy and food security

Considering the necessity of identifying and responding to the current and future challenges of food security in order to lead, collaborate and design the related strategies [30], this paper, by examining the current state of Iran and identifying challenges and threats, seeks to provide strategies to promote the role of agricultural mechanization in agricultural development as a solution to the achievement of food security.

Any attempt to increase agricultural production without considering a proper mechanization strategy would never have a positive outcome [20]. A sustainable agricultural mechanization strategy is a planning strategy that contributes to the goal of sustainable agriculture, and at the same time accepts food self-sufficiency and generates economic and inclusive growth as well as social benefits [19].

Increasing food production along with maintaining natural resources is not an easy task. The second green revolution, which in the second half of the last century was able to produce more than twice the amount of food, is currently not in a good position. The growth rate of major cereal yields (wheat, rice and corn) is declining. Increasing food production requires resource-friendly methods, and this will require the development of new mechanization technology [26]. There are many technology options, but there is relatively scarce evidence of supporting decision making in the form of technology [23]. While one of the major constraints on developing and modernizing production in developing countries is the low level of engineering technology in agriculture [12]. Despite its many values and undeniable benefits, which are briefly mentioned above, mechanization is still considered as an input, such as other inputs, like fertilizers, seeds, and chemical protection products and in the most optimistic way, it is one of the combined management tools aimed at maximizing farmers' productivity and profits [8]. Despite the impact of agricultural

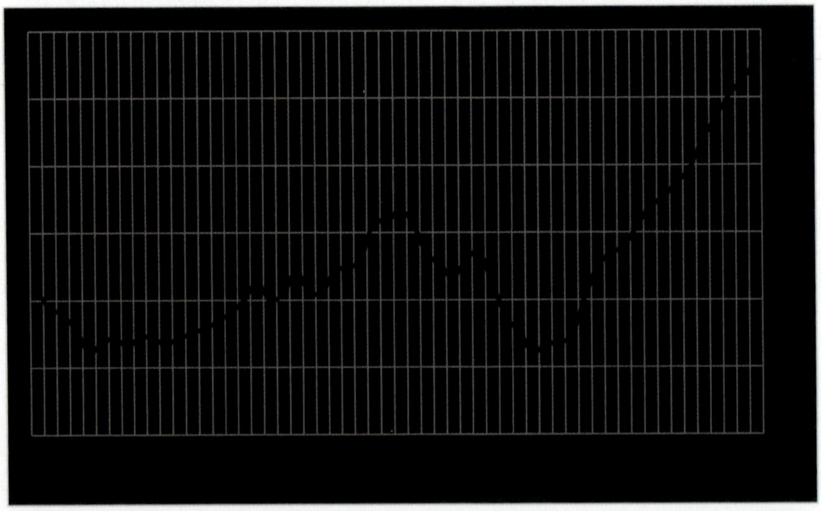

Fig. 2 Agricultural value added. Source: World Bank (OCT 28, 2017)

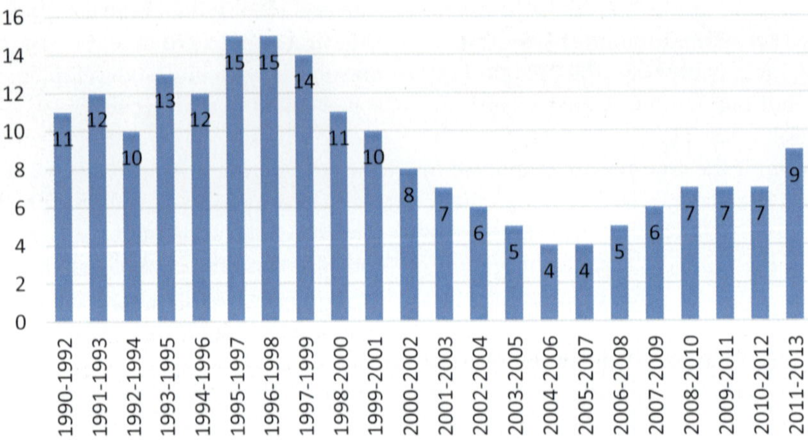

Fig. 3 Value of food imports

mechanization on the agricultural production process, due to the lack of comprehensive research on the extent of the effect of mechanization components on the production and its fields of application, there were always differences of opinion and doubts about the role of mechanization in the agricultural sector. This has prevented the stabilization and improvement in agricultural mechanization in the agricultural sector and even in the sectors of industry and services in developing countries. For example, the FAO report on the state of mechanization in the African countries [12] indicates that most African countries have not taken serious plans for sustainable mechanization, and efforts have failed to reach mechanization issues due to the micro, rather than macro, approaches. Studies show that crop production is more beneficial in areas where agricultural mechanization is provided [15]. According to the research conducted in mechanized agricultural areas, yields were significantly higher than non-mechanized areas. The use of pesticides was also more effective, and the land and fertilizers were used optimally. This is the need of the world today, achieving food security by preserving natural resources for future generations.

Methods

The purpose of agricultural policy is to develop appropriate and sustainable guidelines for promotion of efficient agricultural practices that guarantee food security, generate employment for citizens, provide raw materials for all agricultural industries, and also to acquire foreign currency. Agriculture can work with other sectors of the economy to achieve faster development, poverty reduction and environmental sustainability [3, 6, 20]. This usually requires upgrading of infrastructure and injection of technology to advance production from the early stages of agriculture to mechanical systems through a suitable political framework [3]. Also, economic development studies have focused on the importance of addressing the fundamental constraints on elimination of poverty in general and improvement in the performance of smallholder farmers. Many proven technologies and planting improvement practices have promise to boost agricultural production and reduce poverty in developing countries [21]. Here is where agricultural mechanization plays its role, and therefore developing countries need to design a new strategy based on mechanization. Various activities involved in the formation of a strategy can be divided into three stages: (1) status analysis; (2) strategy development; and (3) strategy approval [12]. All three stages were conducted in this research. The selection of appropriate policies to promote the role of agricultural mechanization in agriculture to achieve food security is the ultimate goal of this research. Policies create the parameters in which development programs are implemented. The strategy determines what resources should be available for development programs and how to identify, mobilize, deploy and maintain these resources for implementation of these programs [20]. On the other hand, mechanization planning requires a quantitative evaluation of the mechanization index and its impact on agricultural production (yield) and economic factors (cost of cultivation, deployment of harmful and mechanical force and economic advantage) [27]. In order to achieve this goal, considering strengths, weaknesses, opportunities and threats (SWOT), a matrix was created as a political and analytical framework. The framework includes internal and external factors that influence the development of agricultural mechanization and seeks to provide ideas for agricultural development with the help of agricultural mechanization. These factors were prioritized by AHP hierarchical analysis method. At the next stage, the strategies were extracted according to the obtained results. In the final section, strategies were prioritized by TOPSIS method.

Study design and data collection

In this research, data were collected from a combination of primary and secondary sources to analyze the role of agricultural mechanization in agricultural development and the achievement of sustainable food security in developing countries. The preliminary data were taken from the Ministry of Agriculture's official statistics, Statistical Centre of Iran, the Agricultural Mechanization Development Center, as well as the Ministry of Commerce of Iran. Secondary data were collected from the stakeholders during the interviews. Experts were selected using the Sampling Snowball method. This technique is a sequential targeted sampling method that can be used in a multi-criteria decision-making process in which the researcher, after identifying the individuals introduced by an expert, asks them to introduce one or more other specialists. These experts included government employees in the areas of planning, implementation, monitoring and evaluation of mechanization, private sector, research and educational institutions, university professors and Ph.D. students in mechanization. Using questionnaires distributed among them, internal factors (strength and weakness) and external factors (opportunities and threats) were identified. In this study, 22 factors were agreed as internal and external factors. By identifying strategies and internal and external factors, comparative questionnaires were designed to compare the factors together. A pairwise comparison was performed by hierarchical analysis method (AHP). A binary comparison was performed based on scale of 1–9. The principles of this comparison scale are shown in Table 1 [29].

Results

Identifying internal and external factors

According to the results obtained and the findings of previous researches of experts, and based on the designed questionnaire, they identified weaknesses, strengths, opportunities and threats (Table 2).

Twenty-two factors were identified as internal and external factors. Disturbances in the form of macroeconomic factors on one hand and environmental constraints on the other especially the water crisis, are among the most important factors identified along with the status of Iran's agricultural mechanization fleet.

Prioritizing the factors

At this stage of the research, the identified internal and external factors were prioritized by hierarchical analysis method (Table 3).

Based on the findings of this section, the weaknesses were identified as the most effective factor compared to other factors with an importance coefficient of 0.467. The threats were identified as the next effective factor, with

Table 1 The fundamental scale of AHP

Score	Definition
1	Equal importance
3	Weak importance of one over another
5	Essential or strong importance
7	Very strong and demonstrated
9	Absolute importance
2, 4, 6, 8	Intermediate values between adjacent scale values

Table 2 The summary of SWOT analysis results of development agricultural mechanization in Iran

Internal factors	
Weaknesses	**Strengths**
1. Lack of proper distribution and the heterogeneous amount of machinery and technology in the agricultural sector	1. Ability of cultivating strategic products in many parts of the country
2. Disproportion size of agricultural land and existing agricultural machinery	2. Dependence of the production of these products on energy inputs
3. Use of amortized machines and consequently increased waste	3. Growing capacity of country mechanization
4. Lack of correlation between the increase in Indicators of machinery and the reduction of waste	4. Significant correlation between mechanization capacity index and increase in wheat production
5. Gradual growth of the agricultural sector compared to other economic sectors of the country	5. Increasing trend and moving toward increasing and improving energy indicators
External factors	
Threats	**Opportunities**
1. High cost of ownership for beneficiaries	1. Climate diversity
2. High dependence on fossil fuels	2. Suitable market for exporting to countries of the region
3. High cost of exchanging rate and impact on the economy	3. Large number of graduates in agriculture
4. Lack of interest in investing in the manufacturing sector	4. Possibility of modifying seeds and using biotechnology in production
5. Growing trend in the service sector and the downward trend in manufacturing sectors	5. Major factories of agricultural machinery production in the country and the history of decades of manufacturing and exporting machinery
6. Expanding urbanization and lack of interest in engaging in agricultural activities	
7. Water bankruptcy of the country	

Table 3 Pairwise comparison of SWOT factors

	Strength	Weakness	Opportunity	Threat	Importance degree of SWOT factors
Strength	1	0.333	2	0.5	0.160
Weakness	3	1	4	2	0.467
Opportunity	0.5	0.25	1	0.333	0.095
Threat	2	0.5	3	1	0.277
					CI=0.01

an importance coefficient of 0.277, and subsequently, the strengths and opportunities were prioritized, respectively, with importance coefficients of 0.160 and 0.095.

These results indicate that the weaknesses and threats are more important in the current challenges of Iran compared to the strengths and the opportunities, leading Iran to the more defensive or more conservative strategies.

Due to the higher coefficient of importance of weaknesses in comparison to threats, it is necessary to pay more attention to internal factors in the agricultural sector and mechanization of Iran rather than external factors.

Prioritizing the sub-actors

Prioritizing the sub-factors based on internal and external factors was the next step in this study. At this stage, sub-factors were prioritized in four groups and were compared to other sub-factors (Table 4).

The weaknesses SWOT factor has the most importance among internal and external factors, and the most important sub-factor is the lack of proper distribution and the heterogeneous amount of machinery and technology to the agricultural sector. The importance coefficient of this sub-factor is estimated at 0.427. This suggests that despite the high agricultural mechanization coefficient in Iran, distribution of machinery in this country has not been done properly. Also, many new technologies in the agricultural sector have not been properly utilized. The second weakness, which earns the second rank, with an importance coefficient of 0.236, also emphasizes this issue. The disproportion size of agricultural land and existing agricultural machinery is identified as the second most important sub-factor. These sub-factors both

Table 4 Relative priority of each SWOT sub-factors

S	S1	S2	S3	S4	S5			Relative priority
S1	1	5	2	3	4			0.420
S2	0.2	1	0.25	0.5	0.5			0.067
S3	0.5	4	1	2	3			0.264
S4	0.33	2	0.5	1	2			0.148
S5	0.25	2	0.33	0.5	1			0.099
								CI=0.01
W	W1	W2	W3	W4	W5			
W1	1	2	4	5	3			0.427
W2	0.5	1	2	3	2			0.236
W3	0.25	0.5	1	2	0.5			0.109
W4	0.2	0.33	0.5	1	0.5			0.072
W5	0.33	0.5	2	2	1			0.153
								CI=0.01
O	O1	O2	O3	O4	O5			
O1	1	4	2	3	0.5			0.266
O2	0.25	1	0.33	0.5	0.166			0.059
O3	0.5	3	1	2	0.5			0.174
O4	0.33	2	0.5	1	0.25			0.097
O5	2	6	2	4	1			0.401
								CI=0.01
T	T1	T2	T3	T4	T5	T6	T7	
T1	1	5	2	3	4	3	0.5	0.232
T2	0.2	1	0.25	0.33	0.5	0.33	0.142	0.036
T3	0.5	4	1	1	3	2	0.33	0.132
T4	0.33	3	1	1	3	2	0.33	0.121
T5	0.25	2	0.33	0.33	1	0.5	0.166	0.0509
T6	0.33	3	0.5	0.5	2	1	0.25	0.081
T7	2	7	3	3	6	4	1	0.343
								CI=0.01

underline the poor distribution of agricultural machinery in Iran, and suggests that the distribution of machinery across the country is not proportionate to the size of Iran's agricultural land. After the land reforms at the time of the former regime in Iran, the division of agricultural land has been one of the issues that policymakers point to as a barrier to Iranian agricultural development. This problem however can be resolved by planning and subsequent distribution of machinery, appropriate to the demand. The gradual growth of the agricultural sector, in comparison with other economic sectors of the country, has been ranked third, with an importance coefficient of 0.153. In recent years, development of the various economic sectors of Iran has not been coordinated. The service sector in Iran has had the highest development rate, and after the agreement on the nuclear program between Iran and the P5 + 1, the oil sector has been well developed. The agricultural sector and the industrial sector did not have significant development and have a small share of gross domestic product. The two following sub-factors are related to the amount of waste, as well as the contribution of mechanization to reducing this amount. Given the role that mechanization plays in different stages of production, it can have an effective contribution to reducing agricultural waste. The statistics in Iran however, indicate low correlation between mechanization and reduction of agricultural waste. The use of amortized machines, lack of scheduled plans and lack of attention to timeliness costs are among the reasons for this low correlation. Statistics show that the amount of agricultural and food waste in Iran is very high. Correct management of agricultural mechanization can play an important role in reducing waste and, consequently, in achieving food security. Threats, after weaknesses, rank second in terms of importance (Table 3). Among the sub-factors in this group, water bankruptcy, with an importance coefficient of 0.343, has the highest rank. Iran has been in crisis for many years and today the crisis has exceeded the warning threshold. The continuation of this process is a threat not only to the agriculture of Iran, but also to its civilization. After the water crisis, the High cost of ownership for beneficiaries is in the second place, while the high cost of exchange rates in the economy of Iran is in the third place. These sub-factors both will ultimately lead to a low level of investment in the agricultural sector. This factor, with an importance coefficient of 0.121, ranks next. Dependence on fossil fuels, urbanization and development of the service sector and declining trend of manufacturing sectors are among the following sub-factors.

The third factor is the Strengths, with an importance coefficient of 0.160. In terms of importance of the sub-factors, the ability of cultivating strategic products in many parts of the country, with an importance coefficient of 0.420, is ranked first. The growing capacity of the country mechanization is ranked second, and the sub-factor of the significant correlation between mechanization capacity index and increase in wheat production is ranked third. The two following sub-factors are related to energy indicators. Increasing trend and moving toward increasing and improving energy indicators, with an importance coefficient of 0.099, is in the fourth place and the dependence of the production of these products on energy inputs is in the fifth place. The increasing trend of energy indicators on one hand and the dependence of production on these indicators on the other indicate that Iran should maintain the status quo and try to improve these indicators.

In the ranking of opportunities sub-factors in Iran, major factories of agricultural machinery production in the country and the history of decades of manufacturing and exporting machinery, with an importance coefficient of 0.401, ranked first. The country's climate diversity ranked second. Iran is a country with four seasons throughout the year. This feature provides a decent opportunity for policymakers to have more options to manage the food market in the country. The large number of graduates in agricultural, with an importance coefficient of 0.174, is the third most important sub-factor. The large number of graduates gives Iran the opportunity to make its agriculture more scientific, as an example, Iran can take steps in the field of biotechnology and seed improvement. The possibility of modifying seeds and using biotechnology in production is in the fourth place. This sub-factor is reinforced with the help of the former sub-factor, the large number of experts in Iran. The last sub-factor is the suitable market for exporting to countries of the region, which Iran can plan for it as an appropriate opportunity.

Extracting strategies

After studying internal and external factors, strengths, weaknesses, opportunities and threats and their ranking, finally, according to the political, economic, social and environmental structure of the country, eight final strategies were proposed by the experts (Table 5).

The extracted strategies were prioritized using the TOPSIS method. At first, each of the eight strategies was ranked based on the sub-factors (Table 6) and finally they ranked accordingly (Table 7).

Discussion

According to Table 7, Injecting proper financial investment into the agricultural sector, with an importance coefficient of 0.194, was ranked first. Despite its prominent role in achievement of food security, Iran's agricultural sector is considered the weakest among the economic sectors. Low development of this sector over the years and lack of interest in investing in it is alarming. As such policymakers should prioritize strategies to strengthen this sector and support it. Earlier, FAO, in a report reviewing the state of agricultural mechanization in African countries [12], presented strategies to achieve lasting effects, such as political will and commitment at the highest possible level, allocating funds to the agricultural sector and providing appropriate loans to small farmers. These proposals were presented by FAO, due to the weakness of the private sector and the lack

Table 5 Suggested strategies

Suggested strategies	
1. Considering the relative advantages of producing and planning for the development of food and international trade	5. Producing or importing agricultural machinery and new technologies, according to their suitability with the conditions of the country, along with strengthening of the supply chain and maintenance service
2. Energy management, moving toward clean energies and improving energy indicators	6. Investing in agricultural research and development and the use of biotechnology and modern knowledge
3. Modifying consumption patterns by replacing high-yield products in the household basket	7. Modifying Iran's cultivation pattern based on the state of the country's production resources with attention to the state of the country's water resources
4. Injecting proper financial investment to the agricultural sector	8. Employing graduates directly in the agricultural sector

Table 6 Overall priority of each SWOT sub-factors of strategies by TOPSIS analysis

	S1	S2	S3	S4	S5	W1	W2	W3	W4	W5	O1	O2	O3	O4	O5	T1	T2	T3	T4	T5	T6	T7
A1	0.605	0.043	0.296	0.166	0.095	0.598	0.110	0.102	0.238	0.215	0.228	0.05	0.066	0.083	0.038	0.064	0.08	0.036	0.033	0.014	0.022	0.858
A2	0.538	0.097	0.338	0.19	0.142	0.598	0.22	0.154	0.17	0.431	0.101	0.034	0.049	0.055	0.23	0.064	0.090	0.11	0.101	0.056	0.022	0.095
A3	0.201	0.043	0.169	0.095	0.079	0.998	0.441	0.154	0.136	0.647	0.178	0.05	0.133	0.074	0.268	0.581	0.04	0.33	0.304	0.127	0.18	0.286
A4	0.403	0.086	0.296	0.166	0.095	1.397	0.882	0.154	0.306	0.647	0.076	0.039	0.133	0.065	0.345	0.581	0.08	0.293	0.304	0.127	0.045	0.286
A5	0.067	0.032	0.253	0.143	0.047	1.197	0.661	0.051	0.204	0.431	0.127	0.045	0.116	0.027	0.23	0.258	0.02	0.036	0.033	0.042	0.067	0.476
A6	0.336	0.053	0.169	0.095	0.047	0.598	0.992	0.154	0.272	0.215	0.076	0.022	0.116	0.027	0.076	0.516	0.09	0.036	0.033	0.014	0.022	0.858
A7	0.201	0.064	0.338	0.190	0.063	1.197	0.661	0.154	0.238	0.359	0.203	0.017	0.133	0.065	0.191	0.193	0.06	0.22	0.101	0.084	0.203	0.381
A8	0.403	0.043	0.253	0.143	0.047	0.598	0.441	0.205	0.136	0.431	0.152	0.045	0.149	0.055	0.230	0.258	0.03	0.11	0.033	0.042	0.158	0.476

Table 7 Overall priority of each strategy

Strategy	Overall priority	Ranking result
A1	0.130	4
A2	0.076	8
A3	0.099	6
A4	0.194	1
A5	0.136	3
A6	0.117	5
A7	0.163	2
A8	0.081	7

of government intervention agricultural investment and commerce, which seems to be in line with the situation in Iran.

Modifying Iran's cultivation pattern based on the state of the country's production resources with attention to the state of the water resources is ranked second with an importance coefficient of 0.163. The severe water crisis in Iran, or in other words, bankruptcy of water, has put the country in a difficult position. According to this situation, Iran must not only take more water-efficiency measures, but must also reform its cultivation pattern based on the need for water for agricultural products and available water resources. And, if necessary, implement "not to do farming" Project in the required areas.

Producing or importing agricultural machinery and new technologies, according to their suitability with the conditions of the country, along with strengthening of the supply chain and maintenance service is ranked third with an importance coefficient of 0.136. The need for investment in the supply chain of machinery as a short-term solution, along with technical training and proper management for businesses and service and maintenance, are among the suggestions of researchers [26]. The use of machines that are unsuitable for the country's condition, on one hand, and the worn-out fleet of agricultural machinery in many parts of the country on the other, makes it necessary to modernize the fleet and to distribute it more appropriately throughout the country, which is possible due to the existence of tractor and combine factories in Iran and their history.

Considering the relative advantages of producing and planning for development of food and international trade is ranked third with an importance coefficient of 0.130. Food security is one of the responsibilities of the government of each country. One of the most important dimensions of food security is food availability. The government is required to distribute a decent number of calories in the country based on the nutritional needs of its population. This amount of energy will be supplied either through domestic production or through imports. According to this strategy, it is recommended that Iran, to establish this dimension of food security, considers the relative advantages of production as its agenda. If a product or several products have no comparative advantage in production, instead of spending resources for its production, the country should strengthen its international trade by importing the needed products and avoid the problem of this dimension of food security, and at the same time, use its resources appropriately to maximize productivity in production.

The next strategy is investing in agricultural research and development and the use of biotechnology and its new knowledge. This strategy ranked with an importance coefficient of 0.117. Due to the limitation of arable land in Iran, the country must move toward increasing productivity per unit area. New technologies and biotechnology are among the things that make this possible for each country and for this reason, Iran should invest in agricultural research and development. Seed modification, genetic modification, etc., are among the many ways to develop agriculture and achieve sustainable food security. Researchers believe that the development of food security is possible through the development and promotion of technology. These achievements are mainly attributed to increased use of crop improvement methods. These results highlight the role of improved methods for promoting agricultural productivity among small farmers [21]. Researchers are also strategizing with emphasis on technology transfer and planning for the development of agricultural mechanization. It is suggested that the first step is to remove the obstacles that mechanization faces including the fact that small farmers do not have the financial ability to move to mechanized cultivation require government support. The government support at the executive and legislative levels, supporting academic education, providing appropriate technology and services, and moving toward the assembly and production of agricultural machinery are among these proposals [20]. The emphasizing on research and development, the selection and transferring appropriate technologies for the region, optimal use of energy resources, increasing the number of machinery, training, providing loans to farmers, as well as providing tax rebates are among the proposed policies in this area [2]. Iran also can take steps to maximize the productive use of resources using precision farming technology, given the country's water crisis. Modifying consumption pattern by replacing high-yield products in the household basket is the sixth ranked strategy that is related to the dimension of food utilization. Modifying this pattern requires long-term cultural and social planning. Dietary habits in Iran are not only rooted in different cultures in this country, but also closely related to religious beliefs. To achieve this, various government and community-based organizations, such as the media, social networks and NGOs, can play a significant role. Employing graduates directly in the agricultural sector, with a 0.081 significance factor, is the next strategy. Iran has a large number of agricultural graduates and has a major need for agriculture production based on science and technology, which has been addressed in previous strategies. The government can, with proper planning, use this potential to develop its agricultural sector and move toward reducing unemployment. Energy management, moving toward clean energies and improving energy indicators, is the last strategy with an importance coefficient of 0.076. Considering the abundant oil and gas resources of Iran, there seems to be less interest in moving toward such energies in the country. Most of the country's environmental crises are caused by pollution from fossil fuels.

Conclusions

In this research, we identified the weaknesses, strengths, threats and opportunities facing Iran's agricultural development to achieve the main objective of the research. In the next step, these factors were prioritized, with weaknesses and threats having the highest score in prioritizing by hierarchical analysis method. Environmental threats especially water bankruptcy; unfavorable conditions of Iran's arable land; economic problems; lack of proper development in agricultural sector in comparison with other economic sectors; exhaustion and disproportion of the mechanization fleet with conditions of the country were considered as the most important problems facing the country's agricultural development. To achieve sustainable food security, with regards to the identified factors and considering the opportunities and strengths of Iran, with the future in mind, the necessary strategies were presented in this paper. Strategies are classified in the category of defensive policies with respect to the importance coefficients of weaknesses and threats compared to the other factors. Government support policies in the agricultural sector; reforming the country's planting pattern according to the climatic conditions; taking into account the relative advantage of agricultural production as well as modernizing Iran's mechanization fleet, investing in agricultural research and development and moving toward modern knowledge; given the large number of agricultural graduates in Iran; the production or import of agricultural machinery and modern technologies, according to their suitability with the country's conditions and future needs; along with strengthening of the supply chain and maintenance services were among these strategies that were prioritized by the TOPSIS method.

Authors' contributions

The authors collaboratively designed the study, performed the literature review and analysis, and wrote and edited the manuscript. All authors read and approved the final manuscript.

Author details

[1] Department of Agricultural Mechanization, Science and Research Branch, Islamic Azad University, Tehran, Iran. [2] Department of Environment, Tehran, Iran.

Acknowledgements

Not applicable.

Competing interests

The authors declare that they have no conflict of interest to the publication of this article. There are no patents or copyrights relevant to this work. There are no relationships or activities to disclose that could be perceived to have influence the submitted work.

Funding

The authors received no financial support for the research and/or authorship of this article. No financial relationships exist with any entities that could be perceived to influence the submitted work.

References

1. Abi Said M, Hakimi A, Babar Z, Bush R, Chaaban J, Crist JT, Mundy M. Food security and food sovereignty in the middle east. Washington, DC: Georgetown University; 2012.
2. Ahmad T. Agricultural mechanization research and development in the Islamic Republic of Pakistan. Beijing: United Nations Centre for Sustainable Agricultural Mechanization; 2015.
3. Akinbamowo RO. A review of government policy on agricultural mechanization in Nigeria. J Agric Ext Rural Dev. 2013;5(8):146–53.
4. Alemu MH, Olsen SB, Vedel SE, Kinyuru JN, Pambo KO. Can insects increase food security in developing countries? An analysis of Kenyan consumer preferences and demand for cricket flour buns. Food Secur. 2017;9(3):471–84.
5. Amongo RC, Larona ML. Agricultural mechanization policies in the Philippines. Beijing: United Nations Centre for Sustainable Agricultural Mechanization; 2015.
6. Byerlee D, Janvry AD, Sadoule E, Townsend R, Klytchnikova I. Agricultral for development. Washington, DC: The World Bank; 2008.
7. Clapp J. Food self-sufficiency: making sense of it, and when it makes sense. Food Policy Issue. 2017;66:88–96.
8. Clarke L. Strategise for agricultral mechanization development. Rome: Agricultural Support System Division, FAO; 2000.
9. FAO. The state of food insecurity in the world (economic crises—impacts and lessons learned). Rome: FAO; 2009.
10. FAO. FAO statistical pocketbook. Rome: FAO; 2015.
11. FAO and UNIDO. Agricultural mechanization in Africa ... Time for action: planning investment for enhanced agricultural productivity. Rome: FAO; 2008.
12. Fonteh MF. Agricultural mechanization in Mali and Ghana: strategies, experiences and lessons for sustained impacts. Rome: FAO; 2010.
13. Graaf JD, Kessler A, Nibbering JW. Agriculture and food security in selected countries in Sub-Saharan Africa: diversity in trends and opportunities. Food Secur Issue. 2011;3:195–213.
14. Grebmer KV, Bernstein J, Prasai N, Amin S, Yohannes Y, Towey O, Nabarro D. Global hunger index getting to zero hunger. Washington, DC: IFRIP; 2016.
15. Haruna IM, Junior JA. Mechanization practice: a tool for agricultural development in nigeria: a case study of Ifelodun local government area of Kwara State. Int J Basic Appl Sci. 2013;2(3):98–106.
16. Joffre OM, Castine SA, Philips MJ, Senaratna Sellamuttu S, Chandrabalan D, Cohen P. Increasing productivity and improving livelihoods in aquatic agricultural systems: a review of interventions. Food Secur. 2017;9(1):39–60.

17. Lehane S. The Iranian water crisis. Nedlands: Future Directions International; 2014.
18. Mesgaran MB, Madani K, Hashemi H, Azadi P. Iran's land suitability for agriculture. Sci Rep. 2017;7:7670.
19. Mrema G, Soni P, Rolle RS. A regional strategy for sustainable agricultural mechanization: sustainable mechanization across agri-food chains in Asia and the Pacific region. Bangkok: FAO; 2014.
20. Negrete JC. Strategies for technology transfer of agricultural mechanization in Mexico. Int Res J Adv Agric. 2015;1(1):1–11.
21. Pan Y, Smith SC, Sulaiman M. Agricultural extension and technology adoption for food security: evidence from Uganda. New York: Institute for International Economic Policy; 2015.
22. Ritzema RS, Frelat R, Douxchamps S, Silvestri S, Rufino MC, Herrero M, van Wijk MT. Is production intensification likely to make farm households food-adequate? A simple food availability analysis across smallholder farming systems from East and West Africa. Food Secur. 2017;9(1):115–31.
23. Rosegrant MW, Koo J, Cenacchi N, Ringler C, Robertson R, Fisher M, Sabbagh P. Food security in a world of natural resource scarcity, the role of agricultural technologies. Washington, DC: IFPRI; 2014.
24. Schmitz A, Moss CB. Mechanized agriculture: machine adoption, farm size, and labor displacement. AgBioForum. 2015;18(3):278–96.
25. Simalenga T. Entrepreneurship in mechanized agriculture technology-oriented operations. Agric Mech J (AMA). 2000;31(3):61–88.
26. Sims B, Kienzle J. Mechanization of conservation agriculture for smallholders: issues and options for sustainable intensification. Environments. 2015;2015(2):139–66.
27. Singh G. Estimation of a mechanisation index and its impact on production and economic factors a case study in India. Biosys Eng. 2006;93(1):99–106.
28. Sundaram KJ, Dawe D, Haen DH, Stamoulis K, Wiebe K. The state of food insecurity in the world 2012: economic growth is necessary but not sufficient to accelerate reduction of hunger and malnutrition. Rome: FAO; 2012.
29. Triantaphyllou E. Multi-criteria decision making methods: a comparative study. Dordrecht: Springer; 2000.
30. Walport M. Global food security strategic plan 2011–2016 updated November 2013. Global Food Security; 2013.
31. Woertz E. Food security in Iraq: results from quantitative and qualitative surveys. Food Secur. 2017;9(3):511–22.

6

Characterisation of food security and consumption patterns among smallholder livestock farmers in Botswana

Sirak Bahta[1*], Francis Wanyoike[2], Hikuepi Katjiuongua[3] and Davis Marumo[4]

Abstract

Background: Achievement of food security is an important objective in the National Development Plan of Botswana. This study set out to characterise food security situation and food consumption patterns among livestock keepers in Botswana. The study also sought to determine the effect of various factors (including livestock-related issues) on nutrition/food consumption patterns of livestock farmers in Botswana. Data used in this study come from a survey of smallholder livestock producers conducted by the International Livestock Research Institute and its partners in Botswana in 2013. A food consumption score (FCS) is a qualitative measure of food security constructed using data collected at a household level, and to investigate the influence of various factors on food security and consumption patterns, both logit and OLS regression were used.

Results: Results show that purchase was the chief source of foodstuffs (including main staples) in these households. About 20% of households claimed that they had suffered lack of food at least once during the preceding 12 months prior to the collection of data used in this analysis. Nearly all of them reported having had to endure going to bed hungry. Application of the World Food Programme (WFP) FCS methodology of food security analysis supported this claim. Based on the FCS, about 9% of households were food insecure. This category constituted 6% that ranked poorly and 3% that were on the borderline according to the WFP's definition of food security. These findings justify continued efforts to ensure realisation of food security among livestock keepers in Botswana.

Conclusion: The results from this study justify continued, but well-targeted efforts to ensure realisation of food security among livestock keepers in Botswana. The significant effect that livestock keeping has on food security implies that promotion of the livestock activities, especially in the rural areas, could foster alleviation of malnutrition, especially among the poor.

Keywords: Food security, Nutrition, Livestock keepers, Botswana

Background

Although Botswana has one of the highest per capita levels of income in Africa, food security at both the household and national levels remains a challenge—a situation mainly precipitated by the structure of the economy and arid climates. Over 70% of Botswana's population resides in the rural areas [1], and majority (70%) relies on traditional/subsistence agriculture for their livelihoods [1]. Commercial farming is not a prominent economic activity in Botswana [1, 2]. Overall, the agricultural sector accounts for only 1.9% of the national income compared to the industry, and service sectors account for 28.7 and 69.4%, respectively [1, 3]. The country is characterised by high inequality with about 20% of the population living below the poverty datum line (PDL), while 30% is either unemployed or underemployed [4].

The diminished share of agricultural production in Botswana's national economy is mainly attributable to the rapid growth of other sectors notably mining (especially diamonds), tourism and manufacturing [5]. However, available data also show that productivity in the

*Correspondence: s.bahta@cgiar.org
[1] International Livestock Research Institute (ILRI), Gaborone, Botswana
Full list of author information is available at the end of the article

agricultural sector has stagnated over many years with levels in the widespread traditional/subsistence sub-sector often being many times lower than in the relatively tiny commercial sub-sector. For example, in 2013, maize yields in the traditional sub-sector averaged only 22 kg per hectare planted compared to 675 kg in the commercial sub-sector and the trend was the same for millet, beans and sorghum. By the same token, commercial livestock producers outperformed their counterparts in the subsistence/traditional sub-sector in terms of high offtake and low mortality rates of animals [2].

The main factors accounting for the poor performance of the Botswana's agriculture sector are numerous. First, Botswana's climate is largely arid to semi-arid and crop farming is confined to only a small area of about 2500–3800 sq km on the eastern and northern margins of the country [6]. Second, crop production is mainly rain fed and yields tend to be poor due to low and erratic rainfall [1]. Third, the rate of adoption of modern farming methods and practices by farmers in the traditional/subsistence sector is low leading to low yields [5, 2]. Other factors include poor soils and also animal and plant diseases and pests [5]. Amidst this poor performance of the agricultural sector, the national policy on agricultural development shifted from food self-sufficiency to food security in 1991 with the focus being on access to food at affordable prices irrespective of source, that is, domestic or imported [7]. This policy shift saw the removal of trade-restricting border measures in the grain industry, leading to increases in food imports. Currently, the country imports about 90% of the national food supply [6].

Despite the policy change, challenges still remain in ensuring that all persons and households have access to food at all times. For example, during an analysis of the impacts of rising international food prices for Botswana, BIDPA [7] reported that food prices tended to be highest in the rural areas already disadvantaged by relatively low levels of income and high rates of unemployment. Similarly, according to FAO [8], over 30% of the people in the country suffered chronic undernourishment. Consistent with the United Nations (UN) former development goal, the millennium development goal (MDG), on hunger and poverty eradication, realisation of food security is an important objective in Botswana's National Development Plan (NDP 10). Similarly, consistent with the UN sustainable development goals (SDGs), the recently ratified National Development Plan (NDP 11) targets to reduce the number of people living under the poverty line by 50% by the year 2020 and increase daily dietary energy supply (DES) from 2150 to 2600 kcal per person by 2010, and further to 2700 by 2016 and thereafter. Besides, efforts aimed at increasing household food production spelt out in the NDP and the government has initiated numerous programmes to combat malnutrition [9].

Largely due to climatic reasons and tradition, the agricultural sector in Botswana is dominated by livestock farming with smallholder livestock producers accounting for about 70% of the agricultural GDP [10]. Not much is, however, known about these smallholder livestock farmers food security situation. Past studies of Botswana's food security studies have shallowly investigated food security. For example, Acquah et al. [6] investigated food security in Gaborone using a fair sample size and different types of food security scores. Similarly, Moreki et al. [11] and Legwaila et al. [12] have, respectively, investigated the role of village chicken and traditional plants in food security. Limitations of these studies include that they either failed to empirically determine the effects of various factors on food security, and/or the studies are focused on a very narrow segment of the population.

The current study, set out to characterise the food security patterns and food security situation among the subsistence livestock keepers, employed a rigorous empirical method to address these shortcomings using farm-level cross-sectional survey data collected under the auspices of a development research project.[1] The survey was implemented in three districts (Southeast, Chobe and Central) of Botswana. The study was motivated by the current lack of knowledge on whether the efforts employed to combat food insecurity were bearing fruits. The study also sought to determine the empirical effects of various factors (including livestock-related issues) on nutrition/food consumption patterns.

The remaining part of the paper is organised as follows: section two presents the methodological approach used in the study and the description of the characteristics of the surveyed households. Section three provides the results of the econometric analysis followed by section four which provides a detailed discussion of the findings. Finally, section five presents some conclusions and policy implications of the findings.

Methodological approach

Data used in this study come from a survey of smallholder livestock producers conducted by the International Livestock Research Institute (ILRI) and its partners in Botswana in 2013. During the survey, a random sample of 600 livestock keepers were interviewed using a structured questionnaire which included a module on

[1] The Smallholder Livestock Competitiveness Project is funded by the Australian Centre for International Agricultural Research (ACIAR) and implemented by the International Livestock Research Institute (ILRI) in partnership with the Botswana Ministry of Agriculture's Department of Agricultural Research (DAR).

frequency of consumption of various food stuffs during the past 2 weeks. Using this survey data and an adaptation[2] of a methodology for measuring food security among communities by the World Food Program (WFP) [13, 14], the study computed a food consumption score (FCS) for each household which was then used to generate a hierarchical ranking of the households in terms of their food security status.

A FCS is a qualitative measure of food security constructed using data collected at a household level. Unlike quantitative[3] measures of food security which are more tedious and costly to assemble [13], qualitative measures are often used by organisations because the information required to construct them is less time-consuming and costly to collect [14, 15]. The procedure on construction and use of FCS in food security analysis is outlined in WFP VAM [14].

First, the various foodstuffs consumed in a community are categorised into eight standard food groups: main staples, pulses, vegetables, fruits, meat and fish, milk, sugar, and oil. Second, for each household, using data on frequency of consumption of the various food stuffs, a food group consumption frequency score (FGCFS) is computed by summing up the number of days the various foodstuffs in the food group have been consumed. In this study, the FGCFS was allowed to have a maximum value of 14 consistent with the 14-day data used in contrast to the 7-day data recommended by WFP VAM (op cit.) who consequently propose a truncating FGCFS value of 7. Therefore, in this study, cases where the computed FGCFS fell above 14 were changed to 14.

Next, the FGCFS was multiplied by assigned weights[4] [14] based on respective food group's nutrient content to generate weighted food group scores (WFGS). The WFGS were then summed across the food groups obtaining the FCS. On the basis of the FCS, hierarchical categories of households by their food security status were generated, namely households with poor, borderline and adequate food consumption score. In the current study, a FCS of 42 or less was considered poor, while 70.5 or above was considered adequate. Borderline FCS included those scores between 42.5 and 70. These cut-off points were double than those recommended by WFP VAM (op cit.) because of the 2-week food consumption frequency data utilised in this study.

To investigate the influence of various factors on food security and consumption patterns, both logit and OLS regression were used. The logit model is commonly used in empirical analysis of phenomena where the dependent variable always takes a value of either 1 or zero (0) [16, 17]. In the current study, the logit model was used to investigate the influence of various factors on the probability of a household falling among those categorised as food insecure. The logit model was specified as follows:

$$P_i = F(x_i'\beta) = {}^{1}/\left[1 + \exp\left(-x_i'\beta\right)\right] \qquad (1)$$

where Pi is the probability of a household being in the category of those with poor FCS, β is a vector of unknown parameters and Xi is a K x N matrix of farm and farmer characteristics hypothesised to exert influence on food security in households. These explanatory variables (see Table 1) included age, gender (male = 1 and 0 otherwise), years of formal education, main occupation (crop farming, livestock keeping, formal employment, business and retired) of the household head; size of household including number of adults and children; geographical location, that is, district where the farm is located (Chobe, Central or South East); access to grazing and cropping land; numbers of livestock (cattle and small ruminants) kept; and claim by respondents that the household had experienced lack of food in the preceding 12 months before they were interviewed for data used in this study.

A potential drawback in solely using a logit model to analyse the determinants of household food security is that it disregards variability in FCS across households with similar hierarchical ranking in terms of food security. Thus, an OLS model was also estimated to determine the effects of various factors on the estimated household food security status, represented by the FCS. The OLS model was specified as follows:

$$Y_i = \beta X_i + e_i \qquad (2)$$

where Y_i is the computed FCS in the ith household, X_i is a K x N matrix of explanatory variables as used in the logit model (see Table 1), β is a vector of unknown parameters to be estimated and e_i is a 1 × N vector of unobservable random error term assumed to be normally distributed with mean = 0 and variance = σ^2.

Table 1 also shows the expected effect of the various explanatory variables on food security. A positive sign implies that the variable enhances food security and vice versa for a negative sign. The choice of explanatory variables was informed by economic theory, objectives, and the literature on food security and nutrition. Ownership

[2] Instead of the 7 days food frequency consumption data commonly used in the WFP methodology, we used 14 days data.

[3] Quantitative measures of food security use data collected primarily at individual level to calculate dietary energy and nutrient intakes, which are then compared with nutrient requirements.

[4] The multiplying weights used in the generation of the weighted food group scores are based on interpretation by scientists about the caloric density, macro- and micronutrient content, and also actual quantities of the different food groups typically eaten. The highest weights are attached to foods with relatively high energy, good quality protein, and a wide range of micronutrients that can be easily absorbed.

Table 1 Summary of the explanatory variables used in the estimated logit model

Explanatory variable	Code	Type of variable	Expected sign
X_1 = Age of the household head	Number of years	Continuous	–
X_2 = Gender of the household head	1 if male, 0 if female	Dummy	+/−
X_3 = Years of formal education	Number of years	Continuous	+
X_4 = Main occupation of the household head	1 = Crop farming; 2 = Livestock keeping; 3 = Formal employment 4 = Non-farm business 5 = Retired	Dummy	+/−
X_5 = Size of household	Number of people	Continuous	+
X_6 = Geographical location of farm (basis of comparison = South East)	1 = Chobe district; 2 = Central district; 3 = South east district	Dummy	+/−
X_7 = Access to grazing	1 if yes, 0 otherwise	Dummy	+
X_8 = Access to cropping land	Number of hectares	Continuous	+
X_9 = Livestock units kept	No. of livestock units	Continuous	+
X_{10} = Household had suffered food unavailability during the last year	1 if yes, 0 otherwise	Dummy	+/−

of resources (including land and different types of livestock) was expected to have a positive effect on food security due to their effect on food production and household income. This is consistent with findings by [18] in Mozambique. Likewise, in line with results from a cross-country study by [19] which demonstrated strong negative relationship between lack of education and food insecurity, years of schooling of household head was hypothesised to have a positive effect on food security.

Gender of household head was included in the regression model as a dummy variable with a value of 1 if the head of the household was male and 0 otherwise. Although the available literature showed no universal relationship between gender of household head versus poverty and vulnerability [20], gender classification of households in this analysis differs with common practice in many past studies. In this study, families where husbands lived away were categorised as "male"- instead of "female"-headed as in many past studies. This leaves households of single women (unmarried with children, divorced and/or widowed) which tend to be relatively poorer [20] and therefore more likely to be food insecure as our basis of comparison. For this reason, gender of household head was expected to have a positive effect on food security.

The econometric modelling also features household size in terms of numbers of "adult persons" and "children" as explanatory variables consistent with the findings by Garret and Ruel (op cit) which demonstrated that family size and composition may affect food security status of the household.

A higher number of adult family members may be associated with availability of more labour resource which can be sold or used for own food production. Therefore, farm size was expected to impact positively on food security. Conversely, household with high numbers of children may be forced to devote fewer resources to purchasing food in order to attend to other needs of children including health, clothing and education. A negative relationship was therefore expected between high numbers of children versus household food security.

The inclusion of dummy variables for each geographical location of farm households as explanatory variables represents an attempt to account for the community-level variables that may influence food security in these households. Such community-level variables may relate to physical environment and economic characteristics of the study sites (Garret and Ruel, op cit). Dummy variables were included for Chobe and Central Districts, while South East District was used as a basis for comparison. Chobe and Central Districts are relatively more rural than South East District which is the smallest district in Botswana and is located adjacent to the capital city, Gaborone [21]. Chobe District, situated in the northern part of Botswana, receives the highest and least variable amount of rainfall, 550–650 mm [22].

Characteristics of the surveyed households

Table 2 presents a summary of descriptive statistics on the surveyed households. Fifty-nine observations were dropped from the analysis due to the problem of missing values for some variables. Results are only shown for the 541 observations utilised in the quantitative analysis of factors influencing food security. The three surveyed districts were well represented in the 541 observations with

Table 2 Selected farm and farmer characteristics in the surveyed households (*N* = 541)

	Mean	SD	Min	Max
Gender of household head(1 = male, 0 = otherwise)	0.79	0.41	0	1.00
Age (yrs) of household head	60.44	13.44	26.00	96.00
Years of schooling of household head	4.90	4.77	0	18.00
Number of adults in the household	3.52	2.14	1.00	21.00
Number of children in the household	1.11	1.45	0	9.00
Size of crop land (Ha)	6.27	14.27	0	270.00
Number of heads of cattle kept	34.60	55.26	0	474.00
Number of shoats kept	4.84	20.55	0	250.00
Claim that a household had experienced food unavailability during the past 12 months (1 = yes, 0 = otherwise)	0.20	0.40	0	1.00
District				
Central district	0.35	0.48	0.00	1.00
Chobe district	0.28	0.45	0.00	1.00
South east	0.37	0.48	0.00	1.00
Main activity of the household head				
Crop farming	0.53	0.50	0.00	1.00
Livestock keeping	0.79	0.40	0.00	1.00
Formal employment	0.26	0.44	0.00	1.00
Business	0.09	0.28	0.00	1.00
Retired	0.29	0.45	0.00	1.00
Access to grazing land	0.91	0.28	0.00	1.00
Access to cropping land	0.75	0.43	0.00	1.00

37% of the observations being in South East, 35% in Central and 28% in Chobe Districts.

About 80% of the households were male-headed, implying that female-headed households accounted for about 20% of the sample. The implied percentage of female-headed households in this study is low compared to 47–67% reported in studies by Moepeng and Tisdell [23] and Central Statistics Office Botswana [24]. Recall that in the current study households where men lived away were categorised as male rather than female-headed since it is very likely that the power of decision-making continues to reside with the male, a demographic head [25].

Heads of households were elderly (average age = 60.4 years) with rather low levels of formal education (mean = 4.9 years of schooling). Consistent with data from Statistics Botswana [26], households tended to be small with a mean of 4.6 persons each and these included 3.5 adults and 1.1 children. Majority of heads of households had livestock keeping and crop farming as their main economic activities (79 and 52%, respectively). Other commonly cited occupations included formal employment and business (26 and 9%, respectively).

Majority of the respondents indicated that they had access to grazing and cropping land (91 and 75%, respectively). Cattle were the predominant species of livestock kept. On average, a household kept about 35 heads of cattle compared to only 4.8 small ruminants (sheep and goats). This discrepancy in numbers of cattle versus small ruminants kept is rather large when compared with data from Statistics Botswana [26] which shows that the country has a total of about 2.1 million heads of cattle and 1.8 million sheep and goats. This discrepancy owes to the fact that the study purposively targeted smallholder households where a cattle rearing was the dominant activity. Most of the small stock animals are reared on the Western and South Western part of Botswana, which are not part of the survey. In addition, as it is shown in Table 2, the large values of standard deviation associated with numbers ofcattle and small ruminants owned per household show that there is wide variation in livestock ownership in the study area.

A few variables exhibited significant variation across districts. On average, households in the South East District had significantly fewer children (0.78) than in Central and Chobe districts (1.23 and 1.41, respectively). Likewise, size of cropping land and number of herds of cattle kept tended to be lower in the South East District (means = 2.85 ha and 18.43 heads) compared to the other two districts (mean (Central) = 6.52 ha and 31.64 heads; mean (Chobe) 9.76 ha and 54.49 heads). This could be due to the high population density per square kilometre in South East district, a small district closer to the capital city, Gaborone [27]. Rearing of sheep and goats was comparatively more common among households in Chobe (mean = 11.49 animals per household) than the South eastern and Central district (0.52 and 1.93 animals per household, respectively).

Results

Food consumption patterns

Table 3 shows the consumption patterns of various foodstuffs among the surveyed livestock producers. Despite widespread reliance on agriculture as a source of livelihood, purchase was the chief source of foodstuffs (including main staples) among the surveyed livestock keepers. Households that cited own production as a source of food represented less than 30% of the surveyed households for all the food commodities. This result confirms findings by BIDPA [7] which showed that households in Botswana spent the highest proportion of their income (24%) on food. The most widely and regularly consumed main staples included maize, sorghum and wheat (79–80% of households at an average frequency of 7.8–9.8 days in a fortnight). Vegetables were also widely (78% of households) and regularly consumed (on average 8.1 days in a fortnight).

Table 3 Food consumption patterns among livestock producers in Botswana

	% of households ($N = 541$)	Number of days consumed	% of respondents citing their source of various food commodities consumed ($N = 541$)						
			Own production	Hunting/ gathering/ fishing	Cash Purchase	Gifts (friends/ relatives)	Bought on credit	Food assistance	Others
Millet	7	4.6	22		71	7			
Sorghum	86	9.8	8	0.2	90	1	0.2	0.4	
Maize (include pap)	87	9.2	2		97			0.8	
Wheat flour (include bread)	79	7.8		0.4	98	0.2	0.2	0.9	
Local grains	4	7.4			95				
Rice	66	4.1			97	1		1	1
Fruits	51	6.1	5	5	89	1			
Vegetables	78	8.1	13	1	85	1		0.2	0.2
Beans, peas, lentils, or nuts	57	3.8	18	0.3	70	5		5	2.3
Eggs	28	4.6	18		81	1			1
Poultry: chicken, duck, other poultry	13	6.1	3		96	1			
Beef	13	6.0	6		94				
Pork	23	3.6	28		62	9			1
Lamb, mutton	33	6.8	4	17	77	1			
Goat meat	62	10.5	22	0.3	76	0.3		0.3	
Fresh or dried fish or shellfish	79	12.1	2	1	97			0.4	
Milk, cheese, yogurt, or other milk product	84	12.8	1	0.2	98	0.4	0.2		
Oils and fats	66	6.2	18	1	80	1			
Sweets, sugar, honey	72	7.9	7	0.2	88	3		0.7	1
Other condiments: coffee, tea, milk in tea	88	12.8	0.2	0.2	99	0.2	0.2	0.2	

Fish was the leading source of animal proteins and accounted for 79% of the surveyed households. Majority of them (about 80%) consumed fish on a daily basis. Goat meat was the second best source of animal proteins and represented about 62% of the surveyed goat meat was consumed during the 2-week period preceding the study compared to eggs, poultry meat, beef, pork, lamb and mutton (which accounted for 13–33% of the households). Consumption of milk and other dairy products was widespread among the selected districts. A larger proportion (84%) of the households consumed milk and dairy products on a daily basis in the fortnight considered. Surprisingly, in a country where rearing of cattle was widespread, purchase still remained the most frequently cited source of milk and dairy products consumed (98% of surveyed households) as opposed to own production (< 1% of milk/dairy product consumers). This could be attributed to the wide availability of small stores which sale pasteurised milk and the awareness of communities on quality and safety issues that emanated from the continuous efforts of the Ministry of Health in informing societies about food-borne diseases associated with raw milk consumption. Added factor could also be the non-dairy nature of the cattle breeds owned by these households. The arid to semi-arid climate of Botswana is mostly suited for beef cattle breeds which produce minimal milk.

Food security status

Besides the use of food consumption frequency data to assess the food security situation among livestock producers in Botswana, livestock keepers were also explicitly probed about food availability in their households during the past 12 months prior to the household survey. Results suggested food insecurity in some households with about 20% of the respondents, claiming that they had suffered lack of food at least once during the past year and nearly all of them having had to endure going to bed hungry (Table 2).

Results of the food consumption frequency analysis confirmed a presence of food insecurity in some of the households. On the basis of the FCS, about 9% of households were found to be food insecure. This category of households included 6% of households that ranked poorly and 3% that were on the borderline according to the WFP's definition of food security. The estimated number of food insecure households in the South East District was relatively low (5%) compared to Chobe and Central districts (10 and 13%, respectively). This geographical distribution of the food insecure households probably reflects the effect of urbanisation on food availability. The households in the South East District, which is more urbanised, were relatively less poor and thus less food insecure than those in Chobe and Central districts.

The overall estimate of the number of food insecure households based on the FCS methodology was low (9%) compared to the number of households who claimed to have suffered hunger during the previous 1 year (20%). This wide discrepancy may be attributable to a couple of reasons. First, the food availability situation in some of the affected households may have changed between the time when they suffered hunger and the period when data used for this study were collected. Additionally, past studies such as [13] have demonstrated a tendency for the FCS methodology to underestimate the prevalence of food insecurity. Nevertheless, the method is still considered to be generally reliable and is widely used [13–15].

Table 4 presents the results of the effects of various factors on food security situation in the surveyed households at the time of collection of data used in this study. For the logit model, the dependent variable equals one for households with poor FCS and zero otherwise. The logit coefficients therefore show the effect of the factors included as explanatory variables in the model on the log of the odds of a household being food insecure. Results of the χ^2 test indicated that the model was statistically significant in explaining the variability in the data ($P = 1\%$).

Livestock producers in the Chobe and Central Districts were significantly more likely to have poor FCS ($P = 1$ and 5%, respectively) compared to their counterparts in the SE District. As already noted, this higher likelihood of higher prevalence of food insecurity in the two rural districts is perhaps due to the higher incidence of poverty in the rural areas of these districts where most of the households interviewed reside [1].

As expected, higher number of adults persons in a household lowered the probability of that household being food insecure ($P = 5\%$). However, contrary to *apriori* expectations, households with higher number of children were found to be less likely to be food insecure ($P = 1\%$). This effect of number of children food security in households probably shows the contribution of children in productive activities in their farm households. Alternatively, the result perhaps reflects the impact of nutrition programmes targeting children or households with children which has been identified as one of the reasons why the FCS methodology understates the prevalence of food insecurity [13].

Households that reported that they had experienced lack of food during the past 12 months prior to the household survey were significantly more likely to be food insecure ($P = 10\%$) rendering credence to their claim. Conversely, being a livestock keeper and having access to grazing land mitigated the probability of a household

Table 4 Determinants of food security status in the surveyed households

	Logit (Y = Poor)			OLS (Y = FCS)		
	Coeff.	SE	$P > z$	Coeff.	SE	$P > t$
Constant	− 0.94	1.98	0.64	87.96***	14.45	0.00
Gender of household head (1 = male, 0 = otherwise)	− 0.01	0.54	0.99	− 1.38	4.04	0.73
Age (yrs) of household head	0.01	0.02	0.60	− 0.08	0.17	0.66
Years of schooling of household head	0.03	0.06	0.61	0.63	0.42	0.14
Number of adults in the household	− 0.31**	0.15	0.04	1.96***	0.78	0.01
Number of children in the household	− 0.99***	0.29	0.001	4.26***	1.16	0.00
Size of crop land (Ha)	0.002	0.02	0.92	− 0.02	0.12	0.84
Number of heads of cattle kept	− 0.001	0.01	0.82	0.00	0.04	0.93
Number of shoats kept	0.01	0.01	0.61	0.11	0.09	0.23
Claim that a household had suffered food unavailability during the last year (1 = yes, 0 = otherwise)	0.95*	0.49	0.06	− 9.49**	4.20	0.02
District (control = SE)						
Central	1.82**	0.76	0.02	− 10.40***	4.22	0.01
Chobe	1.98***	0.72	0.01	− 13.26***	4.36	0.00
Main activity of the household head						
Crop farming	0.17	0.57	0.76	2.69	3.88	0.49
Livestock keeping	− 2.75***	0.63	0.00	15.54***	4.51	0.00
Formal employment	− 0.58	0.63	0.36	7.54*	4.39	0.09
Business	− 0.83	1.12	0.46	8.35	6.17	0.18
Retired	− 0.63	0.65	0.33	6.55	4.59	0.15
Access to grazing gland (0,1)	− 2.26**	0.99	0.02	38.17***	7.37	0.00
Access to cropping land (0,1)	1.21	0.84	0.15	− 6.70	4.77	0.16
$N = 541$	$N = 541$					
LR $\chi^2(10) = 83.4$	$F(18, 522) = 6.06$					
Prob > $\chi^2 = 0.00$	Prob > F = 0.00					
Pseudo $R^2 = 0.36$	R-squared = 0.17					
Log likelihood = − 74.2	Adj R-squared = 0.14					
	Root MSE = 37.26					

Statistical significance levels: ***1%; **5%; *10%

being food insecure (P = 1 and 5%, respectively). This reinforces the importance of the livestock rearing activities and the access to grazing land on the livelihoods of the people of Botswana.

The results of the OLS regression of FCS on the various factors hypothesised to influence food security are also presented in Table 4. The OLS model was statistically significant in explaining the variability in the data (P = 1%) although the value of the adjusted R^2 was rather low. The OLS coefficients show the effect of the dependent variables on the values of the food consumption score (FCS). Quite clearly, OLS regression results were highly consistent with the results of the logit regression. Households in the Central and Chobe Districts had significantly higher values of FCS (P = 1% in both cases). Higher numbers of adult persons and also children had a positive effect on the FCS (P = 1% for both). Livestock keeping and access to grazing land had significantly positive effects of the FCS (P = 1% in both cases). In addition, FCS was significantly higher in households where the head was formally employed (P = 10%). Households that reported that they had experienced lack of food during the last 12 months year posted significantly lower FCS rendering credence to their claim.

Discussion

Although farming is the principal source of livelihood among livestock keepers in Botswana, most of them rely on purchased foodstuffs including main staples and livestock products such as milk for home consumption. As most food commodities available in food stores in Botswana are imported, this makes local people potentially vulnerable when world food commodity prices rise as has been the case recently. This calls for close monitoring of the situation by the government in order to ensure that poor households do not suffer when world and subsequently domestic food commodity prices rise. Moreover, even as the government opens up to food commodity imports in pursuit of food security, it should not disregard promotion of local production and consumption of agricultural commodities in cases where this is technically and economically feasible.

An appreciable number of households in Botswana suffer food insecurity, and the problem is more prevalent in rural districts (up to 13% of households) than in urban areas.[5] The estimated prevalence of food insecure households using the FCS in this study can be regarded as conservative because as past studies have shown the methodology has a tendency to underestimate the prevalence of food insecurity. Despite this drawback, the methodology is generally accepted as reliable and is widely used. Moreover, with 20% of households, claiming that they had endured going to bed hungry during the preceding 12 months prior to the collection of data used in this study, the FCS estimates of food insecurity prevalence seem credible. This reasoning is also supported by empirical results from this study which show that households which claimed to have suffered lack of food during the past 12 months were significantly more likely to have poor FCS and therefore be food insecure.

Findings from the econometric analysis demonstrate important relationships between food security and a number of factors. Households with livestock keeping as their main activity were less food insecure. This justifies government investments that could make the livestock sector more competitive and attractive for people/households to invest in. For example, much of research and investment in the past has been accorded to the beef subsector, while small ruminants and other livestock species have received relatively less attention. It may be highly beneficial for the government to explore how the rearing and marketing of small ruminants could be enhanced so as to widen the opportunities available in the livestock sector. Note that rearing of different species of animals provides a way of optimising land use besides diversification of sources of income for farmers. Likewise, ways through which sustainable access to grazing land by poor farmers' should be explored as this clearly enhances food security in households. Moreover, better livestock market opportunities might replicate to higher household income, hence food security, although past studies ([28, 29]) have showed the relation between herd size and gross margin in Botswana is not straightforward. In this study, the number of cattle and shoats kept by the households which resulted in non-significant effect on FCS was largely expected to impact the food security status of the households. Bahta and Malope [30] found that farmers with large cattle herds are more efficient in terms of profit. However, their study also acknowledged that the relationship between herd size and gross margin in Botswana is a complex one and they further indicated other previous studies ([28]; BIDPA 2006) that have shown that farmers who own more than 200 cattle have the highest gross margin, while farmers with less than 20 TLU in Bahta et al. [28] and with 150–200 TLU cattle herd size in BIDPA (2006) experienced a negative gross margin. Similarly, ownership small stock (shoats), which are widely used as a source of food security in many sub-Saharan Africa, are less marketable in Botswana as there are not any formal or informal small stock markets.

The aforementioned together with the cultural tendency of livestock overstocking as symbols of wealth and prestige could have left rural households food in secure regardless the number of cattle or shoats they possess.

Findings from the current study also give clues on how targeting of safety net interventions to mitigate food insecurity should be done. With food insecurity being more prevalent in the rural areas, more attention should be placed there. Households with a higher number of adult persons are less likely to be food insecure and should therefore attract less attention. Although households with higher number of children were less food insecure, this probably reflects the effects of ongoing efforts that target children in ensuring food security.

Conclusions and recommendations

This study set out to characterise food consumption patterns and food security situation among livestock keepers in Botswana. The study also sought to determine the effect of various factors (including livestock-related issues) on nutrition/food consumption patterns. The study utilised data from a survey conducted by International Livestock research Institute (ILRI) and its partners on 600 livestock keepers in 2013. Based on WFP's method of measuring food security in communities, we computed a food consumption score (FCS) which was then used to generate a hierarchical ranking of households in terms of their food security status. The results of the FCS were used to ground truth farmers' claims about their food security situation. Regression analysis including both logit and OLS models was utilised to determine the effect of various factors on food security status of farm households.

It was found that purchase is the chief source of foodstuffs (including main staples) among livestock keepers in Botswana. Fish and goat meat were the leading source of animal proteins in households. About 20% of the surveyed households claimed that they had suffered lack of food at least once in the preceding 12 months prior to

[5] According to Acquah et al. [6], the food insecurity among poor Gaborone inhabitants is about 62%. However, since the study purposefully targeted poor households, the value cannot represent the rate of food insecurity in urban area.

the household survey with nearly all of them having had to endure going to bed hungry. The FCS confirmed this claim by the survey respondents. On the basis of the FCS, about 9% of households were found to be food insecure. This included 6% that ranked poorly and 3% that were on the borderline according to the WFP's definition of food security. Households that reported that they had experienced lack of food during the 12 months preceding study posted significantly lower FCS rendering credence to their claim.

The results from this study justify continued, but well-targeted efforts to ensure realisation of food security among livestock keepers in Botswana. The significant effect that livestock keeping has on food security implies that promotion of the livestock activities, especially in the rural areas, could foster alleviation of malnutrition, especially among the poor. Perhaps the government and the people of Botswana could save some money by encouraging consumption of locally produced foodstuffs where this is technically and economically feasible.

Authors' contributions

Conception and design of the work proposal, collection of data, interpretation of data and drafting the manuscript article were performed by SB and FW. Analysis and the initial result interpretation were conducted by FW. Revising the manuscript critically for intellectual content and the final was conducted by SB, HK and DM. All authors read and approved the final manuscript.

Author details

[1] International Livestock Research Institute (ILRI), Gaborone, Botswana. [2] International Livestock Research Institute (ILRI), Nairobi, Kenya. [3] World Bank Group (WBG), Addis Ababa, Ethiopia. [4] Botswana University of Agriculture and Natural Sciences (BUAN), Gaborone, Botswana.

Acknowledgements

The assistance and collaboration of local district authorities, and extension agencies in the South East, Central and Chobe districts of Botswana are gratefully acknowledged, as is the co-operation of farm households selected for the survey.

Competing interests

The authors declare that they have no competing interests.

Funding

This study was conducted as part of the Competitive Smallholder Livestock in Botswana project, funded by the Australian Centre for International Agricultural Research (ACIAR) and implemented by the International Livestock Research Institute and the Botswana Ministry of Agriculture's Department of Agricultural Research.

References

1. UNDP. Botswana study guide, strategic information and developments. Washington DC: International Business Publications. 2012; 1.
2. Statistics Botswana. Annual agricultural survey report 2013. Gaborone: Statistics Botswana; 2014.
3. Botswana. The world factbook. Central Intelligence Agency, 2014. http://www.cia.gov/library/publications/the-world-factbook/geos/bc.html.
4. World Bank. Botswana social protection assessment. Geneva: World Bank; 2013.
5. MFDP. Ministry of finance and development planning, national development plan (NDP) 10. Gaborone: Government Printers; 2010.
6. Acquah B, Kapunda S, Legwegoh A, Gwebu T, Modie-Moroka T, Gobotswang K, Mosha A. The state of food insecurity in Gaborone. In: Crush J (ed) Urban Food Security Series Cape Town: Urban Food Security Network. pp.1–38; 2013.
7. BIDPA. Rising global food prices: causes and implications for Botswana. BIDPA briefing paper 2008.
8. FAO. Biofuels: prospects and opportunities. Rome: FAO; 2008.
9. MFDP. Ministry of finance and development planning, national development plan 11. Gaborone: Government Printers; 2016.
10. ILRI, ACIAR, Botswana MOA. Competitive smallholder livestock in Botswana: report of the second project partners meeting on in Gaborone, Botswana; 2014.
11. Moreki JC, Dikeme R, Poroga B. The role of village poultry in food security and HIV/AIDS mitigation in Chobe District of Botswana. Livest Res Rural Dev. 2010;22.
12. Legwaila GM, Mojeremane W, Madisa ME, Mmolotsi RM, Rampart M. Potential of traditional food plants in rural household food security in Botswana. J Hortic For. 2011;3(6):171–7.
13. WFP and FAO. Measures of food consumption—harmonizing methodologies. Rome: WFP; 2008.
14. Wfp VAM. Food consumption analysis: calculation and use of the food consumption score in food security analysis. Rome: WFP; 2006.
15. Gina K, Berardo A, Papavero C, Horjus P, Ballard T, Dop M, Delbaere J, Brouwer ID. Proxy measures of household food consumption for food security assessment and surveillance: comparison of the household dietary diversity and food consumption scores. Public Health Nutr 2010: 13(12):2010–18.
16. Ben-Akiva M, Bierlaire M. Discrete choice methods and their applications to short-term travel decisions. In: Hall R, editor. Handbook of transportation science. International series in operations research and management science. Dordrecht: Kluwer Academic; 1999. p. 23.
17. Green W. Econometric analysis. 7th ed. Edinburgh Gate: Pearson Education Limited; 2012.
18. Garret J, Ruel M. Are determinants of rural and urban food security and nutritional status different? Some insights for Mozambique. World Dev. 1999;27(11):1955–75.
19. De Muro P, Burchi F. Education for rural people and food security: a cross country analysis. Rome: Food and Agriculture Organization (FAO); 2007.
20. Mrkic S, Johnson T, Rose M. World's Women 2010: Trends and Statistics. New York: United NationsDepartment of Economic and Social Affairs; 2010.
21. Chobe District Council. Chobe District Development Plan 6: 2003-2009. Government Printer, Gaborone;2003.
22. South East District Council. South East District Development Plan 6:2003-2009. Government Printer,Gaborone; 2003
23. Moepeng P, Tisdell C. The socio-economic situation of female heads and poor heads of households in rural botswana: a village case study. Social Economics, Policy and Development, the University of Queensland, Brisbane. Working Paper 48; 2008.
24. Statistics Botswana. Household income and expenditure survey 2002/03. Gaborone: Government Printer; 2004.
25. Joshi S. Female household-headship in rural Bangladesh: incidence, determinants and impact on children's schooling. New Haven: Yale University; 2004.
26. Statistics Botswana. Botswana core welfare indicators survey 2009/10. Main report 2013; 1.

27. Statistics Botswana. Population and housing census 2011 analytical report. Gaborone: Statistics Botswana; 2011.
28. Bahta S, Baker D, Podisi B, Marobela O. Competitive smallholder livestock in Botswana: results of a livestock value chain survey in the Central district of Botswana. Gaborone, Botswana: International Livestock Research Institute (ILRI); Nairobi, Kenya; 2013.
29. BIDPA. Final Report on the Viability and Long-term Development Strategy for the Livestock (Beef) sub-sector in Botswana. Botswana Institute for Development and Policy Analysis, Gaboron; 2006.
30. Bahta S, Malope P. Measurement of competitiveness in smallholder livestock systems and emerging policy advocacy: an application to Botswana. Food Policy. 2014;49:408–17.

Edible backyards: climate change and urban food (in)security in Africa

Bright Nkrumah*

Abstract

Background: Food insecurity is continuously seen as a major problem in Africa, and as a challenge confronting millions of Africa's urban population. Although some city residents produce their own food and, therefore, are food secured, others continuously find it difficult to access arable land in order to keep animals or cultivate crops. Climate change, in addition, seems to worsen the food insecurity situation of the region's urban population, especially in the face of drought and flooding waters. To this end, this study recommends that urban agriculture (UA) is well positioned to mitigate the adverse effect of climate change and improve food security.

Results: The study found that the current problem posed by climate change to Africa's urban food security is expected to worsen. Many urban populations are at risk of becoming victims of climate change through severe food supply problems caused by floods, droughts and hailstorms that affect food production in the hinterlands. African urban population has a role to play in finding appropriate solution to sustain food security, especially through UA.

Conclusion: The study concludes that African governments should endeavour to create the necessary conducive environment to encourage urban residents to engage in animal husbandry and food production.

Keywords: Climate change, UA, Adaptation, Vulnerability, Africa

Background

Each year, hundreds of Africans migrate from rural to major cities in search of a better life [1]. Most of these cities continue to serve as platforms that provide greater economic opportunities to several youths, and has, thus, triggered mass rural–urban migration. This rapid migration according to Davis represents a new era for the African continent, considering that for the first time in history, more people reside in urban instead of rural areas [2]. More people in urban areas imply that more food is needed. While some African countries produce sufficient food capable of feeding their entire population, not everyone has adequate access [3]. One reason which accounts for this problem is that many urban dwellers are increasingly finding it difficult to access healthy and nutritious food due to major disruptions to food supplies [4]. For in many parts of African cities, flooding may occur many times in a month, mostly through rainfall which does not reach defined channels, or even overflows from channels. The impact of these flooding on urban population does not only cause local damage to urban agriculture (UA) and transport, but has an indirect effect of disrupting movement of food supplies from the hinterlands to urban areas [5]. This situation poses a serious threat to the ever-growing urban population, especially in the face of climate change.

According to Cribb, one key factor which could exacerbate Africa's food insecurity is the (ab)use of the region's best agricultural land for urban development (power grids, transport, recreational centres and shopping malls) [6]. Ultimately, besides population growth, infrastructural developments generate substantial proportions of gas emissions. It was against this backdrop that Adelekan observes that West Africa's foremost manufacturing city, Lagos is more likely to be hit by continuous flood [7]. She provides a number of explanations for this projection throughout her work, but expresses it best when she classifies 'uncontrolled expansion of the built-up area' and

*Correspondence: bright.nkrumahup@gmail.com
Department of Political Studies, University of the Witwatersrand, Private Bag 3, Wits 2050, South Africa

'urban development as being a large creator of risk for much of the urban population, most especially the urban poor who live in more hazardous physical and human environments along the coast' [7]. Thus, by extension, local manifestations of climate change are more likely to seriously compromise the vulnerability of urban resident since the viability and productivity of existing food production systems in many parts of Africa will be affected [3]. The concept of vulnerability, according to Kelly and Adger, implies the 'ability and inability of individuals and social groupings to respond to, in the sense of cope with, recover from or adapt to, any external stress placed on their livelihoods and well-being' [8]. Inherent in this definition is the identification, adoption and implementation of policy-relevant, robust recommendations and conclusions relevant to immediate and long-term needs of the people concerned. To this end, the next section looks at Africa's increasing urbanisation, vulnerability and adaptation of UA to climate change. The paper will, however, begin by first setting out the meaning of two important themes which run concurrently throughout this paper: food (in)security and climate change.

Discussion

Africans beliefs and concerns about climate change and food (in)security

It is important to recall the words of English economist, Thomas Malthus, in his 1789 *Essay on the principle of population*, where he projected that population growth would ultimately surpass food production [9]. Given that majority of Africa's population are now living in and/or migrating to urban centres, they are becoming increasingly disconnected from direct food sources. Their food (in)security situation is worsened when food cultivated in cities is further threatened by extreme weather conditions and climatic variability [4].

The phrase food (in)security was arguably first coined in the 1970s. Since then, different institutions and scholars have ascribed different meanings to it. According to one such institution, food security is attained when every individual has 'physical and economic access at all times to adequate food or means for its procurement' [10]. Even though this definition is widely accepted, the responses, scale and causes of food *in*security continuous to attract significant debate. In somewhat stark contrast to the above definition, Hendriks provides an expansive definition of this term. She notes that food security is mainly absence of hunger or where there is availability of 'food supply, access, adequacy, utilisation, safety' [11]. The concept of food security, in sum, has informed scholarly and policy debate at the local, national and international levels over the last four decades [3].

On the other hand, Climate change may be simply defined as a change in the pattern of weather, and related changes in land surfaces and oceans [12]. These changes may be triggered by either human influences (such as land use or changes in the composition of the atmosphere) or natural processes (including internal variability in the climate system or changes in the Sun's radiation). The Intergovernmental Panel on Climate Change has projected that by 2050, rainfall in Southern Africa and sections of the Horn of Africa will decline by about 10% [13]. Equatorial countries such as Kenya, Uganda and Cameroon are projected to be warmer by about 1.4 °C. The Sahara and semi-arid parts of southern Africa are also projected to be warmer by as much as 1.6 °C, whereas sea-surface temperatures in the open tropical oceans surrounding the continent will rise by about 0.6-08 °C (less than the global average) by 2050 [13]. As a result, Abebe notes that climate change will aggravate several existing challenges facing urban food production since it will trigger a more extreme weather conditions in Africa, which may be the major cause of famine or inadequate production of food [14].

While traditional agricultural practices in general are likely to be heavily affected by the effects of changing patterns of vector-borne diseases, both subsistence and commercial farming are more likely to be impacted by more severe storms, very hot days and variable rainfall patterns [14]. Thus, the effect of climate change on Africa's food production could be grave, especially due to livestock and crop losses from drought and flood, as well as unpredictable weathers and seasons [3]. For instance, plants such as cabbage and wheat will find it difficult absorbing nitrates into their tissues with the rises in atmospheric carbon dioxide [3]. This situation compromises the nutritional quality of foods since the lack of nitrogen leads to reduction of protein content in plants. It is, however, important to indicate that '[i]nevitably, the effects of climate change will hit hardest on the urban poor, since they are often located in the most vulnerable parts of the cities and have the lowest capacity to adapt to such changes [15]. Action on climate change is, therefore, crucial given that in light of the recent warming rate of 0.18 °C per decade, the global mean temperature currently exceeding 1 °C above the pre-industrial level, will within 30 years from now go beyond the aspirational limit of 1.5 °C set by the United Nations Paris Agreement [16]. As a reaction to this growing concern, some Africans have started exploring innovative ways of farming in suburbs, as well as developing new methods towards local agricultural production and distribution, otherwise known as *UA* [3]. Against this backdrop, the next section turns to answer the question, 'what is UA' and what are its prospect in mitigating the effects of climate change.

Smart food city: the environmental and social dimensions of urban agriculture

Different scholars have provided different definitions of the meaning of UA. It is important to state, however, that while definitions differ, they all embrace the notion that UA encompasses a wide range of activities carried out at many different levels, from the citywide to the domestic level [3]. The very concept of 'urban' agriculture, according to Peres, is contradictory at best considering that crop production and animal keeping are activities that occur in the hinterlands and not in suburbs [17]. This position was quickly refuted by Bartling who asserts that there has been a proliferation of urban food and animal husbandry, and there has been indeed a continued existence and expansion of a wide variety of food production activities in post-colonial Africa [18]. Prain adds that as in other areas of the world, livestock raising and crop cultivation have long history in urban Africa [19]. Truly, every city in Africa has a history of UA where animals were kept and some resilient farmers managed to access and cultivate crops on those strips of land in the urban setting.

In defence of UA, Mougeot notes that there is the need to provide a stronger theoretical basis for UA to enable African urban residents appreciate its relevance towards urban resilience and food security [20]. The approach of Mougeot is to bring together a variety of theoretical building blocks in order to frame an overarching definition of urban food (and animal) production. These blocks encompass: the relations between urban and peri-urban systems, product destinations, the scale of food production, the type(s) of animal and food items produced, the locations where they are produced, and the types of economic activity involved. He, thus, proffers a more consistent and comprehensive definition of UA as.

> *an industry located within or on the fringe of a town, a city or a metropolis, which grows or raises, processes and distributes a diversity of food and non-food products, (re-)using largely human and material resources, products and services found in and around that urban area, and in turn supplying human and material resources, products and services largely to that urban area [20].*

UA can, thus, be defined as a form of agriculture that is integrated into the urban economic and ecological system. What distinguishes an urban from rural agriculture is that the former is embedded in and interacts with the urban ecosystems. Four distinguishable features of UA are: the use of its final products, method of production, size or scale and location of the farming activity. Table 1 provides a detail account of these four features.

As shown in Table 1, UA in Africa transcends community or private gardens. It encompasses activities such as (peri)urban farms, hybrid agricultural activity which incorporates gardening and farming, animal keeping, poultry, commercial bee and hobby, edible landscaping, institutional and demonstration gardens [3]. Other types of African UA are small scale gardens often run by schools, churches, civil societies and community groups. Besides providing a platform for social interaction, these gardens provide participants with opportunities for physical activity, community pride, and culturally appropriate food (including access to fresh vegetables and fruits). In very simple ways, UA is a socioeconomic tool which provides platforms for social networking, while providing employment opportunities for the unemployed. Some affluent suburban residents, on the other hand, are motivated to cultivate crops based on desire to reduce the distance their food has to travel before consumption, and as a means of supplementing their income [3].

The diversity of UA makes it uniquely adaptive to Africa especially as it can be adapted to the needs of diverse stakeholders and a wide range of urban situations. It can be practised in different locations and under varying political and socioeconomic settings. Since it comprises a variety of livelihood systems (spanning from commercialised agriculture to subsistence

Table 1 Key features of UA

Location	Operational method	Size	Final products
Walls	Outdoor and indoor animal rearing	Windowsills	Plants for therapeutic
Rooftops	Aquaculture sites (ocean, canals	Balconies	Animals and plants for consumption
Nature strips	Rivers and Ponds	Large contiguous parcels of land	Ornamental use
Road verges	Raised bed		Medicinal use
Peri-urban land zoned for rural enterprises	Aquaponics		Environmental services
Community or public managed land	Hydroponic venues		Social capital
Industrial or residential areas	Greenhouses		Ecological services
Private properties (balconies, yards and gardens)	In-soil cultivation		Re-usable waste products

production and processing, many urban dwellers can tend home gardens for household food and income without complex technology or huge capital injection [3].

In sum, UA provides more efficient utilisation of organic waste, reductions in greenhouse gas emissions (as a result of production of more food from local sources), increased job creation for new urban farmers, and health benefits from producing one's food. Even though these apparent advantages may well be correct, these claims remain merely as statements of the possible, especially as very few studies have attempted to measure with some degree of certainty these claims. Nonetheless, in light of the enormous potential of UA to mitigate climate change, it is important to turn our attention to some of the barriers which militate against urban farmers and/or suburban residents in their attempt to start or expand their farms [3].

Cities in floods and droughts: diagnosing barriers to urban agriculture

In contemporary urban Africa, virtually all human activities are somewhat regulated by government legislation or institutions. Without the requisite licence, approval or permit, people cannot undertake commercial activities, change land use or erect buildings. And where such authorisation has been withheld -or even granted- a disgruntled neighbour will eventually lodge a complaint with the relevant authority that such an activity is causing great discomfort. Consequently, in many suburbs (including Accra, Pretoria, Addis Ababa and Lagos), the practices and scope of UA has been subjected to an intricate web of legislations triggered by a string of concerns, mainly relating to public health and neighbourliness. Even though some of these regulations are often enforced by municipal government officials, the most influential regulators of UA have probably been (city, environmental, town and urban) planners [21]. To be specific, urban planners (in less developed and developing African countries) have, for many years, treated UA as activities deserving to be proscribed and at the very least regulated. As Morgan notes, admittedly in rather general terms, 'urban planners in Africa have been part of the problem of food insecurity because, until recently, they saw it as their professional duty to rid the city of UA [21].

Usually, urban farmers' proposal for the relaxation of laws governing the rearing of animals (such as chicken and goats) on their properties have often been met with rather discouraging replies from metropolitan governments, which brings to bear the constant debates whether 'rural' practices can be implemented in urban areas [22]. A member of a local government once decried '[w]ould you like to live next door to someone who's got chickens clucking around plus the potential smell and everything else that comes with it [23]?' With some city dwellers entertaining this rather abstract notion that suburbs are no places for keeping farm animals (including pigs and sheep), urban farmers stand the chance of being ostracised or shun by their neighbours, since they may be seen as '[t]urning the city into a zoo' [23]. In stark contrast to many cities in the world where UA is shaping current debates on the nature of sustainable and resilient urban suburbs against the backdrop of contemporary challenges (mostly climate change and peak oil), UA in Africa is at times perceived as not compatible with present-day visions of an ideal city. Thus, the practice of food production in urban areas tends to be low [24]. For the foreseeable future, if African cities are to be more sustainable and resilient in the face of several existential threats, such as climate change, then states must give priority to urban farming, especially as it: (i) has the potential to help mitigate the factors causing climate change; (ii) requires fewer energy inputs; (iii) symbolises a form of localised food production and consumption. Food production in cities undoubtedly holds a lot of prospects in ensuring sustainable forms of urban life in addition to building urban resilience. Nonetheless, before this could be achieved, urban food production must be incorporated into a more comprehensive initiatives of urban planning and, also, into broader food systems for sustainability and resilience.

Conclusion

This paper explored animal husbandry and food production, and, in particular, UA as an important socioeconomic practice, both in terms of serving as a source of food (and income) for urban household and as a means of mitigating the adverse effect of climate change in Africa. The paper provided a brief overview of the changing nature of Africa's climate and food production, then engaged with UA's contributions to social and environmental sustainability and its role in fostering urban food security. To this end, the study makes an important contribution to the field of agriculture and food security by suggesting that before UA can contribute to Africa's adaptation and resilience to climate change, African leaders must: (i) de-regulate laws which hinder people from planting crops or keeping animals; (ii) provide support to church and school gardens, not forgetting to provide expert advice on gardening; (iii) gather and supply organic waste for composing and fertilising; (iv) recognising the importance of communal gardening spaces when assessing applications for development projects in residential areas; and (v) provide support for temporary land use for suburban farmers.

Abbreviation
UA: urban agriculture.

Authors' contributions
The author read and approved the final manuscript.

Acknowledgements
I will like to thank Prof Lawrence Hamilton, Wits University, for his encouragement, advice and support.

Competing interests
The author declares that he has no competing interests.

Funding
No payment was sought or received for writing this paper.

References

1. FAO. Food security, nutrition and livelihoods. 2018. http://www.fao.org/fcit/nutrition-livelihoods/en/ Accessed 3 Mar 2018.
2. Davis J. Food security and community gardening in the Ashburton, Ashwood & Chadstone neighbourhood renewal area. Victoria: MonashLink Community Health Service Ltd; 2010; p. 3.
3. Nkrumah B. Mobilisng for the realisation of the right to food in South Africa. Dphil thesis, University of Pretoria, South Africa; 2017.
4. West J. Can linking small- and large-scale farmers enhance adaptive capacity? Evidence from Tanzania's Southern Agricultural Growth Corridor. In: Inderberg TH, Eriksen S, O'Brien K, Sygna L, editors. Climate change adaptation and development transforming paradigms and practices. London: Routledge; 2015. p. 139–60.
5. Douglas I. Flooding in African cities, scales of causes, teleconnections, risks, vulnerability and impacts. Int J Disaster Risk Reduct. 2017;26:34–5.
6. Cribb J. Seeking answers to the food crisis. Partners in Research for Development. 2008: 58.
7. Adelekan IO. Vulnerability of poor urban coastal communities to flooding in Lagos, Nigeria. Environ Urban. 2010;22:433.
8. Kelly PM, Adger WN. Theory and practice in assessing vulnerability to climate change and facilitating adaptation. Clim Change. 2000;47:328.
9. Malthus TR. An essay on the principle of population. 1798: pp 6–7.
10. CESCR. General Comment No. 12, E/C.12/1999/5. para. 1.
11. Hendriks SL. The challenges facing empirical estimation of household food (in) security in South Africa. Dev Southern Africa. 2005;22(1):103.
12. Tumushabe JT. Climate change, food security and sustainable development in Africa. In: Oloruntoba SO, Falola T, editors. The Palgrave Handbook of African politics governance and development. New York: Palgrave Macmillan; 2017. p. 853–68.
13. IPCC. The Regional Impacts of Climate Change. http://www.ipcc.ch/ipccreports/sres/regional/index.php?idp=11 Accessed 15 June 2018.
14. Abebe MA. Climate change, gender inequality and migration in East Africa. Wash J Environ L Policy. 2014;4(1):104.
15. De Zeeuw H, Dubbeling M. Cities, food and agriculture: challenges and the way forward. RUAF Foundation. 2009: 5.
16. Brown A. Climate change and Africa. Nat Clim Change. 2015;5:811.
17. Pires V. Planning for urban agriculture planning in Australian Cities. State of Australian Cities. Proceedings of the State of Australian Cities. 2011: 1.
18. Bartling H. A chicken ain't nothin' but a bird: local food production and politics of land use change. Local Environ. 2012;17(1):24.
19. Prain G. The Institutional and regional context in African Urban Harvest: Agriculture in the Cities of Cameroon, Kenya and Uganda. In: Prain G, Karanja N, Lee-Smith D, editors. African urban harvest: agriculture in the cities of Cameroon, Kenya and Uganda. London: Springer; 2010. p. 1.
20. Mougeot LJA. Urban agriculture: definition, presence and potentials and risks. Cities Feeding People Series; 2000; 31: 99.
21. Morgan K. Feeding the City: the challenge of urban food planning. Int Plan Stud. 2009;14(4):344.
22. Ogato GS, Amare KA, Geneletti BD. Towards mainstreaming climate change adaptation into urban land use planning and management: the case of Ambo Town, Ethiopia. In: Leal Filho W, Belay S, Kalangu J, Menas W, Munishi P, Musiyiwa K, editors. Climate change adaptation in Africa: fostering resilience and capacity to adapt. London: Springer; 2017. p. 61–85.
23. Burton P, Lyons K, Richards C, Amati M, Rose N, Des Fours L, Victor Pires, Rochelle Barclay. Urban food security, urban resilience and climate change. National Climate Change Adaptation Research Facility. 2013: 38.
24. Asrat P, Simane B. Adaptation benefits of climate-smart agricultural practices in the Blue Nile Basin: empirical evidence from North-West Ethiopia. In: Leal Filho W, Belay S, Kalangu J, Menas W, Munishi P, Musiyiwa K, editors. Climate change adaptation in Africa: fostering resilience and capacity to adapt. London: Springer; 2017. p. 45–59.

8

Government input support on *Aus* rice production in Bangladesh: impact on farmers' food security and poverty situation

Md. Taj Uddin and Aurup Ratan Dhar*

Abstract

Background: *Aus* rice began to lose its importance as farmers slowly started shifting to cultivation of irrigated *Boro* rice encouraged by its higher yields. The government of Bangladesh has launched an incentive programme for small and marginal farmers with an attempt to rejuvenate *Aus* rice cultivation. The aim of this study is to assess the impact of government input support on *Aus* rice production in Bangladesh and its impact on food security and farmers' poverty situation. A total of 120 (i.e., 60 supported and 60 non-supported) farmers from two upazilas of Mymensingh district were selected. Data were analysed with a combination of descriptive statistics, mathematical and statistical techniques.

Results: The results show that rice cultivation occupied the major portion of farmers' total cropped area. The government of Bangladesh had provided with input (both cash and kind) support in order to boost *Aus* rice (UFSHI and NERICA variety) production. Cropping intensity of supported farmers' was increased by 39.7%, whereas it was increased by only 1.4% for non-supported farmers. Productivity of *Aus* rice was 138.0 and 100.0% in stare of supported and non-supported farmers, respectively. Average per capita daily calorie intake of the households was still below the national average level of 2122 kcal. Poverty in terms of deprivation of health education, employment, housing, mobility and income was decreased, and overall livelihood circumstances were improved through government input support on *Aus* rice production.

Conclusions: The study concludes that the motivation provided with by the government through input support on *Aus* rice production was highly appreciated by the farmers. Though per capita daily calorie intake by the farmers was still under the national average level, supported farmers were more food secure than non-supported farmers. Several dimensions of poverty were decreased, and overall socioeconomic status was improved. The study recommended that scientific and technical training programmes should be arranged to raise farmers' awareness about Aus rice production for enhancing their livelihood.

Keywords: Government, Input support, *Aus* rice, Food security, Poverty

Background

Bangladesh is the 4th largest country in the world with respect to rice area and production [1]. At present, the net cultivable area is about 8.50 M ha and net cultivated area is 7.45 M ha [2]. Rice is the staple food for the people of Bangladesh. It grows in all the three crop growing seasons of the year and occupies about 77% (11.42 M ha) of the total cropped area of about 14.94 M ha. Rice alone constitutes about 93% of the total food grains produced annually in the country [3]. *Aus* rice has been contributing to food production in addition to other two rice crops (*Aman* and *Boro*) until mid-1980s. The production seasons of *Aus*, *Aman* and *Boro* rice are *Kharif*—I (April–July), *Kharif*—II (July–November) and *Rabi* (November–April), respectively. *Aus* rice began to lose its importance as farmers slowly started shifting to

*Correspondence: aurup971@gmail.com
Department of Agricultural Economics, Bangladesh Agricultural University, Mymensingh 2202, Bangladesh

cultivation of irrigated *Boro* rice encouraged by its higher yields. The acreage during the *Aus* season dwindled below 10 lakh ha now, from 30 lakh ha in early 1980s. Rice output during the *Aus* season slumped to almost half at 17 lakh tonnes from over 30 lakh tonnes in early 1980s. At the same time, the acreage and production of *Boro* rice shot up several folds. One of the reasons behind the falling of *Aus* rice production is non-availability of land as its seed sowing or transplanting period falls during the March and April months when *Boro* rice still remains in the field. Farmers are not getting enough time and land to grow *Aus* rice [4].

Increased production of rice during the *Boro* season has also dampened farmers' interest to stay in the cultivation of *Aus* rice, which grows on an upland situation and requires less irrigation because of monsoon during its growth period. Farmers do not like to grow *Aus* rice as they get higher output from *Boro* rice. The government wants to use the advantage of less irrigation requirement of *Aus* rice with an objective to cut dependency on the underground water for irrigation and arrest the fall in the water level in the aquifer. *Aus* rice can create new scope by scaling up production and shed dependency on the underground water [5].

Governments in both developed and developing countries intervene in agriculture with a view to achieving a wide range of economic and social objectives. Incentives on agricultural production are a common element in agricultural development, the aim of which is to promote adoption of new technologies and thus increase agricultural productivity [6]. The reasons for government intervention include self-sufficiency, employment creation, support small-scale producers for adopting modern technologies and inputs, reduce price instability and improve the income of farm households. This intervention can take a number of forms such as price support programmes, direct payments and input support to influence the cost and availability of farm inputs like credit, fertilizers, seeds, irrigation water, etc. Of all domestic support instruments in agriculture, input support is the most common. The provision of input support is linked to the achievement of self-sufficiency in food production and food security. Input support can bring economic benefits to society but can also be a major cause of negative environmental externalities when they promote excessive use of fertilizers, agrochemicals and irrigation water.

However, the government of Bangladesh has launched an incentive programme with a cost of Tk. 336.2 million for small and marginal farmers in the country with an attempt to rejuvenate the cultivation of *Aus* rice. Besides, the government has allocated Tk. 19.6 million for enhancing sweet pumpkin and *dhaincha* production. Under the programme, a total of 231,363 small and marginal farmers will get necessary seed and chemical fertilizers at the free of cost for cultivating local high yielding variety (HYV) and NERICA (a stress tolerant African variety) of *Aus* rice in 49 drought prone districts across the country. These necessary inputs have already been moved to different stations across the country, and the farmers are getting these inputs for cultivating the *Aus* rice.

Importance of such modality has been portrayed in a good number of studies. A taciturn effort has been made here to appraise the previous research studies which are: Salam et al. [7] assessed the strategic plan for increasing *Aus* and *Aman* rice cultivation in Bangladesh and identified *Aus*- and *Aman*-based cropping pattern appears to be quite prospective; Parvin [8] focused on cash subsidy support to the farmers by the government for increasing farm level production in Bangladesh and found that marginal farmers have enjoyed more advantages in terms of cash benefits (12.0% surplus) than small and medium farmers; Barkat [9] conducted a quantitative analysis of fertilizer demand and subsidy policy in Bangladesh and found that fertilizer distribution switched from market mechanism to government machinery apparently without any rigorous evaluation study of the system; Chirwa et al. [10] identified the factors influencing access to agricultural input subsidy coupons in Malawi and showed that poor and elderly-headed households were less likely to receive fertilizer coupons and receive less of the subsidized fertilizers, whereas households with larger land parcels and semi-commercialized farmers were more likely to receive coupons and acquire more subsidized fertilizers; Sharma and Thaker [11] examined trends in fertilizer subsidy and its distribution system between farmers and fertilizer industry across different farm sizes in India and highlighted that reduction in fertilizer subsidy has adverse impact on farm production and income of small and marginal farmers as they do not get benefit from higher output prices but do benefit from lower input prices.

It is evident from the above discussion that no empirical study has been conducted yet in Bangladesh on impact of government incentives like input support on *Aus* rice production. Thus, there exists a scope to identify the impact of government input support on increasing *Aus* rice production in Bangladesh. The research will present an updated status of the distribution system of government input support in producing *Aus* rice with its intensification and productivity, and evaluate the impact of government input support on food security and poverty reduction of the *Aus* paddy producing farmers, and hence improve livelihood status of the small and marginal farmers. The specific objectives are: (1) to identify the farmers' socioeconomic status and distribution system of government input support for producing *Aus* rice; (2) to analyse the crop intensification and productivity of *Aus* rice through government input support; (3) to address the factors affecting adoption of cultivating *Aus* rice with government support; and (4) to evaluate the impact of government input support on *Aus* rice farmers' food security and poverty reduction.

Methods

Study areas and sample size

The study was conducted at different villages of two upazilas (Mymensingh Sadar and Trishal) of Mymensingh district. For investigation, two categories of farmers were targeted: supported farmers (who received input support from the government) and non-supported farmers (who did not receive input support from the government). Following purposive sampling technique, 30 supported farmers and following stratified random sampling technique, more 30 non-supported farmers (i.e., a total of 60 farmers) were selected from each of the study areas. Thus, a total of 120 farmers were interviewed for this study to collect field level data. Questionnaire survey, focus group discussions (FGD) and key informant interviews (KII) were also performed for collecting the primary data. The questionnaire was concerned with farmers' socioeconomic profile, distribution system of government input support, information on crop intensification, profitability and productivity of *Aus* rice production, food security status and poverty indicators of the farm households, and problems and suggestions regarding *Aus* rice production. Secondary data sources like reports, publications, handouts, etc., relevant with this study were also examined.

Analytical techniques

Descriptive statistics

Descriptive statistics like sum, averages and percentages were calculated to identify the farmers' socioeconomic status and distribution system of government input support for producing *Aus* rice.

Crop intensification index (CII)

Crop intensification index was constructed to measure the cropping intensity in a given cropland per year [12]. The following formula was used for calculation:

$$\text{Cropping intensity} = (\text{Area}_{\text{GC}} \div \text{Area}_{\text{NC}}) \times 100$$

where Area_{GC}, gross cropped area (ha); Area_{NC}, net cropped area (ha).

Profitability analysis

Profitability of *Aus* rice production was measured in terms of gross return, gross margin, net return, benefit–cost ratio (undiscounted) and profitability ratio. The formula needed for the calculation of profitability is discussed as follows.

Gross return (GR)

Gross return was calculated by multiplying the total volume of output by the price in the harvesting period [13]. The equation was as follows:

$$\text{GR} = X_{\text{mp}}P_{\text{mp}} + X_{\text{bp}}P_{\text{bp}}$$

where X_{mp}, yield of main product (kg/ha); P_{mp}, price of main product (Tk./kg); X_{bp}, yield of by-product (kg/ha); P_{bp}, price of by-product (Tk./kg).

Gross margin (GM)

Gross margin was calculated by the difference between gross return and total variable cost. The following equation was used to calculate GM:

$$\text{GM} = \text{GR} - \sum C_{\text{v}}$$

where GR, gross return (Tk./ha); $\sum C_{\text{v}}$, total variable cost (Tk./ha).

Net return (NR)

Net return was calculated by deducting all costs (variable and fixed) from the gross return. The following algebraic form of NR was used for estimation:

$$\text{NR} = \text{GR} - \sum C_{\text{v}} - \sum C_{\text{f}}$$

where GR, gross return (Tk./ha); $\sum C_{\text{v}}$, total variable cost (Tk./ha); $\sum C_{\text{f}}$, total fixed cost (Tk./ha).

Benefit–cost ratio (BCR)

Benefit–cost ratio (BCR) is a relative measure which is used to compare the return per unit of cost. BCR was estimated as a ratio of gross return to gross cost. The formula used for calculating BCR (undiscounted) was as follows:

$$\text{BCR} = \text{GR} \div \text{GC}$$

where GR, gross return (Tk./ha); GC, gross cost (i.e., $\sum C_{\text{v}} + \sum C_{\text{f}}$) (Tk./ha).

Profitability ratio

Profitability ratio was estimated by dividing supported farmers' net return by non-supported farmers' net return [14], the formula of which was as follows:

$$\text{Profitability ratio} = \text{NR}_{\text{S}} \div \text{NR}_{\text{NS}}$$

where NR_{S}, net return of supported farmers (Tk./ha); NR_{NS}, net return of non-supported farmers (Tk./ha).

Average productivity index (API)

Aus rice productivity was measured using average productivity index (API). The following formula was adopted from [15] for calculation:

$$\text{API} = \sum_{i=1}^{n}\left(\frac{X_i - \bar{X}_i}{\text{SD}(X_i)}\right)^2 \sum_{i=1}^{n}\left(\frac{Y_i - \bar{Y}_i}{\text{SD}\,(Y_i)}\right)^2$$

where X_i, average crop yield in the cropped area (quintal/ha); $\bar{X}_i$, average crop yield in the entire region (quintal/ha); SD (X_i), standard deviation of crop yield in the cropped area (quintal/ha); Y_i, average harvested extent in the cropped area (quintal/ha); $\bar{Y}_i$, average harvested extent in the entire region (quintal/ha); SD (Y_i), standard deviation of harvested extent in the cropped area (quintal/ha).

The productivity grade was determined from the productivity range which is represented in Table 1.

Logit model

In order to identify the factors influencing adoption of *Aus* rice cultivation by the farmers with government support, the following logistic regression analysis (i.e., logit model) was used [16]:

$$K_i = \ln\left[P_i \div (1 - P_i)\right] = \beta_0 + \beta_1 X_1 + \beta_2 X_2 + \beta_3 X_3 + \beta_4 X_4 + \beta_5 X_5 + \beta_6 X_6 + \beta_7 X_7 + \beta_8 X_8 + E_i$$

where P_i is the probability of adoption and non-adoption of *Aus* rice production with government support, $P_i = 1$ indicates adoption and $P_i = 0$ indicates non-adoption.

Dependent variable: K_i, probability of adoption of *Aus* rice production with government support.

Independent variables: X_1, household size (no.); X_2, educational level of household head (years of schooling); X_3, age of household head (years); X_4, farm size (ha); X_5, farm income (Tk.); X_6, non-farm income (Tk.); X_7, extension contact ($P_i = 1$ indicates having extension contact, and $P_i = 0$ indicates having no extension contact); X_8, farming experience (years of farming); β_0, intercept; β_1–β_8, regression coefficients of the dependent variables; E_i, error term.

The marginal probabilities of the key determinants of adopting *Aus* rice production with government support were estimated based on expressions derived from the marginal effect of the logit model given as follows:

$$\text{d}K/\text{d}X = \beta_i\{P_i(1 - P_i)\}$$

where β_i, estimated logit regression coefficient with respect to the *i*th factor; P_i, estimated probability of farmers' adoption status.

Table 1 Range and grade of productivity. *Source*: Adopted from Dharmasiri (2009)

Range of productivity	Grade of productivity
87.5% and above	Very high
62.5–87.5%	High
37.5–62.5%	Medium
12.5–37.5%	Low
Below 12.5%	Very low

Measurement of food security

To identify the food security status of the farmers, two stages of analyses were done. At first, a food security index (Z) was constructed and food security status of each household was determined based on the food security line using the recommended daily calorie intake [17]. Households whose daily per capita calorie intake amounted up to 2122 kcal were regarded as food secure and those below 2122 kcal were regarded as food insecure. The mathematical representations were as follows:

$$Z_i = Y_i \div R$$

where Z_i, food security index for *i*th households (1 = food secure households and 0 = food insecure households, i.e., $Z_i = 1$ for $Y_i \geq R$; and $Z_i = 0$ for $Y_i < R$); Y_i, daily per capita calorie intake of *i*th households; R, daily per capita calorie required for *i*th households; $i = 1, 2, 3, \ldots, 120$.

Based on the household food security index (Z), food shortfall/surplus index (P) and the head count ratio (H) were calculated. Food shortfall/surplus index was calculated as:

$$P = \frac{1}{M}\sum_{i=1}^{m} G_i$$

where P, food shortfall/surplus index; M, number of food secure households (for food surplus index) or food insecure households (for food shortfall index); G_i, per capita calorie intake deficiency (or surplus) faced by *i*th households where, $G_i = [(Y_i - R) \div R]$.

The head count ratio (H) measures the percentages of the households that are food secure or insecure which is defined as:

$$H = M \div N$$

where H, head count ratio; M, number of households that are food secure (for food surplus index) or food insecure (for food shortfall index); N, number of sample households.

German correlation sensitive poverty index (GCSPI)

The GCSPI is an index designed to measure the intensity of poverty [18]. It comprises six equally weighted poverty dimensions; health (weighted indicators: health condition and health impairments), education (weighted indicators:

schooling and graduation), employment (weighted indicators: activity status, minimum wage and time poverty), housing (weighted indicators: housing condition, amenities and living space), mobility (weighted indicators: transport and crime) and income (weighted indicators: households' monthly income). The following formula was used to address the intensity of poverty:

$$\text{Intensity of poverty} = \sum d(w) \times 100$$

where d, households deprived of the indicators; w, weighted score of the indicators.

Problem confrontation index (PCI)

The researcher identified the major problems faced by the farmers in producing *Aus* rice with government input support. An overall score of the problems faced by supported and non-supported farmers was computed by adding their scores of the problems in all 10 selected problems. Each farmer was asked to indicate the extent of difficulty caused by each of the problems by checking any of the four responses such as 'frequently', 'occasionally', 'rarely' and 'not at all', and weights were assigned to these responses as 3, 2, 1 and 0, respectively. The scores of problem confrontation index (PCI) for each selected problem were computed through using the subsequent formula [19]:

$$\begin{aligned}\text{PCI} = {} & \left(P_{\text{frequently}} \times 3\right) + \left(P_{\text{occasionally}} \times 2\right) \\ & + \left(P_{\text{rarely}} \times 1\right) + \left(P_{\text{not at all}} \times 0\right)\end{aligned}$$

where $P_{\text{frequently}}$, number of responses indicating the problem occurred frequently; $P_{\text{occasionally}}$, number of responses indicating the problem occurred occasionally; P_{rarely}, number of responses indicating the problem occurred rarely; $P_{\text{not at all}}$, number of responses indicating no problem at all.

The problems were ranked according to farmers' PCI score which denoted the severity of their responses.

Results and discussion

Demographic information of the farmers

The demographic information of the farmers is represented in Table 2. It is seen that average household size of the supported farmers was 5.0 (whereas 68.0% members were male and 32.0% were female) and non-supported farmers was 6.0 (whereas 65.0% members were male and 35.0% were female). In terms of farmers surveyed, 72.0 and 76.0% were male, and 28.0 and 24.0% were female in case of supported and non-supported farmers, respectively. Average farm size of the supported and non-supported farmers was 0.67 and 0.62 hectare, respectively. Majority of the farmers (62.8% supported and 59.1% non-supported) were under the age group of 15 to 55 years which is considered as active and working group. Most of the supported farmers (41.0%) in the study areas could put sign only, whereas majority of the non-supported farmers (45.0%) were illiterate. About 80.0% supported and 82.0% non-supported farmers were engaged in different agriculture, and other income generating activities (IGA) in the study areas (Table 2).

Land tenancy arrangements of the farmers

Table 3 reveals the land tenancy status of three types of farmers (i.e., small, medium and large) within supported and non-supported categories. It is seen that most of the supported and non-supported farmers were small farmers (76.0 and 68.0%, respectively). Average farm size of supported small and medium farmers was 0.26 and 1.08 ha, which was 0.21 and 1.03 ha in case of non-supported small and medium farmers, respectively. The major share of small and medium farmers' cultivable land was rented/leased-in (76.9 and 50.0%, and 52.4 and 49.1% for supported and non-supported farmers, respectively). Though 30.6 and 22.2% cultivable land was rented/leased-out by supported and non-supported medium farmers, respectively, neither supported nor non-supported small farmer rented/leased-out any cultivable land.

Table 2 Demographic information of the farmers. *Source*: Field survey, 2017

Particulars	Farmers' categories	
	Supported	Non-supported
Average household size (no.)	5.0 (male: 68.0%; female: 32.0%)	6.0 (male: 65.0%; female: 35.0%)
Average sex distribution (% of farmers)		
Male	72.0	76.0
Female	28.0	24.0
Average farm size (ha)	0.67	0.62
Average age (years)		
Below 15 years	12.8	14.3
15–55 years	62.8	59.1
Above 55 years	24.4	26.6
Literacy rate (% of farmers)		
Illiterate	32.0	45.0
Sign only	41.0	32.0
Primary and above	27.0	23.0
Occupational status (% of farmers)		
Agriculture only	20.0	18.0
Agriculture and others	80.0	82.0

Table 3 Farmers' land tenancy arrangements. ***Source*****: Field survey, 2017**

Farmers' categories	% of farmers	Average farm size (ha)	Land tenancy arrangements (ha)		
			Own	Rented/leased-in	Rented/leased-out
Supported					
Small farmers (< 1.00 ha)	76.0	0.26	0.06 (23.1)	0.20 (76.9)	–
Medium farmers (1.01–3.00 ha)	24.0	1.08	0.21 (19.4)	0.54 (50.0)	0.33 (30.6)
Large farmers (above 3.01 ha)	–	–	–	–	–
Non-supported					
Small farmers (< 1.00 ha)	68.0	0.21	0.10 (47.6)	0.11 (52.4)	–
Medium farmers (1.01–3.00 ha)	32.0	1.03	0.26 (24.1)	0.53 (49.1)	0.24 (22.2)
Large farmers (above 3.01 ha)	–	–	–	–	–

Figures in the parentheses indicate percentages of average farm size

Farm area under crop production

Farm area under crop production picturises the total cropped area of supported and non-supported farmers which were 0.48 and 0.43 ha, respectively (Table 4). Rice cultivation constituted 68.8 and 69.8% of total cropped area of supported and non-supported farmers, respectively, where non-rice cultivation was 27.0 and 25.5%, respectively. It is also seen that fruits and agroforestry occupied 4.2% area in case of supported farmers and 4.7% area in case of non-supported farmers. *Boro* rice constituted the highest portion (29.2%) of total cropped area for supported farmers where *Aman* rice constituted the highest portion (27.9%) of total cropped area for non-supported farmers. Being less productive, farmers' attention over *Aus* rice production was less than *Aman* and *Boro* rice (16.7 and 16.3% in case of supported and non-supported farmers, respectively).

Distribution system of government input support

The government of Bangladesh has launched an input support programme in order to boost *Aus* rice cultivation and motivate the farmers towards *Aus* rice production. Under the programme, each farmer will get 5 kg seed, 20 kg urea, 10 kg DAP, 10 kg MoP and Tk. 400 in cash for cultivating UFSHI breed of *Aus* rice on 0.13 ha (1 bigha) of land. For NERICA cultivation, each farmer will get 10 kg seed, 20 kg urea, 10 kg DAP, 10 kg MoP and Tk. 800 in cash for cultivating NERICA breed of *Aus* rice on 0.13 ha (1 bigha) of land (Table 5).

Farmers' perceptions about government input support on *Aus* rice production

To appraise the government input support on *Aus* rice production, the researchers made discussion with the farmers of the study areas. Following the discussion, 10 (ten) statements were selected for this study. Each farmer was asked to indicate his/her opinion regarding the pros and cons of the input support. The percentages of farmers sharing their opinions on the selected statements are represented in Table 6.

It is reflected that 90.0% supported farmers were agreed with the statement that the support given by the government was not necessary to be paid back (Table 6). 85.0% supported farmers gave positive opinion about increase in income which was followed by 75.0% supported farmers' agreement with the statement that poor landless farmers could get support from the government. Disagreement of supported farmers for the statements that farmers might not be helped through this programme in short duration and field level GO-NGO motivational activities were found in case of 55.0 and 46.7% farmers, respectively.

For non-supported farmers, 86.7% were agreed with the statement that political affiliation was required to get input support from the government. 83.3% non-supported farmers were agreed with the statement that support given by the government is not necessary to be paid back which was followed by 81.7% non-supported farmers agreement with the statement that poor landless farmers could get support from the government. On the contrary, 40.0% non-supported farmers were disagreed with the statement that farmers might not be helped through this programme in short duration (Table 6).

Crop intensification analysis

Cropping intensity is explained as the number crops grown in a given cropland per year. It measures the productivity of per unit gross cropped area in a year (Bhaskar 2009). The whole process is named as crop intensification. Intensification of crop production in the study areas is demonstrated in Table 7. It is seen that before government input support, gross and net cropped area of supported and non-supported farmers were 0.36 and 0.22 ha; and 0.39 and 0.23 ha, respectively. Cropping intensity was found as 163.6 and 169.6% in case of supported and

Table 4 Farm area under crop production. *Source*: Field survey, 2017

Crops	Farmers' categories			
	Supported		Non-supported	
	Cultivated area (ha)	% of total cropped area	Cultivated area (ha)	% of total cropped area
Rice crops				
Aus	0.08	16.7	0.07	16.3
Aman	0.11	22.9	0.12	27.9
Boro	0.14	29.2	0.11	25.6
Total rice cropped area	0.33 (68.8)		0.30 (69.8)	
Non-rice crops				
Vegetables	0.03	6.2	0.03	6.9
Cash crops	0.02	4.2	0.01	2.3
Spices	0.02	4.2	0.02	4.7
Pulses	0.03	6.2	0.03	6.9
Oil seeds	0.03	6.2	0.02	4.7
Total non-rice cropped area	0.13 (27.0)		0.11 (25.5)	
Fruits and agroforestry	0.02	4.2	0.02	4.7
Total cropped area	0.48	100.0	0.43	100.0

Figures in the parentheses indicate percentages of total cropped area

non-supported farmers, respectively. On the contrary, after government input support, gross and net cropped area of supported and non-supported farmers were 0.48 and 0.21 ha, and 0.43 and 0.25 ha, respectively. Cropping intensity of supported and non-supported farmers was estimated at 228.6 and 172.0% after government input support. Supported farmers' cropping intensity was increased by 39.7%, whereas non-supported farmers' cropping intensity was increased by only 1.4%.

Table 5 Distribution system of government input support conditions per farmer. *Source*: Upazila Agriculture Office, 2017

Items	Amount	Conditions/remarks
Cash, Tk.	400	For UFSHI (local HYV) cultivation in 0.13 ha (1 bigha) of land
Kind, quantity		
Seed, kg	5	
Fertilizers, kg		
Urea	20	
DAP	10	
MoP	10	
Cash, Tk.	800	For NERICA (African variety) cultivation in 0.13 ha (1 bigha) of land
Kind, quantity		
Seed, kg	10	
Fertilizers, kg		
Urea	20	
DAP	10	
MoP	10	

The results implied that before government input support, supported and non-supported farmers could grow crops for nearly 1.6 and 1.7 times per year in a particular cropland whereas after government input support, supported farmers could grow crops for nearly 2.3 times per year in a particular cropland, but it was 1.7 times in case of non-supported farmers (Table 7). The result is quite similar with [20] where the authors found variability in crop intensity from one farming household to the next was higher among high-intensity households than those of low-intensity households.

Profitability of *Aus* rice production

Profitability of *Aus* rice production was measured in terms of gross return, gross margin, net return, benefit–cost ratio and profitability ratio. Variable and fixed costs were taken into deliberation to estimate the total cost of production. Variable costs included human labour, power tiller, seeds/seedlings, fertilizers, and herbicides and insecticides; and fixed cost included land use cost and interest on operating capital.

Total variable cost of supported and non-supported farmers was Tk. 21,605 and Tk. 31,426 per ha, respectively (Table 8). Total variable cost of supported farmers was lower than non-supported farmers because they were provided with essential seed/seedlings, fertilizers and cash for *Aus* rice cultivation by the government at free of cost. Total fixed cost was estimated at Tk. 8286 and Tk. 8876 per ha for supported and non-supported farmers, respectively. Total cost of *Aus* rice production in case of supported and non-supported

Table 6 Farmers' perceptions about government input support on *Aus* rice production (% of farmers). *Source*: Field survey, 2017

Statements	Farmers' categories					
	Supported ($n = 60$)			Non-supported ($n = 60$)		
	Agree	Neutral	Disagree	Agree	Neutral	Disagree
Diversified food production is increased	58.3	11.7	30.0	40.0	35.0	25.0
Nutritional security of the family members is assured	48.3	30.0	21.7	53.3	20.0	26.7
Farmers income can be increased	85.0	5.0	10.0	80.0	6.7	13.3
Farmers may not be helped through this programme in short duration	28.3	16.7	55.0	36.7	23.3	40.0
Poor landless farmers can get support from the government	75.0	3.3	21.7	81.7	10.0	8.3
Farmers can get quality seed from the government	45.0	18.3	36.7	51.7	28.3	20.0
Support given by the government is not necessary to be paid back	90.0	6.7	3.3	83.3	10.0	6.7
Political affiliation is required to get input support from the government	70.0	21.7	8.3	86.7	8.3	5.0
Income generation from non-farming activities is not included in the approach	48.3	26.7	25.0	48.3	20.0	31.7
Field level government and non-government organizations motivate the famers to adopt the programme	33.3	20.0	46.7	36.7	28.3	35.0

Table 7 Crop intensification index (CII). *Source*: Authors' estimation, 2017

Government input support stages	Farmers' categories					
	Supported			Non-supported		
	Gross cropped area (ha)	Net cropped area (ha)	Cropping intensity (%)	Gross cropped area (ha)	Net cropped area (ha)	Cropping intensity (%)
Before	0.36	0.22	163.6	0.39	0.23	169.6
After	0.48	0.21	228.6	0.43	0.25	172.0
Change in cropping intensity (%)	–	–	39.7	–	–	1.4

farmers was Tk. 29,891 and Tk. 40,302 per ha, respectively.

Table 8 represents that gross return from *Aus* rice cultivation was Tk. 49,920 and Tk. 51,220 per ha for supported and non-supported farmers, respectively. Gross margin and net return of supported and non-supported farmers were estimated at Tk. 28,315 and Tk. 19,794; and Tk. 20,029 and Tk. 10,918 per ha, respectively. BCR of *Aus* rice production was found as 1.67 and 1.27 in stare of supported and non-supported farmers, respectively, which means that by investing Tk. 100 per ha in *Aus* rice production, supported farmers received Tk. 167 in return, whereas non-supported farmers received Tk. 127. Profitability ratio of supported and non-supported farmers was 1.83 which was significant at 1% probability level and it demonstrated that while non-supported farmers earned Tk. 100 per ha from *Aus* rice production, supported farmers earned Tk. 183 per ha with the input support provided by the government.

Measurement of *Aus* rice productivity

Productivity of *Aus* rice in the study areas was estimated using average productivity index which is represented by Table 9. It is seen that average yield of supported and non-supported farmers in the cropped area was 27.8 and 27.5 quintal per ha, respectively, where average yield in the entire region was 28.2 quintal per ha. On the other hand, average harvested extent of supported and non-supported farmers was 27.6 and 27.2 quintal per ha, respectively, where average harvested extent in the entire region was 27.9 quintal per ha. Average *Aus* rice productivity was estimated at 138.0 and 100.0% in case of supported and non-supported farmers, respectively. Although productivity of *Aus* rice was high for both categories of farmers, supported farmers' one was 38.0% higher than non-supported farmers (Table 9). The results are relatively similar with [21] where the authors found improvement in smallholder maize profitability and productivity through input subsidy scheme.

Table 8 Profitability of *Aus* rice production (Tk./ha). *Source*: Authors' estimation, 2017

Particulars	Farmers' categories	
	Supported	Non-supported
Variable costs		
Human labour	17,700	17,340
Power tiller	3705	3705
Seed/seedlings	–	4940
Fertilizers		
Urea	–	2470
DAP	–	1383
MoP	–	988
Total	–	4841
Herbicides and insecticides	200	600
(1) Total variable cost	21,605	31,426
Fixed costs		
Land use cost	6990	6990
Interest on operating capital	1296	1886
(2) Total fixed cost	8286	8876
(3) Total cost	29,891	40,302
Return from *Aus* rice production		
(4) Gross return	49,920	51,220
(5) Gross margin (4 - 1)	28,315	19,794
(6) Net return (4 - 3)	20,029	10,918
(7) Benefit–cost ratio (BCR) (4 ÷ 3)	1.67	1.27
(8) Profitability ratio	1.83*** (0.007)	

Figure within the parenthesis indicates percentages of farmers; *** indicates significant at 1% probability level

Factors affecting adoption of *Aus* rice cultivation with government support

A logit model was used to identify the factors influencing adoption of *Aus* rice cultivation by the farmers in the study areas with government support. Eight independent variables were identified as major determinants of adopting *Aus* rice cultivation by the farmers in this study. The estimated equation was as follows:

$$K_i = 5.134 + 0.029X_1 + 1.103X_2 - 0.271X_3 + 0.660X_4 + 1.580X_5 + 0.004X_6 + 0.000X_7 - 0.001X_8$$

Three out of eight independent variables included in the model were found significant in explaining the variation in adopting *Aus* rice cultivation by the farmers with government input support. These variables were farm size, farm income and extension contact (Table 10).

Household size

The result of marginal effect shows that household size had a positive value of dK/dX and it was 0.002. It indicated that if household size is increased by 1 unit, the probability of adopting *Aus* rice cultivation with government support will be increased by 0.002 times.

Educational level of household head

Educational level of household head had a positive value of dK/dX which was 0.006, and it meant that if educational level of household head is increased by 1 unit, the probability of adopting *Aus* rice cultivation with government support will be increased by 0.006 times.

Age of household head

The result of marginal effect shows that age of household head had a negative value of dK/dX and it was 0.008. It implied that if the age of household head is increased by 1 unit, the probability of adopting *Aus* rice cultivation with government support will be decreased by 0.008 times.

Farm size

The result of marginal effect shows that farm size had a positive value of dK/dX and it was 0.016, which was statistically significant at 10% level of probability. It demonstrated that if farm size is increased by 1 unit, the probability of adopting *Aus* rice cultivation with government support will be increased by 0.016 times. The reason was that with a large farm size, farmers could

Table 9 Average productivity index (API). *Source*: Authors' estimation, 2017

Particulars	Farmers' categories	
	Supported	Non-supported
Average yield in the cropped area (quintal/ha)	27.8	27.5
Average yield in the entire region (quintal/ha)	28.2	28.2
Standard deviation of yield in the cropped area (quintal/ha)	0.5	1.0
Average harvested extent in the cropped area (quintal/ha)	27.6	27.2
Average harvested extent in the entire region (quintal/ha)	27.9	27.9
Standard deviation of harvested extent in the cropped area (quintal/ha)	0.2	0.5
Average productivity (%)	138.0	100.0

Table 10 Estimates of coefficients and marginal effects of logit model. *Source*: Authors' estimation, 2017

Variables	Coefficient (β)	SE	z	P > |z|	95% CI		dK/dX
Constant	5.134	2.967	1.730	0.084	0.690	10.942	–
Household size (X_1)	0.029	0.031	0.942	0.295	– 0.093	0.028	0.002
Educational level of household head (X_2)	0.103	0.079	1.293	0.175	– 0.263	0.048	0.006
Age of household head (X_3)	– 0.271	0.174	– 1.557	0.382	– 0.493	– 0.189	– 0.008
Farm size (X_4)	0.660*	0.367	1.801	0.091	1.008	0.430	0.016
Farm income (X_5)	1.580**	0.682	2.316	0.043	– 1.616	– 1.058	0.016
Non-farm income (X_6)	0.004	0.003	1.519	0.153	0.010	0.000	0.000
Extension contact (X_7)	0.000**	0.000	2.433	0.036	0.000	0.000	0.000
Farming experience (X_8)	– 0.001	0.002	– 0.618	0.579	– 0.005	0.003	– 0.000

**, *Significant at 5 and 10% probability level, respectively

cultivate less productive *Aus* rice in a noticeable amount of cropland keeping cultivation of *Aman* and *Boro* rice in other cropland.

Farm income

Farm income had a positive value of dK/dX which was 0.016, and it was statistically significant at 5% level of probability. It indicated that if farm income is increased by 1 unit, the probability of adopting *Aus* rice cultivation with government support will be increased by 0.016 times. The reason was that farmers earning a healthy amount of monetary income from other farming enterprises could take a risk of profit or loss from *Aus* rice cultivation.

Non-farm income

Non-farm income had a positive value of dK/dX, and it was 0.000. It meant that if non-farm income is increased by 1 unit, the probability of adopting *Aus* rice cultivation with government support will be increased by 0.000 times.

Extension contact

The result of marginal effect shows that extension contact had a positive value of *dK*/*dX* and it was 0.000, which was statistically significant at 5% level of probability. It implied that the probability of adopting *Aus* rice cultivation with government support for those farmers who have extension contact is 0.000 times higher compared to those farmers who do not have extension contact. The reason was that farmers got influenced and motivated by the extension agents to adopt *Aus* rice cultivation with government input support.

Farming experience

The result of marginal effect shows that farming experience had a negative value of dK/dX which was 0.000. It meant that if extension contact is increased by 1 unit, the probability of adopting *Aus* rice cultivation with government support will increase by 0.000 times.

Average annual income of the farmers

The money income earned by the farmers from different sources is depicted in Table 11. Mainly, there were two sources of money income in the study areas, namely farm income and non-farm income. Farm income included income from crop, livestock, poultry, homestead and agroforestry. Non-farm income included income from small business, wage labour, shopkeeping, van/rickshaw pulling and other sources. It is found that average annual income of supported and non-supported farmers was Tk. 92,092 and Tk. 87,459, respectively. In case of supported farmers, 59.4% income (Tk. 54,727) was from farming activities and 40.6% (Tk. 37,365) was from non-farming activities where these were 54.5% (Tk. 47,665) and 45.5% (Tk. 39,794) in terms of non-supported farmers, respectively. The result is supported by [22] where the authors found that crop farming was the largest source of farm income for the sampled farmers.

Food security status of the households

Food security was estimated from the viewpoint of three perspectives, such as availability of safe and nutritious food, access to food and utilization of food. It is evident from Table 12 that food security index value for food secure supported and non-supported households was 1.09 and 1.03, and for food insecure supported and non-supported households was 0.43 and 0.50, respectively.

Based on the recommended daily calorie intake of 2122 kcal, it is observed that 71.7 and 65.0% households were food secure supported and non-supported; and remaining 28.3 and 35.0% households were food insecure supported and non-supported, respectively. Average calorie intake of food secure supported and non-supported households was 2017.4 and 1984.2 kcal per day which was 1788.4 and 1940.1 kcal in case of food insecure supported and non-supported households, respectively, but these were still lower than the national average level (i.e., 2122 kcal) for both food

Table 11 Average annual income of the farmers

Sources of income	Farmers' categories			
	Supported		Non-supported	
	Tk./year	Percentage of total income	Tk./year	Percentage of total income
Farm income				
Crop	30,569 (55.8)[a]	59.4	26,629 (55.9)[a]	54.5
Livestock	12,357 (22.6)[a]		9749 (20.5)[a]	
Poultry	2551 (4.7)[a]		3017 (6.2)[a]	
Homestead and agroforestry	8000 (14.6)[a]		7420 (15.6)[a]	
Others	1250 (2.3)[a]		850 (1.8)[a]	
Total farm income	54,727 (100.0)[a]		47,665 (100.0)[a]	
Non-farm income				
Small business	12,575 (33.7)[b]	40.6	11,972 (30.1)[b]	45.5
Wage labour	8119 (21.7)[b]		10,041 (25.2)[b]	
Shopkeeping	8790 (23.5)[b]		7990 (20.1)[b]	
Van/rickshaw pulling	5044 (13.5)[b]		6796 (17.1)[b]	
Others	2837 (7.6)[b]		2995 (7.5)[b]	
Total non-farm income	37,365 (100.0)[b]		39,794 (100.0)[b]	
Total income	92,092	100.0	87,459	100.0

[a] Figures in the parentheses indicate percentages of total farm income; and

[b] Figures in the parentheses indicate percentages of total non-farm income

Table 12 Food security indices of the farm households. *Source*: Authors' estimation, 2017

Households' categories	Food security indices	Index values	
		Food secure households	Food insecure households
Supported	Food security index (*Z*)	1.09	0.43
	Head count index (*H*)	71.7	28.3
	Per capita daily calorie availability	2017.4	1788.4
	Food shortfall/surplus index (*P*)	0.21	− 0.16
Non-supported	Food security index (*Z*)	1.03	0.50
	Head count index (*H*)	65.0	35.0
	Per capita daily calorie availability	1984.2	1910.1
	Food shortfall/surplus index (*P*)	0.16	− 0.20

secure and insecure supported and non-supported households. The value of food surplus index in stare of food secure supported and non-supported households was 0.21 and 0.16, respectively, which indicated that households of both categories had superfluous food for crisis period, where the value of food shortfall index for food insecure supported and non-supported households was − 0.16 and − 0.20, respectively, indicating a situation of food shortage and no surplus food at the dilemma period for households of both categories (Table 12). The result is quite similar with [23] where the authors observed that average per capita daily calorie intake of the *char* dwellers was below the national average level of 2122 kcal.

German correlation sensitive poverty index (GCSPI)

To calculate GCSPI, poverty situation and livelihood condition of the farm households were evaluated on the basis of six poverty dimensions: health (weighted indicators: health condition and health impairments), education (weighted indicators: schooling and graduation), employment (weighted indicators: activity status, minimum wage and time poverty), housing (weighted indicators: housing condition, amenities and living space), mobility (weighted indicators: transport and crime) and income (weighted indicators: households' monthly income). The proportion of deprived supported and non-supported households was 29.0 and 34.8%, respectively; and the proportion of privileged supported and non-supported

Table 13 German correlation sensitive poverty index (GCSPI). ***Source*****: Authors' estimation, 2017 and HIES, 2010**

Indicators	Households' categories				Weights
	Supported ($n = 60$)		Non-supported ($n = 60$)		
	No. of households deprived (√) or privileged (×) based on the indicators				
	√	×	√	×	
Health					
Subjective health condition, either poor or bad	22/60	38/60	27/60	33/60	1/12
Lack of physical and mental health condition due to health impairments	9/60	51/60	8/60	52/60	1/12
Education					
Less than 9 years of schooling	33/60	27/60	41/60	19/60	1/12
Neither graduation nor training qualification	45/60	15/60	54/60	6/60	1/12
Employment					
Unemployed status of activity	0/60	60/60	0/60	60/60	1/18
Working with below minimum wage	13/60	47/60	9/60	51/60	1/18
Working hour at least 8 h	28/60	32/60	36/60	24/60	1/18
Housing					
In urgent need of complete renovation to avoid the danger of breaking down	5/60	55/60	13/60	47/60	1/18
Lack of comfortable amenities	44/60	16/60	56/60	4/60	1/18
Living space below minimum requirement (45 sq. metre)	10/60	50/60	17/60	43/60	1/18
Mobility					
No personal vehicle available and public transport more than 20 min away	20/60	40/60	17/60	43/60	1/12
Insecure or dangerous neighbourhood	6/60	54/60	0/60	60/60	1/12
Income					
Monthly household income below breadline (11,479 Tk.)	3/60	57/60	8/60	52/60	1/6
Score of the households	0.290	0.710	0.348	0.652	–
Intensity of poverty (%)					
Deprived (√) households	29.0		34.8		
Privileged (×) households	71.0		65.2		

Score of deprived supported households = (22/60 × 1/12) + (9/60 × 1/12) + (33/60 × 1/12) + (45/60 × 1/12) + (0/60 × 1/18) + (13/60 × 1/18) + (28/60 × 1/18) + (5/60 × 1/18) + (44/60 × 1/18) + (10/60 × 1/18) + (20/60 × 1/12) + (6/60 × 1/12) + (3/60 × 1/6) = 0.290; score of privileged supported households = (38/60 × 1/12) + (51/60 × 1/12) + (27/60 × 1/12) + (15/60 × 1/12) + (60/60 × 1/18) + (47/60 × 1/18) + (32/60 × 1/18) + (55/60 × 1/18) + (16/60 × 1/18) + (50/60 × 1/18) + (40/60 × 1/12) + (54/60 × 1/12) + (57/60 × 1/6) = 0.710; scores of deprived or privileged non-supported households were calculated accordingly; percentage of deprived supported households = 0.290 × 100 = 29.0; percentage of privileged supported households = 0.710 × 100 = 71.0; percentages of deprived or privileged non-supported households were calculated accordingly

households was 71.0 and 65.2%, respectively (Table 13), the households were deprived or privileged based on all the indicators of a single dimension or at a combination of the indicators across dimensions. The reason for a better livelihood condition of supported farmers was that farmers could save money from irrigation and other energy inputs (i.e., the inputs provided with by the government) and further employ it in other income generating activities. This result is quite similar with [24] where the authors observed improved rural poverty status through agricultural input subsidy programmes.

Problem confrontation index (PCI)

Problem confrontation index (PCI) is a mathematically problem ranking index which is defined as a set of objects which must satisfy a number of problems or limitations. It represents the entities in a problem as a homogeneous collection of finite limitations over variables in a specific area. The farmers of the study areas were asked to give their opinion on 10 selected problems which were identified during data collection period and after computing the PCI scores, the problems were ranked according to their PCI score. The computed PCI scores of the 10 problems ranged from 93 to 139 (against a possible range from 0 to 180) for supported farmers and 98 to 119 (against a possible range from 0 to 180) for non-supported farmers which were arranged in rank order according to their PCI scores as shown in Table 14.

Majority of the farmers opined that the price of the output was very low in response to their production costs.

Table 14 Problem confrontation index including ten (10) selected problems. *Source*: Authors' estimation, 2017

Identified problems	Farmers' categories											
	Supported ($N = 60$)						Non-supported ($N = 60$)					
	Frequently (3)	Occasionally (2)	Rarely (1)	Not at all (0)	PCI	Rank order	Frequently (3)	Occasionally (2)	Rarely (1)	Not at all (0)	PCI	Rank order
Insufficient support	17	26	12	5	115	4	23	17	14	6	117	3
Lack of proper distribution	24	16	14	6	118	3	16	16	18	10	98	10
Lack of proper training	15	13	22	10	93	10	15	17	20	8	99	9
High price of seed and fertilizer in the market	28	21	8	3	134	2	22	19	15	4	119	1
Lack of good quality seed and fertilizer	19	19	17	5	112	5	17	16	21	6	104	8
Low price of output	34	16	5	5	139	1	23	17	15	5	118	2
Lack of storage of product during harvesting	16	22	15	7	109	8	25	14	13	8	116	4
Lack of transportation facility	20	17	17	6	111	6	19	18	13	10	106	7
Less technical support from different organizations	18	19	15	8	107	9	19	17	17	7	108	6
Lack of scientific knowledge and method	20	14	19	7	110	7	21	18	14	7	113	5

Calculation of PCI score for the problem of lack of insufficient support-PCI score of supported farmers = $(17 \times 3) + (26 \times 2) + (12 \times 1) + (5 \times 0) = 115$; PCI score of non-supported farmers = $(23 \times 3) + (17 \times 2) + (14 \times 1) + (6 \times 0) = 117$; PCI scores for rest of the problems were computed accordingly

The PCI score of this problem was 139 and 104 ranking by 1st and 2nd in terms of supported and non-supported farmers, respectively. High price of seed and fertilizer in the market was another major problem faced by the farmers in the study areas. The PCI score of this problem was calculated at 134 and 119, which was ranked as 2nd and 1st problem along with the statements of supported and non-supported farmers, respectively. Most of the farmers stated about lack of proper distribution as one of the major problems. With the statements of supported and non-supported farmers, the PCI score of this problem was marked as 118 and 113 which was ranked as 3rd and 5th, respectively. Other problems like insufficient support, lack of proper training, lack of good quality seed and fertilizer, lack of storage of product during harvesting, lack of transportation facility, less technical support from different organizations and lack of scientific knowledge and method were ranked as 4th (with PCI score 115), 10th (with PCI score 93), 5th (with PCI score 112), 8th (with PCI score 109), 6th (with PCI score 111), 9th (with PCI score 107) and 7th (with PCI score 110) in case of supported farmers; and 3rd (with PCI score 117), 9th (with PCI score 99), 8th (with PCI score 104), 4th (with PCI score 116), 7th (with PCI score 106), 6th (with PCI score 108) and 10th (with PCI score 98) in case of non-supported farmers (Table 14). The result is fairly comparable with [25] where the author found that lack of information on subsidy programmes was the major problem in Africa.

Conclusions

The study concludes that the motivation provided with by the government through input support on *Aus* rice production was highly appreciated by the farmers. The study exposed that cropping intensity as well as profitability and productivity of *Aus* rice was increased through this programme. It is also seen from the study that three out of eight independent variables had significant influence on adoption of *Aus* rice cultivation by the farmers with government support which were: farm size, farm income and extension contact. Though per capita daily calorie intake by the farmers was still under the national average level, supported farmers were more food secure than non-supported farmers. It is also revealed that farmers' income was increased as a result of minimized input cost. The study also indicates that poverty in terms of deprivation of health education, employment, housing, mobility and income was decreased, and overall livelihood circumstances were improved through government input support on *Aus* rice production. Majority of the farmers stated about low output price, high price of seed and fertilizer in the market and lack of proper distribution as major problems they faced. This study will help and motivate the researchers and policy makers to uncover the critical areas of government forthcoming supports' impact on crop production. Considering the findings of the study, some essential policy recommendations have been arisen which are: motivational and extension services of government should be strengthened and properly implemented to raise farmers' awareness about *Aus* rice production. Also, initiative for scientific and technical training programmes should be arranged by different government and non-government organizations to strengthen safety net programmes for enhancing farmers' food security and reduction in poverty.

Authors' contributions

MTU designed, collected and checked the analysed data; supervised and reviewed the draft manuscript and approved the final manuscript. ARD coordinated the study, collected and analysed the data and prepared the draft manuscript. Both authors read and approved the final manuscript.

Acknowledgements

The authors express earnest gratitude to the respondents and stakeholders of the research areas for their alliance and assistance in conducting surveys and collecting the data and information.

Competing interests

The authors declare that they have no competing interests.

Funding

The authors express gratefulness to Ministry of Science and Technology, People's Republic of Bangladesh, for funding to conduct this research.

References

1. FAO. Food and Agriculture Organization of the United Nations, Rome, Italy. http://www.fao.org/fileadmin/templates/ess/documents/apcas26/presentations/APCAS-16-5.2.5_-_Bangladesh_-_AMIS_in_Bangladesh.pdf (2013). Accessed 05 Oct 2017.
2. BBS. Statistical yearbook of Bangladesh, Bangladesh Bureau of Statistics, Statistics Division, Ministry of Planning, Government of the People's Republic of Bangladesh, Dhaka; 2015.
3. BER. Bangladesh Economic Review, Finance Division, Ministry of Finance, Government of the People's Republic of Bangladesh, Dhaka; 2013.
4. BRRI. Bangladesh Rice Research Institute, Government of the People's Republic of Bangladesh, Joydebpur, Gazipur; 2011.
5. UNB. United News of Bangladesh. Govt launches Tk 336 m incentive to boost *Aus* production. The Financial Express. http://www.thefinancialexpress-bd.com/2016/03/27/23283 (2016). Accessed 02 Oct 2017.
6. Eills F. Agricultural polices in developing countries. Cambridge: Cambridge University Press; 1992.
7. Salam MA, Iftekharuddaula KM, Siddique AB, Rashid MA, Rashid MH, Momin MSI, Kabir MS and Biswas JK. Strategic plan for increasing *Aus* and *Aman* rice cultivation in Bangladesh. In: Conference paper; 2014; p. 1–17.
8. Parvin J. Agricultural input assistance card: direct input subsidy disbursement. Unnayan Onneshan: The Innovators, Centre for Research and Action for Development; 2011; p. 1–9.
9. Barkat A. A quantitative analysis of fertilizer demand and subsidy policy in Bangladesh. Final report submitted to the Manob Sakti Unnayan Kendro (MSUK), Dhaka, Bangladesh. http://fpmu.gov.bd/agridrupal/sites/default/files/Barkat.pdf (2010). Accessed 18 Sept 2017.

10. Chirwa E, Matita M and Dorward A. Factors influencing access to agricultural input subsidy coupons in Malawi. Future Agricultures Consortium, Brighton, Sussex. 2010. http://www.future-agricultures.org/publications/working-papers-document/factors-influencing-access-toagricultural-input-subsidy-coupons-in-malawi/. Accessed 05 Aug 2017.
11. Sharma VP and Thaker H. Fertilizer subsidy in India: Who are the beneficiaries? Working paper, Indian Institute of Management, Ahmedabad, India. 2009. http://www.indiaenvironmentportal.org.in/files/fertilizer%20 subsidy.pdf. Accessed 27 Sept 2017.
12. Bhaskar S. Cropping intensity in India. Knowledge of agriculture. 2009. http://knowledgeofagriculture.blogspot.com/2009/11/cropping-intensity-in-india.html. Accessed 05 Aug 2017.
13. Dillon JL, Hardaker JB. Farm management research for small farmer development. Rome: Food and Agriculture Organization of the United Nations; 1993.
14. Tulsian M. Profitability analysis (a comparative study of SAIL & TATA steel). IOSR J Econ Finance. 2014;3(2):19–22.
15. Dharmasiri LM. Measuring agricultural productivity using the average productivity index (API). Sri Lanka J Adv Soc Stud. 2009;1(2):25–44.
16. Gujarati DN. Basic econometrics. New York: McGraw-Hill; 2003.
17. Babatunde RO, Omotesho OA, Sholotan OS. Factors influencing food security status of rural farming households in north central Nigeria. Agric J. 2007;2(3):351–7.
18. Rippin N. Multidimensional poverty in Germany: a capability approach. Forum Soc Econ. 2016;45(2–3):230–55.
19. Uddin MT, Dhar AR, Rahman MH. Improving farmers' income and soil environmental quality through conservation agriculture practice in Bangladesh. Am J Agric Biol Sci. 2017;12(1):55–65.
20. Islam MM, Uddin MT. Impact of GO-NGO support on crop intensification and food security in Sirajganj *char* areas. Bangladesh J Crop Sci. 2014;25:25–36.
21. Denning G, Kabambe P, Sanchez P, Malik A, Flor R, Harawa R, Nkhoma P, Zamba C, Banda C, Magombo C, Keating M, Wangila J, Sachs J. Input subsidies to improve smallholder maize productivity in Malawi: toward an African green revolution. PLoS Biol. 2009;7(1):2–10.
22. Jannat A, Uddin MT. Farmers' perception about 'one house one farm' project and its impact on enterprise profitability in selected areas of Mymensingh district. Agriculturists. 2016;14(1):43–53.
23. Uddin MT, Dhar AR. *Char* people's production practices and livelihood status: An economic study in Mymensingh district. J Bangladesh Agric Univ. 2017;15(1):73–86.
24. Mason NM, Jayne TS and Mofya MR. A review of Zambia's agricultural input subsidy programs: targeting, impacts, and the way forward. Indaba Agricultural Policy Research Institute (IAPRI), Lusaka, Zambia. https://ageconsearch.umn.edu/bitstream/162438/2/wp77.pdf (2013). Accessed 05 Oct 2017.
25. Dorward A. Rethinking agricultural input subsidy programmes in a changing world. Trade and Markets Division, Food and Agriculture Organization of the United Nations, Rome, Italy. https://www.oecd.org/tad/agricultural-policies/46384527.pdf (2009). Accessed 17 Aug 2017.

Transitory coping strategies of food-insecure smallholder farmer households: the case of Ilu Gelan District, West Shoa Zone, Oromia Reginal State, Ethiopia

Belachew Dessalegn*

Abstract

Background: Over 960 million people in the world are hungry and undernourished. The majority of these people are found in Asia and Africa. Approximately one-third of the people in sub-Saharan Africa are undernourished. The mechanisms pursued by households differ in several aspects within and between households. Coping strategies are short-term, location-specific actions and adjustments against hazard and activities that take place within existing structures. Before coming to the modernization time, every society around the world has attempted to overcome food shortages at household levels. They practice activities to escape them from food insecurity.

Methods: The study was aimed at coping strategies among food-insecure smallholder farmer households in Ilu Gelan District, West Shoa Zone, Oromia Regional State, Ethiopia. A cross-sectional study design and mixed data collection methods were employed. Multistage random sampling technique was employed to select 100 sample households for quantitative data and key informant interview, focused group discussion and observation for qualitative data. Data were analyzed using descriptive and inferential statistics.

Results: A coping strategy index at household level had been calculated, and inferential statistics was used to test the variability of the index by gender of the household head. The mean coping index was 88.54 and 119.14 for males and females, respectively. Using inferential statistics, equality of means was tested. The t value was -5.173 for 98° of freedom, and the mean difference was significant ($p < 0.001$). The study revealed that female-headed households were higher in coping measures and mean of coping strategy index than male-headed households.

Conclusions: In the study districts, smallholder farmer households rely on less preferred and less expensive food items. Coping strategy adopted by poor rural households is a shift to poor, and nutritionally lacking diet leads to health-related problems.

Keywords: Transitory, Coping strategies, Food insecurity, Smallholder farmer, Household

Background

Global estimates of food-insecure populations stand at 825 million [1] to 850 million [2]. Regional estimates of the food-insecure population include 263 million in South Asia, 268 million in China and Southeast Asia, 2012 million in sub-Saharan Africa (SSA), 60 million in South and Central America and the Caribbean, and 50 million in other regions of the world. Contrary to the United Nations' Millennium Development Goals of cutting hunger by half by 2015, the number of food-insecure populations in the world has been on the rise [3]. The stock of food grains in the world in 2007/2008, the lowest in decades, was only 75 million tons for milled rice and 105 million tons for corn in early 2008 [4]. An estimated 75% of the world's poor (those who live on less than < $2/

*Correspondence: belachewdessalegn@yahoo.com
Department of Rural Development and Agricultural Extension, College of Agriculture and Natural Resources, Madda Walabu University, P.O. Box 47, Bale Robe, Ethiopia

day income) live in rural areas and depend directly or indirectly on agriculture [5]. Food prices are rising [6], leading to riots in 30 countries around the globe [7, 8]. Share of family income spent on food is estimated at 10% in the USA, 20% in Brazil, 30% in China, 50% in Kenya and 65% in Bangladesh [8], and 9.7 million people food-insecure Ethiopians require relief assistance to meet basic food needs [9]. Thus, the world poor are under great stress, and an increase in food prices is a threat to global peace and stability [3].

The absence of food security associated with coping strategies increasingly being used. Human beings can struggle to sustain their life when food shortages happen, because food is one of our most basic needs than shelter and cloth. Before coming to the modernization time, every society around the world has attempted to overcome food shortages at household levels. They practice certain activities to escape them from food insecurity. Many years ago, when the world population was much lower than it is now, a man had little difficulty in ordinarily time in growing the food that was needed. At that time, human beings started to lead their life in ancient time by hunting wild animals and gathering fruity crops.

Farmers have developed coping and adaptation strategies to buffer against the adverse effects of climate change and variability [10, 11] by altering their economies and lifestyles with changing circumstances in their environments [12], and the concepts were raised beyond the climate literature [13]. Coping and adaptation to climate change and variability are closely related and interchangeably used in the context of disaster response except that they have different time spans. Coping strategies are autonomous, short-term, location-specific actions and adjustments targeted against a certain hazard and activities that take place within existing structures [14–16]. Coping strategies help to mitigate the negative effects of climate change and variability for the short term, but they are "risk spreading" in nature [11].

On average, agriculture contributes 33% to national income, 70% to full-time employment and 40% to total export earnings in Africa [17]. Most of sub-Saharan Africa relies on agriculture for employment and food security for their economies. Even though agriculture is important for the national economy, it is highly dominated by smallholder farmers who produce under unfavorable conditions characterized by low and erratic rainfall and poor soils [18]. Agriculture in Ethiopia by large is subsistence [19]. Compared to other sectors, agriculture is highly vulnerable to climate change which manifests itself in terms of longer-term trends in the average conditions of rainfall and temperature, interannual variability and the occurrence of droughts, floods, frosts and other extreme events [20]. Ethiopia's economy is dependent on agriculture which is characterized by a low-input, low-output and subsistence production system [21]. It is extremely vulnerable to climate change and variability, resulting in vulnerability of smallholder farmers [22, 23]. Droughts and floods occur frequently in most parts of Ethiopia, indicating how the country is suffering from climate variability and extreme events, and future climate change constitutes a major development challenge [24]. Drought followed by flood is the most common climate-related hazard in Ethiopia [25, 26].

Rainfall variability and associated droughts have been the major causes of food insecurity and poverty traps for many households during the past three decades [27, 28]. In addition, since the water resource of the country is governed by the amount and distribution of rainfall [29, 30], the spatiotemporal variability and declining trend of rainfall have negative impacts on the water resource sector. Climate change is likely to change rainfall patterns, resulting in shorter growing seasons in the future, particularly for subsistence farmers in Africa who rely on rain fed agriculture [31]. Extreme weather events such as droughts and floods are predicted to become more frequent, adding to the global burden of hunger caused by poverty, weak governance, conflict and poor market access [32, 33]. A recent vulnerability mapping in Africa cited Ethiopia as one of the most vulnerable countries to climate change, food insecurity and with the least adaptive capacity [34, 35]. Its geographical and climatic conditions, high dependence on agriculture and weak adaptive capacity were stated as the major reasons for its vulnerability [35, 36]. Sub-Saharan Africa remains the only region in the world where the number of hungry and malnourished populations will still be on the increase even by 2020 [37]. While other regions have improved per capita food availability since the 1970s, food production and availability have perpetually declined in sub-Saharan Africa. It is both a technological and a political/economic challenge and cannot be ignored any longer. Agrarian stagnation in sub-Saharan Africa has defined numerous attempts at transforming subsistence agriculture, even with due consideration to issues related to biophysical constraints and the human dimension challenges [38, 39]. Agriculture accounts about 41.6% of the GDP, employs about 83% of the labor force and contributes around 90% of the total export earnings of Ethiopia [19].

The sector is dominated by about 11.7 million smallholders cultivating about 95% of the national agricultural production, and large farms contributed to only 5% of the total production [40, 41]. This shows that the overall economy of the country depends on smallholder subsistence agriculture. Food production and population statistics in Ethiopia are notoriously unreliable, and all estimates of national food availability and consumption requirements are "guesstimates" at best [42]. Given this limitation of statistics during the late 1980s, 52% of Ethiopia's population consumed less than the recommended daily allowance of 2100 kcal, and Ethiopian agriculture appears to be locked into a downward spiral of low and declining productivity, caused by an adverse combination of agro-climatic, demographic, economic and institutional constraints, trends and shocks. Some observers argue that a "Malthusian crisis" is developing as rapid population growth (almost 3% per annum) is associated with steadily falling landholdings and per capita food production [42]. Between 1960 and 1990, the population doubled from 23 to 48 million, while per capita landholding shrunk from 0.28 to 0.10 hectare, and per capita food output collapsed by 41% from 240 to 142 kg [42].

In Africa, women smallholder famers were one of powerful engines and would play key role in the development process to ensure food security at household level and they made greater contribution to household food security but women smallholders farmers households were the most neglected in the development policies and programs despite its contribution to food security at national levels.

The concept of coping strategies is not new. But, different researchers, scholars, authors and organizations have defined the term coping strategies in different ways. These indicate that there is no comprehensive definition [43]. Devereux defines as a response to adverse events or shocks. Broad definition of coping strategies, namely that "all the strategically selected acts that individuals and households in a poor socioeconomic position use to restrict their expense or earn some extra income to enable them to pay for the basic necessities (food, clothing, shelter) and not fall too far below their society's level of welfare" [44]. Food insecurity is still widespread, especially among developing nations [45]. Over 960 million people in the world are hungry and undernourished [46]. The majority of these people are found in developing countries, most especially in Asia and Africa, and many die of hunger-related diseases (Standing Committee on Nutrition [47]). According to African Food Security Briefs [48], approximately one-third of the people in sub-Saharan Africa are undernourished [45].

The strategies pursued by households differ in several aspects, that is, within the household and between households [49]. Due to varying degrees of wealth among households, different coping behaviors are adopted by households at different poverty levels. However, some coping strategies are common to all households although the extent to which such strategies enable a household to remain afloat depend on the assets at their disposal [43]. Above all, the general tendency is that the lower the household asset status, the more likely the household would engage in erosive responses such as selling off productive assets such as farm implements [50].

Methods

Description of the study area

For the administrative purpose, the Ilu Gelan District is divided into 17 peasants associations and one town. The study area was located at water shed of Gibe River. The population has grown over the years. Ilu Gelan District is located at a 90 km from zonal capital (Ambo). The district capital, Ijaji, is located 215 km from regional capital (Finfine). The district was bounded by Chelia District in north and east, Dano District in the south and Bako District in west (Ilu Gelan Agricultural Offices, 2004 EC). The district is divided into three distinct climatic zones as high land, midland and low land respectively.

The average maximum and minimum temperatures of the district are 32 and 25 °C, respectively. There are four types of soil found in the district: fertile soil, sandy loam soil, clay soil and red soil. Out of the total area of the district Ilu Gelan is covered by red soil 70% and sandy loam soil 10% [51]. Population of the study was smallholder farmer households that live in the rural areas of the district. Currently there are 4073 households in the two selected kebele in the study area [51].

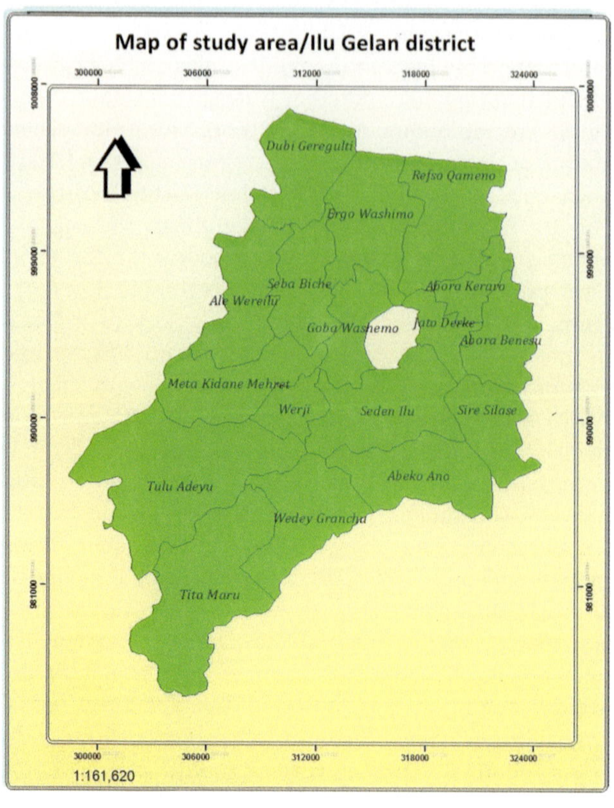

Study design

Household-based cross-sectional survey research study was conducted in Ilu Gelan to identify transitory coping strategies of food-insecure smallholder farmer households. Sampling design was determined by type of universe to be studied, households in this study, sampling unit, sampling frame, size of sample and parameters of interest. Because of the nature of the study, multistage sampling technique was applied in order to increase the reliability and validity of the data.

Sample size determination and method of sampling

Sampling is a technique, which helps us in understanding the parameters or characteristics of the Universe or population by examining only a small part of it. Therefore, it is necessary that sampling technique be reliable [52]. Appropriate sample size depends on various factors relating to the subject under investigation such as the time, cost and degree of accuracy desired. [53]. Nevertheless, the sample size and the sample selection procedure should assure the representativeness of the population. Sample size determination has its own scientific approach.

To analyze coping strategies on food insecurity of the households, the researcher used multistage sampling technique to select sample food-insecure households in the study district.

First stage: The study district, Ilu Gelan, was selected purposively based on; repeated decreasing agricultural production, ever-increasing natural and man-made distresses and its threatening effect on high soil erosion and deforestation in the area, vulnerability to food insecurity to due to climate change and environment failures. Second stage: Two samples have been selected randomly out of 17 kebele. Third stage: The next step was selection of the sample of household head. Finally, 100 sample smallholder food-insecure households were selected by using random simple techniques. The sample size for collecting quantitative data for this research is determined by using Yemane Formula [54]. The study used the following formula to calculate sample size. The researcher has adopts [54] for determining sample size. $n = \frac{N}{1+N(e^2)}$.

The following steps were used to determine sample size derived from the above formula to collect quantitative data using semi-structured interview schedule. Here, n designates the sample size the research uses; N designates total number of households heads; e designates maximum variability or margin of error 8%; 1 designates the probability of the event occurring.

Therefore, $n = \frac{N}{1+N(e^2)}$

$$n = \frac{4073}{1 + 4073(0.08)2} = 100$$

Therefore, total sample size was 100 household head farmers, out of which 50 were selected from Seba Biche and the remaining 50 were from Meta Kidane Mehreta. Selection was made proportionally from total household living in both kebeles. The total sample drawn was 100.

Methods of data collection

This study based on a micro-level, and it is derived from a cross-sectional primary data. The structured household questionnaire was used to collect the data from 100 farm households. The data were collected through household survey. Moreover, focused group discussions and key informants interviews were conducted in the village communities. Data analysis procedures consist of descriptive statistics and coping strategies index (CSI). The CSI is developed by [55] to measure the food security situation. The basic idea of CSI is to combine the frequency and severity of coping strategies. The frequency of coping strategies requires the means of scoring of relative frequency which measures how many days per week a household had to rely on the various coping strategies ranking from "never" to "every day." The severity of coping strategies is measured using focused group discussion via asking the individuals to classify their coping strategies based on their opinion (1 = less severe, 2 = moderate, 3 = severe and 4 = very severe). The means of scoring reflect the severity weight of each coping strategy that household has adopted. Thus, the CSI score is calculated by combining of both "frequency" and "severity" of coping strategies. The result of the CSI score denotes that a household with a higher value is more food insecure compared with a household with a lower value.

The survey gathered data pertaining to the demographic and economic aspects of the households. I also included items related to the causes of food insecurity in the district, as well as access to the market, receipt of food aid and the distance of the village away from the main road. The situation of the stock status was measured through the availability of sufficient sustainability daily rations for household members, the number of meals taken per day and the number of days per week meat or fish eaten in the household. The odds of reporting that one often experienced food shock were also included in the analysis.

Method of data analysis

The data analysis process was carried out after collection of the required information from primary sources.

The data were analyzed by Statistical Package for Social Sciences (SPSS) version 20. SPSS was used to analyze different variables through descriptive statistics such as, frequencies, mean, standard deviation and percentage.

Results

Measuring household food insecurity

In this study, sample households were classified into food secure and food insecure on the basis of the 2100 kilocalorie threshold. Households with daily calorie consumption greater than or equal to 2100 kcal per day were categorized as food secure and those households whose calorie intake fell below this food security threshold grouped as "food-insecure" based on Ethiopian Health and Nutrition Research Institute (EHNRI) recommendation [56].

Coping strategies

To increase longevity and starting some things that makes life, the indicators of coping mechanisms are listed in Table 1. But it is different for male- and female-headed households. In addition to that, the study considered some food security proxy indicators. These are the food consumption score which combines information on meal frequency and dietary diversity and the coping strategy index which has proven to be a good indicator of food security level.

Coping strategies and weighing

According to the information gathered from focused group discussion and key informant interview in the study districts, the use of these coping strategies depends on the most frequently used coping strategies they used. Table 2 shows average severity weight for various coping strategies among selected localities smallholder farmer households in Ilu Gelan, in 2017/2018G.C. The coping strategy index was adopted and modified in the present study [55]. During FGD and key informant interview, the major coping mechanisms in the area were listed and finalized. The revised CARE/WFP list is given in Table 2. There were 19 coping strategies listed. As explained in "Methods" section, weighing of the strategies is very important to ensure the cultural sensitivity of the population. This exercise is done during the FGD and KII. The average weight of each strategy during two FGDs was finalized after the consensus ranking with key informants. For consensus ranking, the individual strategies listed have been ranked into four categories, where 1 and 4 indicate the least and most severe category, respectively, and 2 and 3 indicate intermediate. The weight assigned to each strategy is also shown in Table 2. There was no complete consensus about the ranking except some respondents employing their children to wealthy households for keeping cattle, and selling fire wood and charcoal. However, a quick look will indicate that there was good consensus on withdrawing children from

Table 1 Coping mechanisms by sex of household head. *Source*: Field survey, 2018

Coping mechanisms	Male	Female	Significance
Cuttin down consumption of more nutritious food items (meat, egg, etc.)	2.33	3.65	0.00
Selling of asset and others materials	0.7	1.01	0.227
Selling charcoal and fire wood	0.51	3.0	0.006
Consumption of wild food	2	5	3.50
Engage in petty retailing of items	0.66	3.86	0.00
Seasonal migration/mobility for labor	0.93	0.82	0.71
Selling fire wood and charcoal	1.5	2.49	0.01
Sending out children to work on others' farm	0.82	1.22	0.347
Relying on less preferred and less expensive food	2.29	4.49	0.00
Dropping of children from school	1.10	2.44	0.015
Reduction in non-food expenditures	1.93	3.87	0.000
Daily labor in both rural and urban areas	1.47	0.50	0.001
Purchase of food by credit/relying on borrowing	1.61	1.23	0.026
Skip entire day without food	1.61	0.91	0.015
Reducing number of meals per day	2.57	4.32	0.000
Sending family members to abroad for remittance	0.23	0.18	0.045
Selling liquor ('Arake' and 'Tela')[a]	1.85	0.68	0.001
Sale of timber	0.69	0.22	0.025
Begging	0.31	0.13	0.180

[a] *'Arake'* and *'Tela'* are locally made alcoholic drinks

Table 2 Coping strategies grouped and ranked by FGDs and KII. *Source:* Field survey, 2018

	Coping mechanisms	Focused group discussion and key informative ranking for each individual behavior			
		META FG-1	SIBA FG-2	Average mean	Consensus ranking after KII
1	Cutting down meat consumption	5	3	4.0	4
2	Sending children to urban	5	4	4.5	4
3	Selling of asset and others materials	4.5	3.5	4.0	4
4	Engage in petty retailing of items	3	3	3.0	3
5	Seasonal migration/mobility for labor	3.5	4	3.75	4
6	Borrowing money	4	4	4.0	4
7	Selling fire wood and charcoal	3	3	3.0	3
8	Employing their children for others person	5	4	4.5	4
9	Relying on less preferred and less expensive foods	4	4	4.0	4
10	Dropping of children from school	3	2	2.5	2
11	Reduction in consumption expenditures	4	2	3.0	3
12	Moving from agriculture and becoming daily laborer	3	3	3.0	2
13	Purchase of food by credit	3	5	4.0	4
14	Skip entire day without food	4	3	3.5	4
15	Reducing meals	5	3	4.0	4
16	Migrating (particularly to Middle East and may include vulnerable jobs)	2	1	1.5	1
17	Consuming alcohols	2	5	3.5	4
18	Timbering	3	2	2.5	2
19	Begging	4	5	4.5	4

Table 3 Mean coping strategies index. *Source*: Field survey, 2018

	Mean coping index	Std. deviation	Std. error
Male ($n=61$)	88.54	30.33	3.88
Female ($n=39$)	119.14	26.35	4.22
Total ($n=100$)	100.47	32.39	3.24

school as the most commonly implemented indicator of coping strategy. In general, the consensus ranking should be a whole number that is the most frequent response. Both men and women small holders farmer households use all the coping strategies in the listed above at least part of them, in order to fulfill their food discrepancy. All the relative frequency of using different coping strategies was ranked from "never" to "every day" in a 1–7 scale.

In the study area, different coping mechanisms were employed by both male- and female-headed households. As indicated in Table 1, female-headed households were significantly different and more engaged in coping mechanism of selling fire wood and charcoal and engage in petty retailing of items. The analysis also showed that consumption-based copping mechanism of female-headed household like cutting down meat consumption, relying on less preferred and less expensive food, reduction in consumption expenditure and reducing meals was significantly different and more affiliated when compared with male-headed households and dropping of children from school was more seen in female-headed households. Table 1 indicates that male-headed households were significantly different in coping mechanisms and more affiliated to the coping mechanisms of daily laborer, timbering, skip entire day without food, and alcohol consumption.

Coping strategies index

The coping index was calculated for food-insecure households ($n=100$) using the methodology suggested in "Methods" section. The mean coping index was 88.54 and 119.14 for males and females, respectively. Using inferential statistics equality of means was tested. The t value was -5.173 for 98 degrees of freedom, and the mean difference was significant ($p<0.001$). It implies that coping mechanisms of female-headed households were significantly different from male-headed households (Table 3).

Discussion

Human beings can struggle to sustain their life when food shortages happen, because food is one of our most basic needs rather than shelters and cloth. Before coming to the modernization time, every society around the world has attempted to overcome food shortages at household levels. They practice certain activities to escape them from food insecurity. In the past, when the world population was much lower than it is now, a man had little difficulty in ordinarily time in growing the food that was needed. At that time, human beings started to lead their life in ancient time by hunting wild animal and gathering fruity crops. But it is different from male and female households. In addition to that, the study considered some food security proxy indicators. These are the food consumption score which combines information on meal frequency and dietary diversity and the coping strategy index which has proven to be a good indicator of food security level.

According to information obtained from group discussion and key informant interview, food-insecure households used strategies such as migration, dropping children out of school, cutting down trees and child labor as coping mechanisms. Even though many rural girls move to towns for daily labor, the income they earn does not cover their food and clothing. In the absence of gainful daily labor, some of them are often forced to involve in undesirable things to lead their life (prostitution and robbery). Studies conducted in Ethiopia have indicated that children migrating to urban areas, especially in Addis Ababa, are exposed to high child labor abuse in urban areas and are mostly deprived of their basic rights to education [57]. Another coping strategy adopted by poor rural households is a shift to poor and nutritionally lacking diet which sometimes leads to health-related problems including mental disorder. The downside of the effects of the coping strategies aforementioned has also implications for social norms. For instance, sending children to the streets earn money by begging is a shameful act according to Oromo[1] culture. However, today abject poverty and HIV/AIDS pandemic have forced many children to street begging even among the Oromo. The primary victims of this problem are girls.

As regards the effect of early marriage, parents betroth their daughters at early ages hoping that the latter would provide the former with income they could earn after marriage. This is often the case when the husband is believed to have sufficient income to support his wife's family. Ironically, however, early marriages can end up leading women to engage in less profitable activities like brewing and selling local liquor (e.g., "*Areke*" and "*Tela*" which are locally made liquors). Brewing such liquors is not only less profitable, but it has debilitating effect on their health. Early marriage is also one of the factors accounting for high dropout rates of girls at both elementary and high schools. Dropping out of school is not limited to girls; boys are also forced to quit school and sent out to work on others' farms and earn income for their family. Even worse is that the income that boys earn in this way is too little to even cover their own cost of clothing let alone helping their family buy sufficient food. So, these results were confirmed with the study revealed thatin Table 1 those reported by households surveyed. As the Table 1 displays that the dropping of children from school was one of the coping mechanisms practiced in the study district and significant differences were found between household food insecurity and coping mechanism (at t value 0.015). This was similar to the values reported in Tanzania by Ballard [58], and another option by the same finding was the withdrawal of children, especially girls from school in order to utilize their labor and save money, which, among other things, had ramifications for future literacy levels and the child's participation in the modern economy. Again dropping of children from school was much higher among food-insecure households observed in the surveyed areas (Table 1). So, these values were in agreement with those reported by Brock and Coulibay [59] who pointed out that globally, withdrawing children from school is a short-term strategy that has permanent effects that could make it difficult to reduce food security in the long term. The data show in Table 1 that in the study districts smallholder farmer households strongly rely on less preferred and less expensive food items and statistically strongly significant differences at (at t value 0.000) were observed in the surveyed district (Table 1). These indicated that households have significant differences noted between food insecure and coping mechanisms in Table 1. These values were in line with those values obtained by Ahmed Mohammed Abdulla. This result conforms to findings of a study by [60] conducted among the pastoralists in the Oromia (Yabelo) region of Ethiopia. One-sample t test reveals that there is a statistically significant difference between male-headed and female-headed households in terms of their coping strategies indices.

As given in Table 1, sending family members to abroad, to the Middle East countries in particular, in expectation of future remittance is another coping mechanism practiced by the sample households in the study district and significant values were obtained (t value: 0.045). This study was also consistent with the finding of Abdulla [60]. The study also indicates that the coping strategies practiced by most of the households were reduction in non-food expenditures and daily labor in both rural and

[1] Oromo is the ethnic group residing in the study area.

urban areas as well as purchase of food by borrowing. However, borrowing involves high interest rate in the form of usury. This is particularly true in the case of borrowing from individuals as opposed to formal and microfinance institutions. Due to the high interest rate, very few households borrow from individuals lending money through usury. [61] also found a similar result in the study they conducted in the Southern Nations, Nationalities and Peoples Region of Ethiopia. The study also shows that sale of timber, begging, reducing daily meals and selling productive assets and other materials are coping strategies predominantly pursued by households who had a higher likelihood of being food insecure. So, the results were also in line with those of Bryant [62].

This [62] idea is that distress sale of assets is a major coping strategy adopted by households at times of acute food shortage. In addition, the result is consistent with the study of that of Ekesa in 2008 [36]. According to information obtained from focused group discussion and key informant interview, farm households in the study area often have too little to eat. Lack of enough food was mainly because of insufficient crop harvest and lack of purchasing power. The latter is ascribed to lack of sufficient income, both farm and non-farm/off-farm. Limited purchasing power robs the households of their entitlement to food whenever it is available in the market. Generally, the source of food insecurity among the study participants arises from both problems of availability and access. The FGD and key informant interview show that the major food crops consumed in the study area include cereals, legumes, roots and tubers, bananas, vegetables, fruits, meat and milk products. Cereals crop like sorghum and maize and legumes like beans and soya bean were under produced due to natural and man-made factors. Roots and tubers such as cassava, potatoes and sweet potatoes were consumed as most important food security crops during food shortage, but now they are produced by fewer numbers of households because of environmental factors.

To cope up with the fall in incomes, households sold off assets, one such asset is land. Bryant [63] found a similar finding. In Cambodia, a study conducted revealed that about one in five children whose households were affected by male labor migration had to start working in order to help their mothers [64]. Food is one of the basic needs which are crucial for human beings. We can do nothing without food. However, [65] indicated large amount of food production in the world does not ensure any country's food security. Time to time there are many factors to worsen the food security problem in many parts of the developing world. CARE/WFP [55] identified education level of household head, income source, market and food prices and supply of food commodities as some of the factors underlying food insecurity. In order to cope with food shortages, most households are often compelled to ration the meager food at their disposal, i.e., cutting the amount and number of meals, favoring certain household members over others and skipping whole days without eating [49]. Households can take measures to overcome food insecurity problem. Poor households are the hardest hit by food insecurity when compared with the relatively better-off ones. The most food-secure households are those who achieve adequate access to food while using only a small proportion of available resources, whereas the most food-insecure households are those most at risk who fail to achieve adequate access even by devoting a large proportion of available resources to food [66].

Daily labor in both rural and urban areas, purchase of food by borrowing and petty trade, seasonal migration/mobility for labor and selling fire wood and charcoal were identified as widely practiced coping mechanisms. Withdrawing children from school and sending them out to work on others' farm, relying on less preferred and less expensive food and reducing non-food expenditures and skipping meals were other common coping mechanisms pursued by the households in the study area.

Smallholder farming households in Ilu Gelan pursued a host of short-term coping strategies to overcome the problem of food shortages. The irony, however, is that these coping strategies employed by the households were not effective in reducing household food insecurity in the study area. Thus, even though the coping strategies could ease problem food shortage over a very short period, they were not able to provide long-term solution to the food security problem poor household face. While agriculture plays a major role in the reduction in household food insecurity, the food insecurity problem cannot be solved by promoting agriculture alone. Hence, policies aimed at reducing food insecurity should look beyond agriculture if long-term solution is sought.

Conclusions

The main objective of the study was to identify the coping mechanisms food-insecure households in Ilu Gelan districts adopted in order to grapple with food shortages they face. The data show in the study districts smallholder farmer households rely on less preferred and less expensive food items. Coping strategy adopted by poor rural households is a shift to poor and nutritionally lacking diet leads to health-related problems. A one-sample t test reveals that there is a statistically significant difference between male-headed and female-headed households in terms of their coping strategies indices. The downside of the effects of the coping strategies aforementioned has also implications for health and social norms.

Begging is also one of the coping mechanisms that practiced in the study district, but insignificant differences (at t value 0.180) were observed between food-insecure households and coping mechanisms in the surveyed district (Table 1). These values were disagreement with those result obtained. However, today abject poverty and HIV/AIDS pandemic have forced many children to street begging. The primary victims of this problem are girls. Smallholder farming households in Ilu Gelan pursued a host of short-term coping strategies to overcome the problem of food shortages. The irony, however, is that these coping strategies employed by the households were not effective in reducing household food insecurity in the study area. Thus, even though the coping strategies could ease food shortage problem, over a very short period they were not able to provide long-term solution to the food security problem poor household face. While agriculture plays a major role in the reduction in household food insecurity, the food insecurity problem cannot be solved by promoting agriculture alone. Hence, policies aimed at reducing food insecurity should look beyond agriculture if long-term solution is sought.

Household in Ilu Gelan was engaged in traditional and hand tool agriculture with simple drafting animal. Data obtained from FGD and key informant interview show that the production of crop did not provide sufficient food for household family consumption to impact on household food security status. Low production due to erratic rainfall reduced the availability of crops for household consumption and opportunities for income generation.

Households did not produce sufficient quantities of crops throughout the year due to natural and man-made factors. However, the households to cope with food insecurity relied on reduction in non-food expenditures. Daily labor in both rural and urban areas, purchase of food by credit/relying on borrowing and sending family members to abroad for remittance and obtained from their own production and also food received as gifts from relatives and from non-farm activities were more practiced among food-insecure household.

Recommendations

In view of the findings of the study and the above concluding remarks, the following recommendations are made in order to promote sustainable food security at smallholder farmer's household level in the study district. Gender monitoring and evaluation of the ongoing programs should be carried out to ensure benefits to all and particularly to the vulnerable, female-headed households. Withdrawing children from school was also a common for food-insecure households copping strategy in the study districts. To reduce school withdrawal, the government in collaboration with the World Food Program served school children with feeding program.

In addition to this, implementing disaster and risks management programs to support rural poor and smallholder farmers during natural calamities and other shocks is desperately needed to give prescription for the households. Moreover, encouraging a wise use of natural resources as well as improved technologies, increasing animal and crop diversification and increasing local social net work systems among the community during shocks and stress would be important. Pay attention to policy and strategies of access to use water resources uses both surface water and ground water like modern irrigation technologies and unemployment issues is the big problem. So, households should be focusing on home gardens activities.

Abbreviations

GDP: growth domestic product; WEP: World Food Program; FGD: focused group discussion; HIV: human immune virus; KI: key informants; EHNRI: Ethiopian Health and Nutrition Research Institute; CSI: coping strategies index.

Authors' contributions

This research has been designed and led by BD who structured the concepts, reviewed all of the studies and analyzed both quantitative and qualitative data. He further developed the manuscript, identified and developed important concepts, validated and designed the arguments, conceived and conducted quantitative study and edited the final research. The author read and approved the final manuscript.

Acknowledgements

I convey my deepest thanks to my friends for the completion of this work and energetic encouragement, suggestion and insight and guidance to complete the paper. And I would also like to express my deepest thanks to Gemechu Shale Ogato (Assistant Professor and Ph.D.) at Ambo University as well Sajitha (Ph.D.) at Ambo University for their editing and constructive comments throughout my research work.

Competing interests

The author declares that there are no competing interests.

References

1. Lobell DB, Burke MB, Tebaldi C, Mastrandea MD, Falcon WP, NaylorR L. Prioritizing climate change adaptation needs for food security in 2030. Science. 2008;319:607–10.
2. Borlaug N. Feeding a hungry world. Science. 2007;318:359.
3. Lal R. Soils and food sufficiency. A review. Agron Sustain Dev. 2009;29(1):113–33.
4. USDA-FAS. Agricultural statistics, foreign agric. service, Washington, DC. 2008.
5. FAO. The state of food security in the world. Rome: FAO; 2006.
6. Normile D. As food prices rise, U.S. support for agricultural centers wilts. Science. 2008;320:303.
7. Brown LR. Why ethanol production will drive world food prices even higher in 2008. New York: Earth Policy Institute; 2008.
8. Hoyos C, Blas J. West rethinks strategic threats. Financial Times. http://www.ft.com; 2008. (**consulted on October 29th 2013**).
9. UNOCHA. ECHO Factsheet—Ethiopia, September 2016; 2016.
10. Burton I, Diringer E, Smith J. Adaptation to climate change: international policy options. Prepared for the Pew Center on Global Climate Change; 2006.
11. Cooper P, et al. Coping better with current climatic variability in the rain-

fed farming systems of sub-Saharan Africa: an essential first step in adapting to future climate change? Agric Ecosyst Environ. 2008;126:24–35.
12. Below T, et al. Micro-level practices to adapt to climate change for African small-scale farmers. IFPRI Discussion Paper No. 953. Washington, DC: International Food Policy Research Institute; 2010.
13. Berman R, Quinn C, Paavola J. Identifying drivers of household coping strategies to multiple climatic hazards in Western Uganda: implications for adapting to future climate change. Working Paper No. 149. Centre for Climate Change Economics and Policy; 2013.
14. Ashton PJ. Avoiding conflicts over Africa's water resources. Ambio. 2002;31(3):236–42.
15. Turner B II, Kasperson R, Matson P. A framework for vulnerability analysis in sustainability science. Proc Natl Acad Sci. 2003;100(14):8074–9.
16. Ashraf M, Routray T. Perception and understanding of drought and coping strategies off farming households in north-west Balochistan. Int J Disaster Risk Reduct. 2013;5:49–60.
17. Otsuka K, Larson DF, Hazell PBR. An overview. In: Otsuka K, Larson DF, editors. An African green revolution. Finding ways to boost productivity of small farms. Dordrecht: Springer; 2013.
18. Gebreselassie S. Land, land policy and smallholder agriculture in Ethiopia: options and scenarios. 2006.
19. Girmay G. Smallholder subsistence agriculture: why Ethiopia is still here. Am Res J Agric Orig Artic. 2015;1(3):2015.
20. IPCC. Summary for policy makers. In: Field CB, Barros V, Stocker TF, Qin D, Dokken DJ, Ebi KL, Mastrandrea MD, Mach KJ, Plattner G-K, Allen SK, Tignor M, Midgley PM, editors. Managing the risks of extreme events and disasters to advance climate change adaptation. A special report of working groups I and II of the intergovernmental panel on climate change. Cambridge: Cambridge University Press; 2012. p. 1–19.
21. EPCC (Ethiopian Panel on Climate Change). First assessment report. Working group II agriculture and food security, Published by the Ethiopian Academy of Sciences; 2015.
22. Robinson S, Strzepek K, Cervigni R. The cost of adapting to climate change in Ethiopia: sector-wise and macro-economic estimates Ethiopia Strategy Support Program (ESSP). Working Paper 53; 2013.
23. Wagesho N, Goel N, Jain M. Temporal and spatial variability of annual and seasonal rainfall over Ethiopia. Hydrol Sci J. 2013;58(2):354–73.
24. Bewket W. Climate change perceptions and adaptive responses of smallholder farmers in central highlands of Ethiopia. Int J Environ Stud. 2012;69(3):507–23.
25. Gebrehiwot T, van der Veen A. Climate change vulnerability in Ethiopia: disaggregation of Tigray region. J East Afr Stud. 2013;7(4):607–29.
26. Adnew M, Bewket W. Variability and trends in rainfall amount and extreme event indices in the Omo-Ghibe River Basin, Ethiopia. Reg Environ Change. 2014;14:799–810.
27. Conway D, Schipper E. Adaptation to climate change in Africa: challenges and opportunities identified from Ethiopia. Glob Environ Change. 2011;21(1):227–37.
28. Nyasimi M, et al. Evidence of impact: climate smart agriculture in Africa. CCAFS Working Paper no. 86. Copenhagen: CGIAR Research program on climate change, agriculture, food security (CCAFS); 2014.
29. Cheung W, Senay G, Singh A. Trends and spatial distribution of annual and seasonal rainfall in Ethiopia. Int J Climatol. 2008;28:1723–34.
30. Korecha D. Characterizing the predictability of seasonal climate in Ethiopia. Ph.D. thesis. University of Bergen; 2013.
31. World Bank. World development report, 2008: development and agriculture. Washington: The World Bank; 2008.
32. Beddington JR, Asaduzzaman M, Clark ME, Fernandez Bremauntz A, Guillou MD, Howlett DJB, Jahn MM, Lin E, Mamo T, Negra C, Nobre CA, Scholes RJ, Van Bo N, Wakhungu J. What next for agriculture after Durban? Science. 2012;335:289–90.
33. Intergovernmental Panel on Climate Change (IPCC). Summary for policymakers. In: Parry ML, Canziani OF, Palutikof JP, van der Linden PJ, Hanson CE, editors. Climate change 2007: impacts, adaptation and vulnerability. Contribution of working group II to the fourth assessment report of the intergovernmental panel on climate change. Cambridge: Cambridge University Press; 2007.
34. Stige LC, Stave J, Chan K, Ciannelli L, Pattorelli N, Glantz M, Herren H, Stenseth N. The effect of climate variation on agro-pastoral production in Africa. PNAS. 2006;103:3049–53.
35. Oluoko O, Alice A. Vulnerability and adaptation to food insecurity and poverty in Kenya. Ann Assoc Am Geogr. 2011;101(1):1–20.
36. Eriksen S, O'Brien K, Rosentrater L. Climate change in Eastern and Southern Africa. Impacts, vulnerability and adaptation. GECHS Report 2008:2, ISS University of Oslo; 2008.
37. Rukuni M. Africa: addressing growing threats to food security. J Nutr. 2002;132:3443S–8S.
38. Otsuka K. Why can't we transform traditional agriculture in SubSaharan Africa? Rev Agric Econ. 2006;28:332–7.
39. Vermeer DE. Food sufficiency and farming in the future of west Africa—resurgence of traditional agriculture. J Afr Stud. 1989;10:74–83.
40. Chanyalew D, et al. Ethiopia's agricultural sector policy and investment framework (2010–2020); 2011.
41. Asfaw, et al. Livestock production and marketing. ESSP II Working Paper; 2011.
42. Devereux D, Sussex I. Food security in Ethiopia: a discussion paper for DFID, Addis Ababa, Ethiopia; 2000.
43. Devereux S. Livelihood insecurity and social protection: re-emerging issue in rural development. Dev Policy Rev. 2001;19(4):517–9.
44. Snel E, Staring R. Poverty, migration, and coping strategies: an introduction. Eur J Anthropol. 2001;38:7–22.
45. Akerele D, et al. Food insecurity and coping strategies in South-West Nigeria. Berlin: Springer; 2013.
46. FAO. FAO and emergencies, from prevention to building back better. 2011. http://www.fao.org/emergencies/home/news/emergency-news/emergencydetail/en/item/8894/icode/?uidf=1345. Accessed 30 Dec 2011.
47. SCN. Double burden of malnutrition—a common agenda. 33rd annual session, Geneva, Switzerland, 13–17 March 2006. 2006. http://www.unsystem.org/scn/. Accessed 30 Dec 2011.
48. AFSB. Africa food security brief quarterly bulletin, Issue No. 1: June 2011. 2011. www.afdb.org/fileadmin/uploads/afdb/Documents/Publications/Africa%20Food%20Security%20Brief%20,%20Issue%201.pdf. Accessed 30 Dec 2011.
49. Maxwell D, Watkins B, Wheeler R, Collins G. The coping strategies index: a tool for rapid measurement of household food security and impact of food aid programs in humanitarian emergencies. CARE and Food Programme; 2003.
50. Hoddinott J. Shocks and their consequences across and within household in rural Zimbabwe. 2004. http://www.sarpn.za/document/. Accessed 01 July 2006.
51. Ilu Gelan district Agricultural office. Annual report; 2013/2014.
52. Chandan JS. Statistics for business and economics. New Delhi: Vikas Publishing House Put Lt; 1998.
53. Rangaswamy R. A textbook of agricultural statistics. New Delhi: Wiley Eastern Limited; 1995.
54. Yamane T. Statistics: an introductory analysis. 2nd ed. New York: Harper and Row; 1967.
55. CARE/WFP. The coping strategies index: field methods manual. Nairobi: CARE and WFP. Additional copies and further information are available from both agencies; 2003.
56. Ethiopian Health and Nutrition Research Institute (EHNRI). Food composition table for use in Ethiopia. Part III and IV. Addis Ababa: Research Institute; 2000.
57. Golini A. Migration and urbanisation in Ethiopia with special reference to Addis Ababa-In depth analysis from the 1994 population and Housing census of Ethiopia, CSA, Addis Ababa/Rome; 2001.
58. Ballard R. The context and consequences of migration Jullundur and Mirpur compare. New Community. 1999;11(1):117–36.
59. Brock K, Coulibay N. Sustainable livelihoods Project: Mali country Report'. Manuscript, IDS Sussex; 1999.
60. Abdulla AM. Determinants of household food security and coping strategies: the case of Bule-Hora District, Borana Zone, Oromia, Ethiopia. Yabelo: Yabelo Pastoral and Dry Land Agriculture Research Center; 2015.
61. Tefera T, Tefera F. Determinants of households food security and coping

strategies for food shortfall in Mareko District, Guraghe Zone Southern Ethiopia. J Food Security. 2014;2(2):92–9. https://doi.org/10.12691/jfs-2-3-4.
62. Bryant J. Children of international migrants in Indonesia, Thailand and the Phillipines: a review of evidence and policies. Innocenti Working Paper No. 2005-05, Florence: UNICEF Innocenti Research Centre; 2005.
63. Black R, King R. Editorial introduction: migration return and development in West Africa. Popul Space Place. 2004;10:75–83.
64. Barrett CB. Food security and food assistance programs. Ithaca: Department of Agricultural Resource and Managerial Economics, Warren Hall, Cornell University; 2002.
65. WFP. Food security and vulnerability in selected towns of south nation & nationality region, Ethiopia Vulnerability Assessment and Mapping (VAM), Addis Ababa, Ethiopia September; 2009.
66. Maxwell S, Frankenrerger T. Household food security: concepts, indicators, and measurement. IFAD and UNICEF, Roma; 1992.

The role of crop diversification in improving household food security in central Malawi

Nelson Mango[1*], Clifton Makate[1], Lawrence Mapemba[2] and Mathinda Sopo[3]

Abstract

Background: This paper concerns the role of crop diversification in improving household food security in central Malawi. In this country, the agricultural sector is dominated by smallholder farming and rain-fed food production systems that are facing increasing challenges from land degradation and declining soil fertility. Maize is the staple food crop, and as such, the majority of farmers grow it regardless of land suitability. This has led to what scientists have labeled as "maize poverty trap." In the event of prolonged drought, maize fails thus leaving farmers food insecure. However, research in Sub-Saharan Africa has shown that crop diversification provides smallholder farmers with a diversity of diet, improves their income, and nutrition security. Due to increased cases of malnutrition and food insecurity, in the wake of climate change, government of Malawi has in the past few years intensified extension efforts for crop diversification.

Methods: The study is based on a sample of 271 randomly selected smallholder farming households from central Malawi. It investigates the influence of crop diversification and other household socioeconomic characteristics on the household Food Consumption Score and Household Food Insecurity Access Score. In our analysis, we rely heavily on a combination of ordinary least squares techniques and some descriptive statistics.

Results: Our results show that crop diversification, cattle ownership, access to credit and attaining of education have a positive and significant effect on the household Food Consumption Score. Precisely, crop diversification, cattle ownership and access to credit are all significant at 5% level, while education is significant at 10%. In addition, crop diversification and attaining of formal education by household head were found to have a negative and significant effect on Household Food Insecurity Access Score and were all significant at 1% level.

Conclusion and policy recommendation: Based on our study findings, we conclude that crop diversification is one viable option in smallholder farming that can ensure establishment of resilient agricultural systems that can contribute significantly to household food security. In terms of policy, the results imply that the current efforts by government of Malawi to intensify promotion of crop diversification should remain a priority policy direction due to the continued malnutrition and food insecurity threat. This is particularly so in this era of climate variability that poses an extra burden to farmers.

Keywords: Household food security, Crop diversification, Food Consumption Score, Smallholder farmers, Malawi

*Correspondence: nelsonmango16@gmail.com
[1] International Centre for Tropical Agriculture (CIAT), P.O. Box MP 228, Mt. Pleasant, Harare, Zimbabwe
Full list of author information is available at the end of the article

Background

Introduction

Malawi is largely an agricultural country. An estimated 84% of Malawians live in rural areas where about eleven million people (which is approximately 65% of the population) are engaged in smallholder subsistence farming. Agriculture contributes significantly to the economy. It accounts for more than 33% of gross domestic product (GDP), 90% of exports, over 80% of employment and approximately 65% of raw materials to the manufacturing industry [1]. Furthermore, over 75% of the food consumed in the country is produced by small-scale farmers [1, 2]. This stresses the importance of smallholder farmers and agriculture itself in general in Malawi. Although agriculture is the most important sector of the economy in Malawi in terms of contribution to livelihoods, GDP, employment creation and export earnings, the country is still on the list of the world's poorest countries [3]. According to the UNDP [4] human development report, about 74% of the population in Malawi still lives below the income poverty line of US$ 1.25 a day and 90% below the US$2 a day threshold.

Malawi faces a lot of challenges including that of attaining food security which is one of the objectives of the Agriculture Sector Wide Approach (ASWAP) of 2011 [1]. Most of the challenges, however, are caused by climate change (droughts, floods), declining soil fertility, poor agricultural policies and bad macroeconomic environment. Due to these challenges, smallholder farmers in Malawi are poverty stricken. These challenges affect individual farmers and they put the household food security status at risk. Moreover, farmers are more vulnerable to the overall effects of climate change since they have limited resources to invest in expensive coping strategies [5, 6].

Consequently, this risk encourages smallholder farmers to diversify in crop production in order to stabilize their food stocks and incomes. Crop diversification is based on cultivating more than one variety of crops belonging to the same or different species in a given area. Crop diversification is one way of developing a resilient agricultural system, especially where communities depend largely on agricultural products (food and fodder) for their livelihoods [7, 8]. Crop diversification is also one of the most ecologically, feasible, cost-effective and rational ways of reducing uncertainties in agriculture, especially among small-scale farmers. In addition, crop diversification brings about higher and spatial temporal biodiversity on the farm and increases resilience, i.e., the ability of an ecosystem to return to its original productive state after being disturbed [7]. According to Di Falco and Chavas [9], cultivating several crop species can also help smallholder farmers to manage price and production risks. In moisture-stressed, ecologically fragile agriculture systems crop diversification can also be a viable strategy to increase farm-level crop productivity [9]. All these benefits of diversification contribute to improved yield for the smallholder farmer which will translate to more and a variety of food for consumption and marketable surpluses from production.

Although crop diversification can be a viable option in reducing the risks associated with food insecurity, low incomes from agricultural production and nutrition insecurity among other challenges, only limited research on this subject has been conducted in Malawi to date. However, there is renewed global interest in crop diversification, mainly ascribed to the present rising concerns about loss of biodiversity, environmental and human health [7]. This therefore calls for scientific expertise in carrying out research of the various aspects of crop diversification that could provide alternative and more viable tactics for crop production. However, efforts to promote crop diversification in Malawi are evident from the past and recently. A recent, sound example is the Agricultural Sector Wide Approach (ASWAP) of 2011 which targeted vertical and horizontal crop diversification among other strategies to improve productivity, income, food, and nutrition security in Malawi. Non-governmental organizations (NGOs) and research organizations have been promoting crop diversifications directly and indirectly for the same reasons as well in Malawi. Examples include the International Food Policy Research Institute (IFPRI), Food and Agriculture Organization of the United Nations, United Nations Development Program, Irish Aid, and many other institutions. Most of these institutions contributed to the crafting of the ASWAP and funding it.

This work examines the role of crop diversification in improving food security. It specifically examines the role of crop diversification intensity in smallholder farming in Malawi on food consumption (volume and diversity) and food insecurity coping strategies severity. The study tests the hypothesis that crop diversification has no significant influence on the household Food Consumption Score (FCS) and Household Food Insecurity Access Score (HFIAS). To the best of our knowledge, there is no study that has examined the effects of crop diversification in smallholder farming in central Malawi, especially targeting food consumption and severity of food insecurity coping strategies. This study makes an important addition to the literature on the influence of crop diversification efforts in smallholder farming in Malawi in terms of household food security.

The rest of the paper is organized as follows: Upcoming subsection gives the literature on the role of crop diversification in improving household food security. The second section outlines the research methodology adopted by this study. Third section presents the results and discussions, and the fourth section presents the study conclusions and policy recommendations.

Role of crop diversification in improving yield stability, nutrition diversity and hence food security

Yield stability is one fundamental component targeted by smallholder farmers. Most of the smallholder farmers rely on seasonal yields for food and economic returns. Thus, the implication of fluctuating and/or poor yields can be very profound since it means less food for the family and less income for other household basic needs. Yield stability is an important step toward attaining household food security. To be food secure, a household must have access to adequate food at all times. The household should not risk losing access to food as a consequence of sudden shocks, e.g., economic or climatic crisis. If smallholder farmers invest in crop diversification, this will help them cushion the problem of food insecurity due to the most likely increase in yields, as reported by several previous studies (such as [8, 10]), and bring the yield stability and insurance effect [7] since if one crop fails they can still depend on the other crop. Smithson and Lenné [11] reviewed a total of 100 studies of intra-specific crop mixtures in smallholder farming (mostly grains and legumes) and concluded that yields were often slightly higher compared to pure stands of component cultivars. Moreover, increasing crop diversity by intercropping tobacco, maize, sugarcane, potato, wheat, and broad bean was reported to increase yields for the same season between 33.2 and 84.7% for some combinations [5]. In addition, a meta-analysis study on cereal mixtures by Kiær et al. [12] reported an increase in grain yields of cereals. Thus, with increasing pressures from climate change and other problems constraining smallholder farmers, strategies need to be developed to make our food and farming systems more resilient to the devastating challenges.

The combination of various crops in agro-ecosystems in smallholder farming not only permits more efficient utilization of agro ecological processes, but also provides diversity of human diet and/or improves household income, nutrition, and security. Improving income further improves the purchasing power of the household thereby allowing purchase of other food products. Thus, crop production diversification and consumption habits should include a broader range of crop plant species, in particular those identified currently as underutilized and/or scarce in household food diets ceteris paribus. Crop diversification can as a result contribute significantly to livelihoods, improved health and nutrition, household food security, climate resilience, and ecological sustainability.

Methods

The study aims to attribute food security outcomes of farmers to crop diversification particularly extent of diversification. The study is based on an extract of 271 households interviewed during a baseline survey in Malawi in 2011. The data were collected through questionnaire survey from a stratified random sample of farmers in central Malawi. This section briefly describes sampling methods used to measure food security, crop diversification, and the econometric approaches used to establish causality between food security and crop diversification.

Sampling and data collection

This study uses cross-sectional household data from the baseline survey collected through structured interviews under the auspices of "Increasing smallholder farm productivity, income, and health through widespread adoption of Integrated Soil Fertility Management (ISFM) in the Great Lakes Region and Southern Africa" project—a case study of Malawi. A total of four districts (Lilongwe, Salima, Dowa and Mchinji) from central Malawi were included in gathering this data. The study population was composed of inhabitants in the communities that were selected for the project. In Lilongwe and Salima districts, these included Extension Planning Areas (EPAs) under the Ministry of Agriculture, Irrigation and Food Security, in Dowa these included Area Development Programs (ADPs) under World Vision, and in one part of Lilongwe and one part of Mchinji these included Farmer Associations under National Association of Smallholder farmers of Malawi (NASFAM). Since this was a household level and community survey, the selected participants for the survey were households. Simple random sampling technique was used to select villages from lists obtained from resident district agricultural development officers. Within selected villages interviewed, households were randomly chosen from households lists provided by representative extension personnel using simple random sampling techniques as well. A total of 599 households were then selected for the survey. However, some households were dropped during data cleaning prior to analysis as they did not have sufficient data thereby reducing the sample to 271 households. Data collection for this study was done in December 2011 through face-to-face administration of questionnaires. Data collection involved a household survey using a questionnaire with structured and semi-structured questions. The survey collected information on household demographics, socioeconomic characteristics, crop production and marketing, crop diversification, crop management, input use, food consumption, food insecurity coping strategies, and other farm- and farmer-specific characteristics.

Measuring food security

According to the FAO [13] and Coates et al. [14], food security is a state in which all people at all times have both physical and economic access to sufficient safe and nutritious food to meet their dietary needs and food preferences for a healthy and active life. Households are food insecure when they have uncertain or limited access to food through socially acceptable channels—a problem affecting many households worldwide, Malawi included. In this paper, we use FCS as a measure of household food security in the study sites.

Food Consumption Score (FCS)

To estimate food security, the study employed FCS approach and computed in accordance with guidelines provide by EFSA [15]. FCS was measured based on dietary diversity, food frequency, and the relative nutritional importance of nine different food groups. The FCS is designed to reflect the quantity and quality of people's diet at household level. A composite score is derived from a weighted sum based on the food type and frequency of consumption during a 7-day period. Precisely, dietary recalls questions were used to collect information on the consumption of selected food groups common in Malawi. The interviewees were asked about frequency of consumption over a recall period of past 7 days. FCS was calculated using the formula proposed by EFSA [15]. In the formula, FCS is derived by multiplying the weight for each food group/type by the frequency (number of days) these food groups/types were consumed; the values for all food types consumed during the 7 days recall period were summed up to give the FCS. The formula can be expressed as follows:

$$\begin{aligned} \text{FCS} = {} & a \times f(\text{cereal and or tubers}) + a \\ & \times f(\text{pulse}) + a \times f(\text{milk}) + a \\ & \times f(\text{fruit}) + a \times f(\text{meat and or fish}) \\ & + a \times f\left(\text{sugar}\right) + a \times f\left(\text{vegetables}\right) \\ & + a \times f(\text{oil}) + a \times f(\text{condiments}) \qquad (1) \end{aligned}$$

where FCS = Food Consumption Score, f = frequency of food consumption (number of days for which each food group was consumed during the past 7 days), and a = weighted value representing nutritional value of selected food groups [15]. Food groups were assigned different weights reflecting their nutritional density. The FCS has thresholds consumption categories of poor food consumption (0–21), borderline food consumption ($21 < \text{FCS} \leq 35$), and acceptable food consumption ($\text{FCS} > 35$) [15]. The FCS was adopted as it provides a more accurate measure of the quality of the household diet. Moreover, it accounts for the nutritional value of food in addition to the number of different types of food consumed. However, the FCS bears some weaknesses, mainly because the measure does not consider foods consumed outside the household and it does not provide any information of intra-household food distribution. To some extent, the 7-day recall makes it impossible to consider quantity of food eaten. Despite its weaknesses, FCS is still considered one of the very useful measures of household food security.

Household Food Insecurity Access Score (HFIAS)

The HFIAS is a continuous measure of the degree of food insecurity (access) in the household in the past 30 days. According to Deitchler et al. [16], the HFIAS reflects the three universal domains of household food insecurity that is anxiety about household food insecurity, insufficient quality and insufficient quantity of food supplies. This indicator captures the household's perception about their diet regardless of its nutritional composition [17, 18]. This food insecurity measure focuses on consumption-related strategies and captures the household's behavioral and psychological responses to food insecurity or perceived food insecurity. The HFIAS is based on the assumption that households' experiences of food insecurity cause predictable reactions and responses that can be captured and quantified through a survey and then summarized into a score.

During the survey, the respondents were asked nine occurrence questions that consist of a generally increasing level of food insecurity. The occurrence questions can be summarized as follows: (Q1a) worrying about food adequacy; (Q2a) eating the kinds of less preferred foods; (Q3a) eating limited variety; (Q4a) inability to eat less preferred foods; (Q5a) eating smaller meal than needed; (Q6a) eating fewer meals in a day; (Q7a) failing to get food of any kind; (Q8a) sleeping at night hungry; and (Q9a) going the whole day or night without eating anything. Specifically, the respondents were asked whether a specific condition (Q1a–Q9a) associated with the experience of food insecurity *ever* occurred during the past 30 days. Respondents were asked to either say yes = 1 if event occurred or no = 0 if the event did not occur. Each severity question is followed by a frequency-of-occurrence question, which asks how often a reported condition occurred during the previous 4 weeks. There are three response options representing a range of frequencies (1 = rarely, 2 = sometimes, and 3 = often). The minimum HFIAS is zero and occurs when a household responds 'no' to all questions on the household food insecurity access scale. Alternatively, 27 is the maximum HFIAS and is obtained by summing up of all frequencies on the frequency-of-occurrence questions when a household responds yes to occurrence question and 'often' as frequency of occurrence to the nine frequency-of-occurrence questions. Following the guidelines by Coates et al. [19], the HFIAS is computed as follows:

HFIAS (0–27) = summation of the frequency-of-occurrence during the past 30 days for the nine food insecurity-related conditions

$$\begin{aligned}\text{HFIAS}(0-27) = {} & Q1a*F1+Q2a*F2+Q3a*F3 \\ & +Q4a*F4+Q5a*F5+Q6a*F6 \\ & +Q7a*F7+Q8a*F8+Q9a*F9\end{aligned} \quad (2)$$

At a household level, a high HFIAS shows that a household is very food insecure, while a low score shows that a household is less food insecure.

Measuring crop diversification

In order to measure crop diversification for the particular crops of interest, we employed the Crop Diversification Index (CDI). The CDI is an index of concentration and has a direct relationship with diversification such that a zero value indicates specialization and a value greater than zero signifies crop diversification. With the CDI index, it was then easy to identify those farmers that are practicing crop diversification and those not practicing diversification. The CDI is obtained by subtracting the Herfindahl index (HI) from one (1-HI). Precisely, the CDI is calculated as follows:

$$S_i = \frac{A_i}{\sum_{i=1}^{n} A_i} \quad (3)$$

where, S_i = proportion of ith crop; A_i = area under ith crop; $\sum_{i=1}^{n} A_i$ = total cropped area; and $i = 1, 2, 3, 4, \ldots, n$ (number of crops)

$$\text{But HI} = \sum_{i=1}^{n} S_i^2 \quad (4)$$

$$\text{Therefore CDI becomes } 1-\sum_{i=1}^{n} S_i^2 = 1-\text{HI} \quad (5)$$

In this study, we used nine crops common in smallholder farming in central Malawi to calculate the index. The nine crops included cereals (maize, barley, and rice), pulses (groundnuts, cowpea, soybean, and common bean), root and tubers (sweet potatoes) and vegetables. On vegetables, the household considered vegetable growers were the ones growing at least one vegetable crop per season.

Measuring influence of crop diversification on food security

To ascertain causality between crop diversification and food security outcomes, the study adopts an ordinary least squares (OLS) regression. Since we have CDI (a continuous variable and food security outcomes FCS and HFIAS as dependent variables (all continuous variables), we decided to use OLS regression. According to Isik-Dikmelik [20], it is very sound and correct to use OLS to ascertain influence of a continuous variable on another continuous variable like in our case. The OLS model is specified as

$$Y_i = \alpha_0 + \alpha_1 X_{i1} + \cdots + \alpha_7 X_{i7} + e \quad (6)$$

where, Y_i = household food security outcome (either FCS or HFIAS), X_{i1} = crop diversification, X_{i2} = cattle ownership (1 = yes; 0 = no), X_{i3} = household size, X_{i4} = access to credit (1 = yes; 0 = no), X_{i5} = education of household head (1 = at least primary education; 0 = otherwise), X_{i6} = age of household head and X_{i7} = ownership of a grain storage facility, α_0 = intercept, α_1 to α_7 are coefficients, and e is the error term. Table 1 shows the description of variables used in our analysis.

The linear, double-log, and semi-log functional forms were tried, and the best model was chosen based on the R^2 value, F-value, sign and statistical significance of the variables. The semi-log model was the one chosen because of a superior R^2 value, F-value, sign and statistical significance of the variables. This implies that all the dependent variables were used in log form.

Results and discussion

Descriptive statistics

The descriptive statistics of variables used in our analysis are shown in Table 2. The statistics are based on a sample of 271 farming households from central Malawi.

Food Consumption Score (FCS) and Household Food Insecurity Access Score (HFIAS)

Statistics show that household FCS on average was at 37.01 indicating borderline food consumption. Minimum and maximum FCS was found to be 14 and 87.5, respectively. The distribution of households based on the FCS is shown in Fig. 1.

Table 1 Description of variables used

Variable	Description
FCS	Household's Food Consumption Score
HFIAS	Household Food Insecurity Access Score
CDIIndex	Crop Diversification Index ($0 \le \text{CDI} \le 1$)
Cattle	Cattle ownership (1 = yes; 0 = no)
Hsize	Size of the household
Credit	Access to credit (1 = yes; 0 = no)
Educ_hh	Education level of household head (1 = at least primary education; 0 = otherwise)
Agehh	Age of household head (in years)
Storagefacility	Ownership of a grain storage facility (1 = yes; 0 = no)

Table 2 Summary statistics of variables used for analysis

Variable	Variable description	Mean	SD	Min	Max
FCS	Food Consumption Score of the household	37.01	15.78	14	87.5
HFIAS	Household Food Insecurity Access Score	8.18	5.63	0	27
CDIIndex	Crop Diversification Index ($0 \leq CDI \leq 1$)	0.61	0.16	0	1
Cattle	Dummy variable for cattle ownership (1 = yes; 0 = no)	0.10 (*27*)	0.30	0	1
Hsize	Family size	5.95	2.28	1	13
Credit	Dummy for access to credit (1 = yes; 0 = no)	0.30 (*81*)	0.46	0	1
Educ_hh	Dummy variable for attaining at least primary education (1 = yes; 0 = no) for the household head	0.86 (*233*)	0.06	0	1
Agehh	Age of household head in years	43.60	12.56	20	78
Storagefacility	Dummy variable for ownership of a grain storage facility (1 = yes; 0 = no)	0.99 (*268*)	0.09	0	1

Given in parenthesis are frequencies corresponding to average mean proportions of all dummy variables, i.e., mean proportion for cattle ownership 0.10 corresponds to 27 out of 271 farmers who owns cattle in the sample

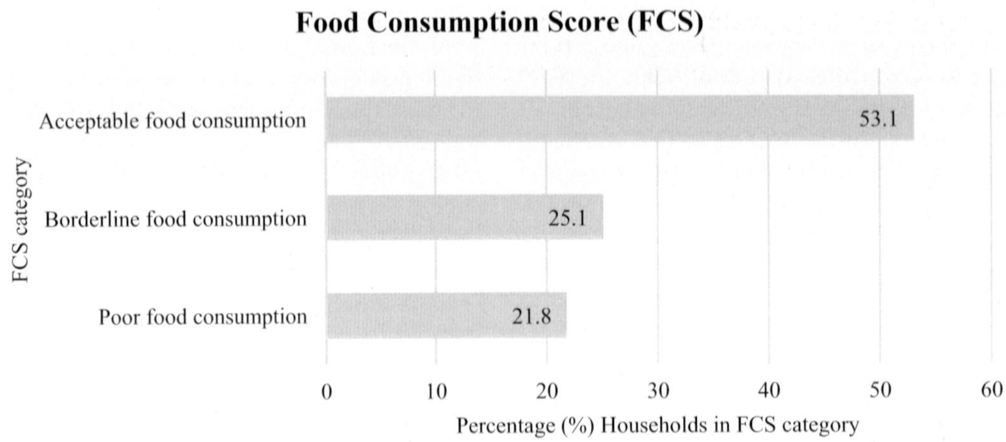

Fig. 1 Distribution of households in central Malawi based on Food Consumption Score

Based on FCS, the distribution shows that the majority of the household (53.1%) are food secure, 25.1% have borderline food consumption, and only 21.8% have poor food consumption. In terms of HFIAS, results show that 22.9% of the households are food secure, 70.5% are mild to moderate food insecure, and only 6.6% are severely food insecure. Distribution of households based on the HFIAS is shown in Fig. 2.

The two measures of food security portray almost a similar trend. The two measures show that at least 75% of the households are moderate to food secure. In trying to relate the two measures of food security, FCS and HFIAS, to crop diversification we found that proportions of crop diversification are higher among borderline food secure to food secure households. However, we found no significant differences in average CDI index by food security category. One notable thing is that average indices by food security status category (both FCS and HFIAS) are above 50% indicating high crop diversification intensity. Table 3 shows the average indices per food security category.

Crop Diversification Index and other socioeconomic characteristics

In terms of diversification, statistics of the CDI show that average within the sample of households is 0.61 with a standard deviation of 0.16. The result implies high crop diversification intensity of the sampled households in central Malawi. In terms of cattle ownership, the sampled households had a very poor ratio of cattle ownership. On average, only 10% of the households owned cattle at the time of the survey. Household sizes at the time of the survey were found to be moderate, about six family members per household. The minimum household size recorded was 1 and maximum of 13 members. Credit access from government and or non-governmental institutions was not common in the group of smallholder farmers sampled in central Malawi. Only 30% of the respondents had accessed credit for use in agricultural production-related activities either from government and/or non-governmental institutions at the time of the survey. The result implies that credit access for financing agricultural

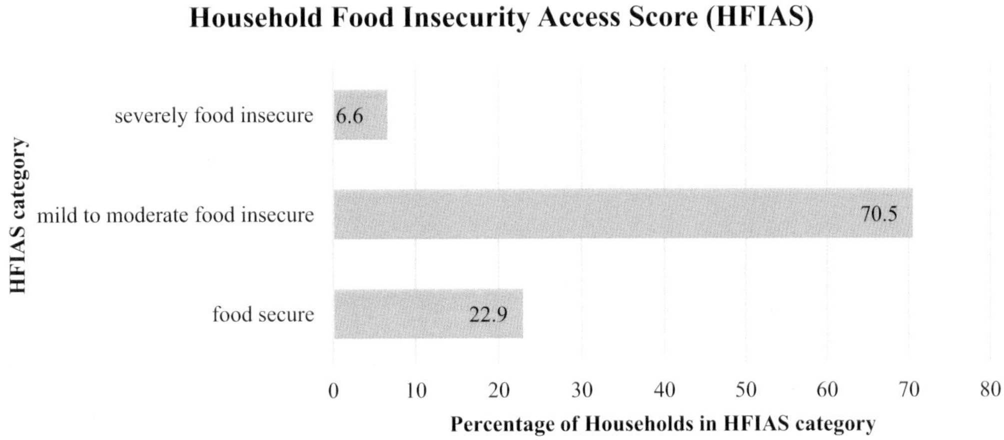

Fig. 2 Distribution of households in central Malawi based on Household Food Insecurity Access Score (HFIAS)

Table 3 Food security status of households and intensity of crop diversification

Food security/insecurity indicator	% of households in category	Average CDI index by category
FCS		
Poor FCS	21.77	0.6
Borderline FCS	25.09	0.61
Acceptable FCS	53.14	0.54
HFIAS		
Food secure	22.9	0.58
Mild to moderate food insecure	70.5	0.58
Severely food insecure	6.6	0.55

activities is still very limited in central Malawi. Education wise, about 86% of household heads had attained at least primary education at the time of the survey. The result implies that the level of education in the study area is good which makes extension messages and any other production- and marketing-related information from various sources easily understood by the smallholder farmers. In addition, the group of farmers was found not to be very old with an average age of 43.6 years. Minimum and maximum age for the group was found to be 20 and 78, respectively. Another characteristic researched on was whether the households own or rent a grain storage facility. Results show that about 99% of the sampled smallholders had access to a grain storage facility at the time of the survey.

Influence of crop diversification on household food security

The ordinary least squares (OLS) regression as mentioned earlier under research methods is used to analyze the effect of diversification on two household welfare indicators, FCS and HFIAS. The results are shown in Table 4. The F-statistics for the two models are both significant at 1%, implying that the variables included in the OLS models are jointly significant in influencing the respective dependent variables HFIAS and FCS. The value of R^2 for the FCS and HFIAS model is 19.4 and 20%, respectively. This means that 19.4% of the variation in FCS and 20% of the variation in HFIAS are explained by the variables included in the two respective models.

Results show that CDI, cattle ownership, access to credit, education, positively influence FCS, while CDI and education have a negative influence on HFIAS.

Crop diversification

Crop diversification as measured by the index was found to have a positive influence on FCS and a negative influence on HFIAS. The coefficient of CDI is significant at 5% and shows a positive influence on household FCS. Households with higher crop diversification intensities are more likely to have diversity in terms of food crops that can be consumed within the household thus justifying the positive relationship. This implies crop diversification improves food consumption in central Malawi. On

Table 4 OLS regression of the welfare effects of crop diversification

Variables	FCS		HFIAS	
	Coef.	$P > t$	Coef.	$P > t$
CDIIndex	0.400	0.047**	− 1.028	0.005***
Cattle	0.270	0.011**	− 0.182	0.370
Hsize	− 0.010	0.460	0.008	0.726
Credit	0.166	0.012**	− 0.006	0.960
Educ_hh	0.095	0.089*	− 0.344	0.001***
Agehh	0.001	0.744	− 0.005	0.324
Storagefacility	0.053	0.872	− 0.660	0.260
_cons	1.111	0.004***	3.664	0.000***
R^2	19.4%		20%	
Adjusted R^2	17%		17%	
F	3.33***		2.95***	
N	271		271	

*Significant at 10%; ** significant at 5%; ***significant at 1%

the other hand, the coefficient of CDI is significant at 1% and shows a negative influence on HFIAS. This means that household with higher crop diversification intensities is less food insecure as compared to those with relatively lower crop diversification intensities. The result implies that crop diversification reduces the severity of food insecurity in central Malawi. Households which have relatively higher number of crop species grown per season are less likely to adopt more desperate food insecurity coping strategies. Therefore, farmers who intensify crop diversification are better off than their counterparts as diversification is positively related to food consumption and negatively related to food insecurity mainly due to the benefits of crop diversification to include, raising farm productivity, income, and reducing production and price risks. Similar studies have found crop diversification to impact positively on food security of the household [5, 7, 8]. More so, the merits of crop diversification in improving food security can manifest through better management of price and production risk [21]. This is possible since growing more than one crop species in a single season gives the farmer options which can ensure him/her manage price and production risks better as compared to less diversified farming enterprises.

Considering that Malawi has a variable climate, crop diversification helps farmers insure against disasters such as floods and drought. Statistics report more than 40 weather-related disasters that occurred in Malawi between 1970 and 2006, including about 16 droughts and flood events [22]. These present serious problems to the smallholder farmer in Malawi since he/she can find difficulties in adopting modern coping strategies to the problem of crop diversification. Furthermore, if smallholder farmers continue to increase acreage of the traditional maize and tobacco at the expense of more drought- and flood-tolerant crops (such as cassava, sweet potato, pigeon peas and other crops) could exacerbate the impact of drought and floods on food security [23]. This therefore justifies the positive influence of crop diversification on FCS and negative influence on HFIAS. Further crop diversification helps farmers exit maize poverty trap in case of adverse weather conditions that do not allow maize to reach maturity stage.

Cattle ownership

Cattle ownership in central Malawi was found to influence the FCS. The coefficient of cattle ownership is significant at 5% and positively related to food consumption. Ownership of cattle is important in improving household food consumption in several ways. The results show that households that own cattle have higher FCSs than those that do not own cattle. First, cattle ownership signifies wealth and in this respect wealthier households have a better purchasing power of a variety of food stuffs unlike poorer households. Secondly, cattle play a vital role as a productivity enhancing input in farming activities as they can provide draft power and farmyard manure. Thirdly cattle also serve as a source of food for the household by producing milk and meat. Cattle ownership is therefore very important in improving food consumption in central Malawi. To support this finding, Bogale and Shimelis [24], found that cattle are an important factor that contributes to food security mainly because: livestock contributes to draft power, subsistence household needs, income and nutritional requirements.

Education

Education of the household head was found to influence both the FCS and HFIAS. The sample of smallholder farmers had 86% of smallholder with at least primary education. The coefficient of education is positively related to FCS and is significant at 5%. The coefficient of education is also negatively related to HFIAS and significant at 1%. These results indicate that households whose heads attained at least primary education have better food security status through better food consumption and their food insecurity status will be less severe compared to those households whose heads never attained school. This is not surprising because educated household heads are more likely to understand extension messages and manage crop diversification efficiently and hence improve their welfare. In addition, education can effectively improve prospects of the farming households to diversify their livelihoods through participation in off-farm formal employment activities. Moreover, education influences access to and use of information. It develops the capacity of farmers to enhance food security. Makombe et al. [25] and Idrisa et al. [26] noted that household head education's level has a positive effect on food security. That is, learned households are more likely to receive information and use it to make informed decisions compared to the less learned ones. Such households are assumed to have better management techniques that can ensure all-year-round supply of diversified and preferred food.

Access to credit

Access to credit was found to influence the household FCS. The coefficient of credit is significant at 5% level and positively related to the FCS. The result indicates that households that have access to credit had higher FCS compared to those that did not have credit access. This may be because households that had access to credit had additional capital available for the farmers to invest in agriculture including crop diversification and could also obtain more income generated by the farming enterprise which could be channeled to improve household consumption. In another sense, access to credit can also improve crop diversification activities through improved access to productivity enhancing inputs such as seed and fertilizers. Successful crop diversification improves food diversity and income, which therefore improves food security within the household. According to Akaakohol and Aye [27], households with access to credit are more likely to have capital to invest in on-farm and off-farm activities which generates more income for the household which is then used to improve food consumption patterns.

Conclusions and policy recommendations

This study has examined the influence of crop diversification on household food security in central Malawi. The study used two measures of household food security, the FCS and the HFIAS. The OLS results show that crop diversification is positively correlated with the FCS and negatively correlated with the HFIAS. This means that households with higher crop diversification intensities are more likely to have a diverse diet and they are also less likely to adopt desperate food insecurity coping strategies. Farming households with more than one crop grown tend to be more secure in terms of food supplies and income and hence are able to cater for the food requirement of their households. Crop diversification hence improves food security through improving food stocks in terms of quantity and variety and also in improving income through sale of crop produced from a variety of grown crop species which then is used to further improve consumption patterns.

Furthermore, cattle ownership and access to credit have positive significant influence on household FCS, while education of the household head positively influences FCS and negatively influences the HFIAS. The results suggest that education of the household head, access to credit, and cattle ownership are crucial factors for food security in central Malawi. We therefore conclude that crop diversification among other factors is a viable option in building resilient and affordable agricultural systems in smallholder farming that can contribute significantly to livelihoods, improved health and nutrition, household food security, and ecological sustainability. This is mainly because crop diversification improves food availability for the household and also income which translates to improved food consumption through food purchases. Crop diversification benefits the farmer mainly in the sense that cultivating several crop species helps to manage both price and production risks, which in the end ensures more food options for the household and income through market participation from the surpluses. Moreover, crop diversification also benefits the smallholder farmer as it can improve farm-level crop productivity.

In terms of policy, results suggest that the government of Malawi need to intensify promotion of crop diversification in smallholder farming, especially to those currently less diversified to improve the food security status of the rural people. Moreover, policies to ensure smallholder farming households' access to credit, education, and draft power are also recommended. For instance, the government in collaboration with the microfinance organizations can work possibility for offering small loans with low interest rates without collateral to smallholder farmers. The loans can be granted on productivity basis to ensure loan repayment by the farmers.

Abbreviations

ADP: Area Development Program; ASWAP: Agricultural Sector Wide Approach; CDI: Crop Diversification Index; CIAT: International Centre for Tropical Agriculture; EFSA: Emergency Food Security Assessment; EPA: Extension Planning Areas; FCS: Food Consumption Score; GDP: Gross domestic product; HFIAS: Household Food Insecurity Access Score; HI: Herfindahl index; IFAD: International Fund for Agricultural Development; MoAFS: Ministry of Agriculture and Food Security; NASFAM: National Association of Smallholder farmers of Malawi; NGOs: Non-governmental organizations; OLS: Ordinary least squares; UNDP: United Nations Development Programs.

Authors' contributions

NM is the principal researcher who conceived and designed the study, took part in designing the questionnaire, participated in the writing of the paper, and wrote the final manuscript. CM did data entry and analysis, and prepared and drafted the manuscript. LM led the designing of the questionnaire, and writing and reviewing the final manuscript. MS participated in the designing of the questionnaire, led the preliminary surveys, interview schedule design, pretesting, and data collection. All authors read and approved the final manuscript.

Author details
[1] International Centre for Tropical Agriculture (CIAT), P.O. Box MP 228, Mt. Pleasant, Harare, Zimbabwe. [2] Department of Applied Economics, Lilongwe University of Agriculture and Natural Resources, Bunda College Campus, P.O. Box 219, Lilongwe, Malawi. [3] International Centre for Tropical Agriculture (CIAT), c/o Chitedze Agricultural Research Station, P.O. Box 158, Lilongwe, Malawi.

Acknowledgements
The researchers acknowledge the International Centre for Tropical Agriculture (CIAT) that commissioned this study, enumerators from, Lilongwe University of Agriculture and Natural Resources, Malawi Ministry of Agriculture, Irrigation and Food Security, and World Vision in Malawi. They also thank all reviewers for their comments and suggestions on earlier drafts of this article. Special thanks to Ms Janine Thorne for editing the paper.

Competing interests
The authors declare that they have no competing interests.

Funding
This research was supported by funding from the International Fund for Agricultural Development (IFAD) to whom the authors express immense gratitude.

References
1. MoAFS. The 2010/2011 farm inputs subsidy programme implementation guidelines. Lilongwe: Ministry of Agriculture and Food Security Government of Malawi; 2011.
2. Chibwana C, Fisher M, Shively G. Land allocation effects of agricultural input subsidies in Malawi. Lilongwe: International Food Policy Research Institute; 2010.
3. IFAD. Rural poverty in Malawi. Lilongwe: International Fund for Agricultural Development (IFAD); 2007.
4. UNDP. Human development report Malawi. Lilongwe: United Nations Development Programme (UNDP); 2009.
5. Lin BB. Resilience in agriculture through crop diversification: adaptive management for environmental change. Bioscience. 2011;61(3):183–93. https://doi.org/10.1525/bio.2011.61.3.4.
6. Makate C, Makate M, Mango N. Smallholder farmers' perceptions on climate change and the use of sustainable agricultural practices in the Chinyanja Triangle, Southern Africa. Soc Sci. 2017;6(1):30. https://doi.org/10.3390/socsci6010030.
7. Njeru EM. Crop diversification: a potential strategy to mitigate food insecurity by smallholders in sub-Saharan Africa. J Agric Food Syst Community Dev. 2013;3(4):63–9. https://doi.org/10.5304/jafscd.2013.034.006.
8. Makate C, Wang R, Makate M, Mango N. Crop diversification and livelihoods of smallholder farmers in Zimbabwe: adaptive management for environmental change. SpringerPlus. 2016;5(1):1–18. https://doi.org/10.1186/s40064-016-2802-4.
9. Di Falco S, Chavas JP. On crop biodiversity, risk exposure and food security in the highlands of Ethiopia. Am J Agric Econ. 2009;91(3):599–613. https://doi.org/10.1111/j.1467-8276.2009.01265.x.
10. Cowger C, Weisz R. Winter wheat blends (mixtures) produce a yield advantage in North Carolina. Agron J. 2008;100:169–77. https://doi.org/10.2134/agronj2007.0128.
11. Smithson JB, Lenné JM. Varietal mixtures: a viable strategy for sustainable productivity in subsistence agriculture. Ann Appl Biol. 1996;128(1):127–58.
12. Kiær LP, Skovgaard IM, Østergård H. Grain yield increase in cereal variety mixtures: a meta-analysis of field trials. Field Crops Res. 2009;114(3):361–73. https://doi.org/10.1016/j.fcr.2009.09.006.
13. FAO (Food and Agriculture Organization): Declaration on world food security. World Food Summit. Rome: FAO; 1996.
14. Coates J, Swindale A, Bilinsky P. Household Food Insecurity Access Scale (HFIAS) for measurement of household food access: indicator guide. Version 3. Washington, DC: Food and Nutrition Technical Assistance Project, Academy for Educational Development; 2007.
15. EFSA. Emergency food security handbook. 2nd ed. Parma: EFSA; 2009.
16. Deitchler M, Ballard T, Swindale A, Coates J. Introducing a simple measure of household hunger for cross-cultural use. Washington, DC: Food and Nutrition Technical Assistance; 2011.
17. Coates J, Swindale A, Bilinsky P. Household Food Insecurity Access Scale (HFIAS) for measurement of food access: indicator guide. Washington, DC: Food and Nutrition Technical Assistance Project, Academy for Educational Development; 2007.
18. Mango N, Zamasiya B, Makate C, Nyikahadzoi K, Siziba S. Factors influencing household food security among smallholder farmers in the Mudzi district of Zimbabwe. Dev South Afr. 2014;31(4):625–40. https://doi.org/10.1080/0376835X.2014.911694.
19. Nyikahadzoi K, Siziba S, Mango N, Mapfumo P, Adekunhle A, Fatunbi O. Creating food self reliance among the smallholder farmers of eastern Zimbabwe: exploring the role of integrated agricultural research for development. Food Secur. 2012;4(4):647–56. https://doi.org/10.1007/s12571-012-0218-8.
20. Isik-Dikmelik, A.: Trade reforms and welfare: an ex-post decomposition of income in Vietnam. World Bank Policy research working paper. Washington DC: World Bank; 2006.
21. Baumgärtner, S., Quaas, M.F.: Managing increasing environmental risks through agro-biodiversity and agri-environmental policies. Working paper series in economics 2008, University of Luneburg; 2008.
22. Action Aid. Climate change and smallholder farmers in Malawi. Lilongwe: Action Aid; 2006.
23. Holden S, Lunduka R. Too poor to be efficient? Impacts of the targeted fertilizer subsidy program in Malawi on farm plot level input use, crop choice and land productivity. Ås: Norwegian University of Life Sciences; 2010.
24. Bogale A, Shimelis A. Household level determinants of food insecurity in rural areas of Dire Dawa, Eastern Ethiopia. Afr J Food Agric Nutr Dev. 2009;9(9):1914–26.
25. Makombe T, Lewin P, Fisher M. The Determinants of food insecurity in rural Malawi: implications for agricultural policy. Lilongwe: Malawi Strategy Support Program (MaSSP); 2010.
26. Idrisa YL, Gwary MM, Shehu H. Analysis of food security status among farming households in Jere Local Government of Borno State, Nigeria. J Trop Agric Food Environ Ext. 2008;7(3):199–205.
27. Akaakohol MA, Aye GC. Diversification and farm household welfare in Makurdi, Benue State, Nigeria. Dev Stud Res Open Access J. 2014;1(1):168–75. https://doi.org/10.1080/21665095.2014.919232.

11

Magnitude and associated factors of household food insecurity in Fedis Woreda East Hararghe zone, Oromia region, Ethiopia

Mahlet Mulugeta[1*], Gebeyaw Tiruneh[2] and Zewdie Aderaw Alemu[3,4]

Abstract

Background: Poverty and food insecurity remain as the major challenges to achieve economic development in Ethiopia. Explaining the household food insecurity situation and identifying associated factors will help in making better decision and resource allocation. This study aimed to determine the magnitude and identify the associated factors of household food insecurity in Fedis Woreda that has been affected by food insecurity for the quite time with a total population estimated to be 133,382 persons, of which the estimated urban population is 26,575 and estimated rural population is 127,877. The Woreda is in food deficit every year where the food gap is covered by food aid amounting to 15–25% for the poorest households of Ethiopia. Cross-sectional study was used to collect data on calorie availability at household level, demographic characteristics, socioeconomic factors and coping strategies from 743 households. Household food insecurity was obtained by comparing the total food calorie available for consumption in the household per adult equivalent to the minimum level of subsistence requirement per adult equivalent of 2200 kcal. Data were analyzed using Statistical Product and Service Solution. Binary logistic regression analysis is used to determine the association between dependent and independent variables.

Results: About 58% of the sampled households in the study area were food insecure. The binary logistics regression analysis identified educational status of the household head [AOR, 95% CI 0.59 (0.38–0.91)], annual farm income [AOR, 95% CI 0.44 (0.287–0.675)], sufficient crop production [AOR, 95% CI 0.45 (0.274–0.748)], dietary diversity [AOR, 95% CI 0.548 (0.382–0.786)] and oxen ownership [AOR, 95% CI 0.454 (0.323–0.639)] were the major factors inversely associated with food insecurity status. On the other hand, large family size [AOR, 95% CI 6.143 (2.40–15.0)], lower than average monthly off-farm income [AOR, 95% CI 1.85 (1.18–2.91)] and small land size [AOR, 95% CI 3.04 (2.04–4.52)] were increased the chance being food insecure household in the study area.

Conclusions: The findings imply that more than half of the community was food insecure where improvement in food security situation needs to promote family planning, undertake different income generating activities and improve access to farmer's education. These areas could provide entry points for policy intervention to reduce food insecurity and create community livelihood opportunities.

Keywords: Food insecurity, Cross section, Household, Ethiopia

Background

Food security, as a concept emerged at the United Nations Food and Agriculture Organization, the World Food conference in 1974 by considering food availability as the central argument [1, 2]. The World Food Summit in 1996 defined food security as a situation "when all people, at all times, have physical, social and economic access to sufficient, safe and nutritious food to meet their dietary needs and food preferences for an active and healthy life" [2].

Food security is a prerequisite for healthy eating and foundational to human and environmental health. It is a basis for the prevention of chronic disease and

*Correspondence: mahletmulugeta3@gmail.com
[1] Food and Agricultural Organization, Addis Ababa, Ethiopia
Full list of author information is available at the end of the article

promotion of healthy growth and development. It is integral to healthy living and environmental health protection [3]. Sub-Saharan Africa is one of the most food insecure regions in the globe [4]. The failure of Sub-Saharan African countries to feed their population has been attributed to climate shocks, mainly drought and the subsequent water scarcity, resource degradation, bad governance and inefficient policies, widespread epidemic, technological stagnation and conflict [5].

Poverty and food insecurity remain as the major challenges to achieve economic development in Ethiopia, and especially in the rural area of the country. This is due to the subsistence nature of Ethiopian agriculture, its mere dependence on rainfall and the existing backward technologies, which have made farmers highly vulnerable to famine and food insecurity. Many Ethiopians live in conditions of chronic hunger with both daily energy supply 1880 kcal/capita/day and 44% prevalence of undernourishment [6]. Hence, the issue of food insecurity and efforts to achieve food security will remain the primary concern of governments and households mainly, those people in lower income or vulnerable groups in the country [7]. The study conducted in 2012 in Shashemene District found in southern Ethiopia [8] indicated that food insecurity in Ethiopia derives directly from dependence on undiversified livelihoods based on low-input and low-output rain-fed agriculture. Ethiopian farmers do not produce enough food even in good rainfall years to meet consumption requirements. Food accessibility was also limited due to a weak subsistence agriculture-based economy, depletion of assets, absence of income diversity and a lack of alternative coping mechanisms.

The depth of food insecurity varies within and between households. The food insecurity status of a household and its members is very sensitive to livelihood stresses and thus changes over time. Empirical evidences have also shown that many factors are responsible for household food insecurity. For instance, a case study conducted in Dire Dawa in 2003 investigated family size, annual income, amount of credit received, irrigation use, age of household head, status of education, cultivated land size, livestock ownership and number of ox owned to be the most determinants of food insecurity [9]. In addition, a study conducted in Oromia [10] has shown family size, number of oxen owned, use of chemical fertilizer, size of cultivated land, farm credit use, total annual income per adult equivalent, food consumption expenditure, livestock owned and off-farm income per adult equivalent to be the major causes of food insecurity. Fedis Woreda or district is one of the severely food insecure Woredas (district) found in the East Hararghe zone in Oromia region as they live on marginal and moisture stressed, and heavily degraded and less productive land. Moreover, increasing population, recurrent drought and resource degradation in the study area have made the food security situation worse. Realizing this issue, many governmental and non-governmental organizations are intervening at least to lessen the adverse effects of the food problem, but there is yet little success.

Therefore, this study determines the magnitude and identifies the determinant factors of food insecurity. Explaining the household food insecurity situation and identifying associated factors will help in bringing an improvement in making better decisions and resource allocation to improve the food insecurity status of the region. Moreover, it will support development practitioners and policy makers have better knowledge as to where and how to intervene in rural areas to bring food security or minimize the severity of food insecurity in the country and in the region in general and in Fedis Woreda in particular.

Methods

Study setting

The East Hararghe zone is located in the Oromia region. It is among the chronically food insecure areas of the country which faces recurrent drought situation. The livelihoods in the zone comprise of agro-pastoralists and pastoralists. There are 19 Woredas in the zone with 14 of them being agro-pastoralists. The rainfall is bimodal with the Kiremt rain being important in the crop-dependent areas. The amount of rainfall varies between 650 and 750 mm, while the average temperature of the Woreda diverges between 25 and 30 °C [11].

Fedis is one of the lowland Woredas that have been affected by food insecurity for the quite time with a total population estimated to be 133,382 persons, of which the estimated urban population is 26,575 and estimated rural population is 127,877. The Woreda is in food deficit every year where the food gap is covered by food aid amounting to 15–25% for the poorest households. The altitude of the Woreda ranges from 1050 to 2118 m above sea level [12]. It is a predominantly Woina Dega livelihood zone. The main sources of income are agriculture (particularly chat and livestock sales), self-employment (firewood sales) and local labor (harvesting and packing chat). Sorghum and maize are grown for home consumption; purchase is an important source of food. The area is considered to be a food deficit area—reflecting small land holdings and erratic rains. Currently, the government is carrying out of the Productive Safety Net Program to support chronically food insecure households.

Source and study population

All households in Fedis Woreda of East Hararghe zone were source population, and all randomly selected households from Fedis Woreda were the study population.

Sample size and sampling procedures

To determine sample size for the study population, the following assumptions were made. The actual sample size for the study was determined using the formula $n = (Z_{\alpha/2})^2 P(1 - P)/d^2$ assuming 5% marginal error and 95% confidence interval at alpha ($\alpha = 0.05$). According to the study conducted on *Food insecurity and Coping strategy in Kersa Woreda*, the prevalence of food insecurity is found to be 70% [13]. As both Woredas have similar livelihood in terms of crop and livestock production system, the prevalence is assumed to be similar [13]. Therefore, using the formula $N = (Z_{\alpha/2})^2 P(1 - P)/d^2 = (1.96)^2\ 0.7\ (1 - 0.7)/(0.05)^2 = 323$. Since two-stage sampling technique was used, the sample size was multiplied by a design effect 2. $N = 323 \times 2 = 646$. By assuming non-response rate and incomplete questionnaire taking 15%, the total sample size was $N = (646 \times 15\%) + 646 = 743$.

A two-stage random sampling procedure was used to select 743 rural households. First, four kebeles (the smallest administrative unit) were randomly selected from 19 kebeles, and then, the sample size was distributed proportionate to the size of each kebele's population. Finally, the 743 households were selected from the corresponding kebeles using a systematic random sampling using sampling interval of every 7th household.

Inclusion criteria—all housholds that are represented by spouse by the time of interview. Exclusion criteria—housholds that are not represented by physically or mentally fit spouse by the time of interview will be excluded from the interview.

Study variables

Dependent variable

In this study, food insecurity at household level was considered as a dependent variable (food secure/insecure).

Independent variables

Household head age, sex, marital status, educational status, family size, total farm land, livestock ownership, off-farm income, household income/farm income and modern farm input utilization were considered as independent variables.

Operational definition

Household food security was measured using household calorie availability. Household calorie availability was computed from each food item consumed and was grouped and adjusted for food processing to obtain the net weekly calorie availability using the Ethiopian Health Research Institute (EHNRI) Food Composition Table [14]. The net weekly calorie availability was divided by seven to obtain the household daily calorie intake. The family size of each household was converted into an adult equivalent family size, which considers age and sex of each family member in the household. The daily net calorie consumption of the household was divided by the adult equivalent family size to obtain the daily calorie availability per adult equivalent of the household.

Severe food insecure households Households whose daily calorie intake per adult equivalent found to be less than 1500 were considered as severely food insecure.

Moderate food insecure household Households whose daily calorie intake per adult equivalent found to be between 1500 and 1800 were considered as moderate food insecure.

Mild food insecure household Households whose daily calorie intake per adult equivalent found to be between 1800 and 2200 were considered as mildly food insecure.

Food secure households Households whose daily calorie intake per adult equivalent found to be greater than 2200 were considered as food secure household.

Inadequate dietary diversity Households who had consumed ≤ 3 types of food groups (namely from cereals, vegetables and vitamin A-rich fruits).

Medium dietary diversity Households who had consumed 4–5 types of food groups (namely from cereals, vegetables, vitamin A-rich fruits and oil).

Adequate dietary diversity Households who had consumed ≥ 6 types of food groups (namely from cereals, vegetables, vitamin A-rich fruits, oil, fish, legumes, nuts and seeds).

Data collection method

A structured survey questionnaire was designed in English and translated into Oromifa, the local language of the region. The questionnaire was tested prior to the data collection process near to the study population area which has similar livelihood and depending on the results of the pretest; it was revised in the lights of suggestions received. The household spouse was the main respondent. The questionnaire encompasses information on demographic characteristics, crop production, asset ownership, such as land use, oxen, access to different services like market, a health post and credit as well as coping strategies employed by the households during food shortage.

Data management and quality control

To collect the data, twelve data collectors and one supervisor were recruited for ten working days, i.e., from April 8, 2014, to April 21, 2014. To assure the data quality, 1-day training was given to twelve data collectors and one supervisor on April 8, 2014. Each data collector conducted two pretests on 24 clients before the actual work started. Necessary information and instruction on the objective; relevance of the study was given to the respondents. Following data collection, the sample household responses were coded and analyzed using Statistical Product and Service Solution (SPSS) version 20. Descriptive statistics, such as mean, percentage and frequency, was used. In order to identify the determinants of food insecurity and assess their relative importance in determining the probability of being food secure or not, binary and multivariate logistic regression was used. In the regression analysis, bivariate binary logistic regression was done between each of the independent and dependent variables to select the most important variables for the multivariate analysis. To identify the independent effects of each predictor variable, variables with *P* value less than 0.2 were considered for the multivariate binary logistic regression analysis. The dependent variable was household food insecurity (HFIS) that is dichotomous taking a value of 1 if the household is food insecure, 0 otherwise.

The information, which identifies the food insecure from the food secure, was obtained by comparing the total food calorie available for consumption in the household per adult equivalent (AE) to the minimum level of subsistence requirement per AE 2200 kcal. A household below this threshold is said to be food insecure ($Zi = 1$), otherwise food secure ($Zi = 0$). Once the group are categorized as food secure and food insecure, the next step is to identify the socioeconomic factors that are correlated with the food insecurity. A variety of models can be used to establish the relationship between the potential determinant factors and food insecurity. The study employed the logit model in line with earlier researchers. Following Bogale study [15], the cumulative logistic probability model can be econometrically stated as:

$$Pi = F(Zi) = \frac{1}{1 + e - (\alpha + \Sigma\ \beta iXi)}$$

where Pi is the probability that an individual is being food insecure given Xi, Xi is a vector of explanatory variables, α and β are regression parameters to be estimated and e is the base of the natural logarithm.

For ease of interpretation of the coefficients, a logistic model could be written in terms of the odds and log of odd. The odds ratio is the ratio of the probability that a household would be food insecure (Pi) to the probability of a household being food secure ($1 - Pi$).

That is, $\frac{Pi}{1-Pi=ezi}$

Taking the natural logarithm of the equation yields:

$$\text{Ln}\frac{Pi}{1 - Pi} = Zi = \alpha + \beta 1x_1 + \beta 2X_2 + \cdots + \beta mX_m$$

If the error term εi is taken into account, the equation becomes:

$$Zi = \alpha + \Sigma\beta^{im} i = 0 = 0\,Xi + \varepsilon i$$

In this study, the explanatory variables used in the model included:

$X1$	Age of household head (age group) 1 = 15–24, 2 = 25–34, 3 = 34–49 and 4 = 50–64
$X2$	Education level of household head (1 = Can't read and write, 2 = Primary school, 3 = Secondary school)
$X3$	Family size (1 = 1–3, 2 = 4–6, 3 = 7–10, 4 = > 10)
$X4$	Land size (0 = < 0.5, 1 = 0.5–1, 2 = > 1)
$X5$	Ox ownership (0 = No, 1 = Yes)
$X6$	Sufficient crop production for the year (0 = No, 1 = Yes)
$X7$	Annual farm income (1 = No income, 2 = 100–500, 3 = 501–1000, 4 = > 1000)
$X8$	Average monthly off-farm income (1 = No income, 2 = 100–500, 3 = 501–1000, 4 = > 1000)
$X9$	Health status, if they have been sick during the past 3 months (0 = No, 1 = Yes)
$X10$	Food aid/remittance received (0 = No, 1 = Yes)
$X11$	Modern agricultural input used (0 = Don't use, 1 = Used improved seed only, 2 = Used fertilizer only, 3 = Used both inputs)
$X12$	Dietary diversity (1 = ≤ 3 food groups, 2 = 4 and 5 food groups, 3 = ≥ 6 food groups

Results

Out of the seven hundred forty-three sample households, 730 (98.3%) were considered for the analysis. In the study, 697 (95.5%) of the households were male-headed and 33 (4.5%) female-headed households. The mean age of the study was 37 year (± 8.768) where the minimum and maximum age was 20 and 60, respectively. Looking at the marital status, 98.8% are married. Result on the education level shows that 79.3% of households can't read and write. The average family size for the overall sample household was 5.69 (± 1.559), where the minimum and maximum size was 2 and 15, respectively (Table 1).

Household food insecurity

Based on the food security cutoff point, i.e., 2200-kcal threshold point as a benchmark, 427 (58.5%) sample households were found to be food insecure. Of the total study participants, 128 (17.5%), 123 (16.8%) and 176 (24.1%) were severe, moderate and mild food insecure households, respectively. In order to combat the seasonal food shortage, households have used different coping mechanisms where borrowing money accounts highest (Table 2).

Factors associated with household food insecurity

The association between each explanatory variable and household food insecurity status was done. Bivariate binary logistic regression analysis between each predictor variable and the household food insecurity status was performed to identify the significant candidate predictor variables that would qualify for the multivariate analysis. Taking the *P* value of < 0.2 as a cutoff, twelve independent variables were fitted to the model. Then, using a forward stepwise regression, eight variables were found to be significant ($P < 0.05$). The eight factors that were retained were the educational status of the household head, family size, farmland size, household's annual farm income, household average monthly off-farm income, ownership of oxen, sufficient crop production over the year and dietary diversity (Table 3).

Households that have obtained better income from their own land are less likely to become food insecure than those households who had no or little income. Other things are held constant; the odds ratio in favor of the probability of being food insecure decreases by 56% as households earning increases by one Birr [AOR 0.440 (95% CI 0.287–0.675, $P < 0.001$)]. However, the presence of an off-farm income in the household has a negative influence on the status of household food insecurity, meaning that households who earn better income are 1.8 times more higher to be food insecure [AOR 1.849 (CI 1.175–2.911, $P = 0.008$)]. Those household who have eaten at least four or five types of food groups per day are 45.2% less likely to be food insecure compared with those who have eaten less food items [AOR 0.548 (CI 0.382–.786, $P = 0.001$)].

Categorization of household based on education exhibited that households that have attended at least primary school are 40.9% less likely to be food insecure compared with those who can't read and write [AOR 0.591 (CI 0.385–.908 $P = 0.016$)].

Households who have a larger family size are six times higher to be food insecure than households with few family sizes [AOR 6.143 (CI 2.402–15.708, $P < 0.001$)].

Regarding the land size, households who had a land size between 0.5 and 1 ha are three times more higher to be food insecure than households that have less than half hectare [AOR 3.037 (CI 2.039–4.521, $P < 0.001$)]. Households who had sufficient production over the years are 54.7% less likely to be food insecure than those who had experienced shortage of crop production (AOR 0.452 (95% CI 0.274–0.748, $P = 0.002$).

Table 1 Sociodemographic characteristics of Fedis Woreda households, 2014

Variable	Category	Frequency	Percent
Sex	Female	33	4.5
	Male	697	95.5
Age in years	15–24	29	4.0
	24–34	242	33.2
	35–49	376	51.5
	50–60	83	11.4
Education	Can't read and write	576	79.3
	Primary school	141	19.3
	Secondary school	10	1.4
Marital status	Married	699	98.8
	Divorced	3	0.4
	Widowed	28	3.8
Family size	1–3	28	3.8
	4–6	482	66
	7–10	218	29.3
	>10	2	0.3

Table 2 Coping mechanism used by Fedis Woreda households for seasonal food shortage (multiple response is possible)

Coping mechanism	Frequency of insecure HH 427 (58.5%)	Frequency of secure HH 303 (41.5)
Borrowed money/food from relatives/neighbors	187 (43.8)	96 (31.7)
Family members went to other places in search of jobs/migration	50 (11.7)	44 (14.5)
Sold asset	28 (6.6)	24 (7.9)
Worked for payment in kind/cash	15 (3.5)	17 (5.6)
Selling chat	181 (42.4)	101 (33.3)

Table 3 Results from multivariable analysis-adjusted for demographic and socioeconomic factors of household food insecurity of Fedis Woreda, East Hararghe zone, Oromia region April 2014

Explanatory variables	Food security status		AOR	95% CI		*P* value
	Insecured HH	Secured HH		Lower	Upper	
Education level						0.024
Can't read and write	357	222	1			
Primary school	67	74	0.591	0.385	0.908	0.016
Secondary school	3	7	0.299	0.061	1.471	0.137
Family size						< 001
1–3	9	19	1			
4–6	254	228	2.652	1.083	6.493	0.033
> 7	164	55	6.143	2.402	15.708	< 0.000
Land size						< 001
< 0.5	67	111	1			
0.5–1	344	178	3.037	2.039	4.521	< 0.001
> 1	16	14	1.748	0.710	4.305	0.225
Oxen ownership						< 001
No	284	153	1			
Yes	143	150	0.454	0.323	0.639	
Annual farm income						0.002
No income	121	52	1			
1–500	7	4	0.663	0.153	2.878	0.583
501–1000	16	13	0.611	0.251	1.487	0.278
> 1000	283	234	0.440	0.287	0.675	< 0.000
Average monthly off-farm income						.029
No income	325	242	1			
1–1000	15	16	1.165	0.518	2.623	0.712
> 1000	87	45	1.849	1.175	2.911	0.008
Sufficient crop production over the year						0.002
No	382	245	1			
Yes	45	58	0.453	0.274	0.748	
Dietary diversity						0.002
Inadequate	322	166	1			
Medium	103	133	0.548	0.382	0.786	0.001
Adequate	2	4	0.219	0.034	1.422	0.112

Those households that had an oxen are 54.6% less likely to be food insecure than those who don't have any [AOR 0.454 (95% CI 0.323–0.639, $P < 0.001$)].

Discussion

In the current study, more than half of the households in Fedis Woreda were food insecure. The finding is less than the study conducted in Kersa Woreda of Oromia region [13] showing 70% food insecurity. The difference could be related to the study period. There is a 2-year difference between these studies where the study conducted in Kersa was in 2012, whereas the current study, i.e., Fedis, was conducted in 2014.

The amounts of annual income of the household obtained from own farm have played a significant role in determining the household food insecurity. Likewise, a study conducted in southern Ethiopia [8] also indicated that those farmers who have better access to different types of farm income are less likely to become food insecure than those households who have little income. Moreover, a study conducted in Dire Dawa also shows a study conducted in Dire Dawa [15] which shows that households that have access to better income opportunities are less likely to become food insecure than those households who had no or little access. Nevertheless before in this study, the presence of an off-farm income in the household has an inverse relation with the household food security, meaning that households who earn better income are more food insecure. This could be explained by the idea of the income which may not have been used for purchase of food item; rather, it might have been used for other expenses like for purchasing alcohol drinks and khat.

Regarding to the crop production, those households who have sufficient production for the years were less likely to be food insecure than households that have experienced shortage of crop production. Considering own farm production being the major source of food, having sufficient production for the year will make the household to have sufficient food. Moreover, it will also support the household to use the income they might get from own farm for purchase of supplementary food items and improve their diet diversity. Another variable that has significance on the household food insecurity is dietary diversity. Those households who have consumed at least four or five types of food groups were less likely to be food insecure. It is also supported by the study conducted in ten countries [16] which reports that increasing the household dietary diversity will significantly improve the energy availability.

Education level of the household head is an important variable mostly assumed to have an impact on food insecurity status of the household. According to this study, households whose head has attended at least primary school were less likely to be food insecure compared with those who can't read and write. It also agrees with the findings of study conducted in Dire Dawa which also shows that the higher the educational level of household head, the more food secure the household is expected to be [17]. The result also coincides with the theoretical evidences that education equips individuals with the necessary knowledge of how to make a living. It is also believed that households who have at least primary education are the ones to be more likely to benefit from agricultural technologies and thus become food secure. However, another study conducted in Dire Dawa shows that the coefficient of education level of the household head was not statistically significant. This may imply that education of household head has not yet enhanced households' capabilities to adopt better production technologies and accept technical advice from extension workers and diversifying their source of income than the illiterate ones which would have reduced the risk of food insecurity among households [17].

Family size is another important variable, which has an influence on household food insecurity. Households who have larger family sizes were six times higher to be food insecure than households with fewer family sizes. The finding is supported in a study conducted in southern Ethiopia [8] which also shows that as the family size increases, food insecurity increases. This could be due to the fact that as family size increases, the amount of food for consumption in one's household increases and, therefore, that additional household member shares the limited food resources. This study showed that households who had a land size between 0.5 and 1 ha were three times more food insecure than households that have less than half hectare.

The relationship between the ownership of oxen and food insecurity turned out to be negative and statistically significant. This is an indication that ownership of ox acts as a hedge against food insecurity in the study area. A study conducted in Oromia region [18] showed that ox ownership is found to have a significant and positive relationship with household food security.

The study focused on identifying factors that are expected to influence household food insecurity in the rural areas of Fedis Woreda. As the unit of analysis is a household, the study lacks macrolevel variables that affect household food security status such as market price and crop diversity. Moreover, qualitative methods were not used which could have enabled to find out additional relevant information. Lastly, there might be a recall bias in estimating the quantity of food consumed during the past week.

Conclusions

In the study area, large proportion of the study participants were food insecure. In the current study, having larger number of family size, owning a large size farmland and earning a better income from off-farm income have increased the chance of being food insecure household. Meanwhile, better education of the household head, earning a better income from own farm, sufficient crop production over the year, dietary diversity and ox ownership have a negative influence on the state of household food insecurity.

Based on the findings and conclusion of the study, the following policy recommendations are forwarded. Policy measures directed toward the provision of better family planning to reduce household family size should be given adequate attention. High yielding improved crop varieties, and the use of modern inputs should be facilitated and strengthened. Government as well as humanitarian agencies such as NGOs and UN agencies should consider and strengthen income generating activities. The regional and federal governments should provide access to education for farmers both formal and informal like Farmers Field School should be strengthened. Government, NGOs and UN agencies should continue promoting the consumption of diversified food and also strengthen nutrition education. Finally, further studies should be conducted on the area of food security considering detailed and accurate information on various variables including political, climatic and weather (rainfall and temperature), topography, natural disasters, ecological conditions and other factors that affect food security. Moreover, some variables which were found to be different compared with other studies like off-farm income and food aid should be further studied.

Abbreviations

AE: adult equivalent; DRMFSS: Disaster Risk Management and Food Security Sector; EHNRI: Ethiopian Health and Nutrition Research Institute; FAO: Food and Agriculture Organization of the United Nations; FS: food secure; FIS: food insecure; GDP: gross domestic production; HFIS: household food insecurity; HH: household; IFPRI: International Food Policy Research Institute; Kcal: kilocalorie; SDPRP: Sustainable Development and Poverty Reduction Programme; SSA: Sub-Saharan Africa; SPSS: Statistical Product and Service Solution; PASDEP: Plan for the Accelerated and Sustained Development to End Poverty.

Authors' contributions

Conception and design of the work proposal, collection of data, analysis and interpretation of data were done by MM, GT and ZA. Drafting the manuscript article, revising it critically for intellectual content and final approval of the version to be published were done by MM, GT and ZA. Finally, before submission for publication, all authors read and approved the final manuscript.

Author details

[1] Food and Agricultural Organization, Addis Ababa, Ethiopia. [2] GAMBY Medical and Business College, Bahir Dar, Ethiopia. [3] Department of Public Health, College of Health Sciences, Debre Markos University, Debre Marqos, Ethiopia. [4] School of Public Health, Addis Ababa University, Addis Ababa, Ethiopia.

Acknowledgements

We would like to thank Mulugeta Hiruy, Muluembet Tekele, Yanet Mulugeta, Rediet Mulugeta, Eshetu Fikadu, Zeleke Dabessa, Fekadu Alemayehu Selamawit for all the support, advice, encouragement you have provided us. We also would like to thank our colleagues Ato Dereje Asaminew who has supported us in data collection and arranged everything. This research would not have been successful had it not been for the cooperation of Ato Seid, all the data collectors and the households (*Fedis Woreda*) who committed their time to provide us all the information we sought and we are truly grateful for that. The last but not the least, we would like to express our gratitude to all our friends, especially Melaku, Sinetshay, Yemserach, Selam and those people who have spent their time directly or indirectly to assist us in this work for their constant encouragement and cooperation.

Competing interests

The authors declare that they have no competing interests.

Funding

No funding of the project.

References

1. Food and Agriculture Organization of the United Nations. Trade forms and food security: conceptualizing the linkages, Rome, Italy; 2003. Cited 17 Feb 2017.
2. FAO. Rome declaration on world food security and world food summit plan of action. World food summit 13–17, Rome; 1996.
3. Ministry of Health. Food security. Healthy eating and food security, British Columbia. 15 February 2014; 2014 http://www.health.gov.bc.ca/healthyeating/.
4. FAO, IFAD, WFP. The state of food insecurity in the world. Meeting the 2015 international hunger targets: taking stock of uneven progress. Rome: FAO; 2015. p. 2015.
5. Food and Agriculture Organization of the United Nations. The state of food insecurity in the world: addressing food insecurity in protracted crises, Rome, Italy; 2010. http://www.fao.org/docrep/013/i1683e/i1683e.pdf. 11 Nov 2013.
6. Van der Veen A, Gebrehiwot T. Effect of policy interventions on food security in Tigray. Northern Ethiopia. Ecology and Society. 2011;16(1):18.
7. UN News Service. Ethiopia: food assistance needs rising for 2010, UN relief wing reports". 3 Feb 2010; 2010. http://allafrica.com/stories/201002031062.html. 20 Nov 2013.
8. Mitiku A, Fufa B, Tadese B. Empirical analysis of the determinants of rural households food security in Southern Ethiopia. The case of Shashemene District. Basic Research Journal of Agricultural Science and Review. 2012;1(6):132–8.
9. Abebaw S. Dimensions and determinants of food insecurity among rural households in Dire Dawa, Eastern Ethiopia: A thesis submitted to the school of graduate studies, Alemaya University; 2003.
10. Tesfaye K. Household food insecurity in Dodota-Sire District, Arsi Zone: coping strategies and policy options. A thesis presented to the school of graduate studies, Alemaya University; 2005.
11. Zenna T. Business and management consultancy (TZBMC), CARE ABDISHE/linking initiatives, stakeholders to achieve gender-sensitive livelihood security (LINKAGES) project end line evaluation report; 2016.
12. Action Contre La Faim Ethiopia report: Final report: nutrition causal analysis, East Haraghe Zone, Fedis and Kersa Wordas, Ethiopia; 2014.
13. Belay S. Food insecurity and copping strategies: a perspective from Kersa district, East Hararghe Ethiopia. Food Science and Quality Management. 2012;5:19.
14. Ethiopian Health and Nutrition Research Institute (EHNRI). Food composition table for use in Ethiopia. Part III and IV. Addis Ababa: Research Institute; 1998.
15. Bogale A, Shimelis A. Household level determinants of food insecurity in rural areas of Dire Dawa, Eastern Ethiopia. African Journal of Food Agriculture Nutrition and Development. 2009;9(9):1914–26.
16. Hoddinott J, Yohannes Y. Dietary diversity as a household food security indicator. Technical Note. 2002;4(4):2002.
17. Aschalew, F. Dimensions and determinant of household food insecurity in Dire Dawa, Ethiopia. MSc Thesis presented to school of graduate studies of Alemaya University, Alamaya; 2009.
18. Alemu HK, Alemu ZG, kundhlande G. Causes of household food insecurity in koredegada Kebeles, Oromia zone, Ethiopia. Agrekon. 2005;44(4):543.

12

Food insecurity of rural households in Boset district of Ethiopia: a suite of indicators analysis

Getachew Teferi Moroda*, Degefa Tolossa and Negussie Semie

Abstract

Background: Despite several efforts made so far to improve the overall food insecurity situation, the challenge is still a major problem in Ethiopia since a long time ago. Hence, the purpose of this study is to examine the food insecurity situation and identify the determinants among the rural households of Boset district. To this end, 397 household heads were selected through systematic sampling technique from six sample *kebeles*. In addition, focus group discussions, key informant interviews, and personal observations were also used to supplement the survey data. Then, the food insecurity status of households was measured with a suite of indicators.

Results: The results revealed that 26.5%, 21.7%, and 41.3% of respondents were highly food insecure through Months of Adequate Household Food Provisioning, Household Food Insecurity Access Scale, and Household Dietary Diversity Score, respectively. On top of these, 56.9%, 46.1%, and 64.0% of the respondents did not have access to water supply, not owned latrine, and dispose waste in an unsafe way, respectively. Furthermore, results from the inferential statistics showed that educational status, farmland size, total annual income, distance from health facilities, and the availability of supporting organizations were positively associated with household food security situation, while access to irrigable land, frequent drought, distance to input/output markets, and distance to road transport were negatively associated.

Conclusion: From the study findings it can be observed that all the dimensions of food (*in*)security should be focused for effective intervention. More specifically, those determinants with both positive and negative associations with food security may deserve the attention of the local authorities. Similarly, there is a need for a reorientation of an approach which is beyond a quick and simple fix. Besides, an integration of efforts between different sectors at both local and national levels is sought to bring a lasting solution to food insecurity.

Keywords: Food insecurity, Determinants, Suite of indicators, Boset, Rural Ethiopia

Background

The issue of where our next meal comes from has bedeviled humankind for much of our existence [1]; thus, food insecurity is a daily reality for hundreds of millions of people around the world [2]. In our today's time, one can observe our world's population living in a situation where there are serious strains. To this end, it was illustrated that "a threefold challenge now faces the world: match the rapidly changing demand for food from a larger and more affluent population to its supply; do so in ways that are environmentally and socially sustainable; and ensure that the world's poorest people are no longer hungry" [3]. Furthermore, it was predicted that "it seems incontrovertible that we are facing a global paradox, such that by 2030, we will have to produce more food with less water to feed approximately another billion people" [4].

A closer look at the Ethiopian context depicts that the food insecurity problem is an issue which deserves special attention to be tackled. Different scholars have depicted that food insecurity has been a daunting

*Correspondence: kennakogt@gmail.com
Center for Rural Development, College of Development Studies, Addis Ababa University, Addis Ababa, Ethiopia

challenge to Ethiopia [5–8]. In recent years, commercial food import and food aid have been accounting for a *significant* proportion of the total food supply in the country [8]. Due to such high dependence on food import, Ethiopia was identified as vulnerable to uncertainties of food import from the international market [9]. Moreover, it was confirmed that, despite the attempts made to improve the food security situation, the *actual* number of people exposed to food shortages in Ethiopia has remained significantly high [10].

Oromia Bureau of Agriculture [11] indicated that the food insecurity situation in Boset district was so precarious. The bureau revealed that due to the fact that Boset district is prone to frequent drought, it was included in the Productive Safety Net Program (PSNP). A district could be included in the PSNP when confirmed by experts that there prevails chronic food insecurity situation. In 2014/2015 there were 58,131 households who were beneficiaries from the PSNP.

Based on the fact that food insecurity is a real challenge for the life and livelihood of Ethiopia, and the study area in particular, it can be justified for a further study so that a solution could be sought. It was mentioned that food security matters immensely, because the consequences of food insecurity can affect almost every facet of society [12]. In addition, a document produced by International Food Policy Research Institute (IFPRI) [13] has stressed that food insecurity situation is a human tragedy on a vast scale, made even more heartbreaking because it is avoidable. Schanbacher [14], on the other hand, mentioned that if food sovereignty's demands are not met, it constitutes a massive violation of human rights. It was also maintained that, in geopolitical terms, deeper food crisis will undoubtedly engender more collective insecurity [15]. Relatedly, taking statements of former US President Obama at the 2012 G8 Summit, it was indicated that food security is "an economic imperative" since a poorly nourished population is a less economically productive one [12].

Given the above-discussed rationale, the concept of food (*in*)security needs to be operationalized so that a common understanding could be reached. This is because food (*in*)security is a dynamic phrase having so many definitions emerging overtime [16–19]. A review of different literature showed that food (*in*)security has evolved from a focus on availability of food, to access and utilization, and to stability in all dimensions [20]. Notwithstanding all the contentions and varying definitions of food (*in*) security over time, now scholars seem to agree on the comprehensive definition given on the World Food Summit of 1996. We have adopted the following definition in this paper:

> *Food security exists when all people at all times have physical or economic access to sufficient, safe and nutritious food to meet their dietary needs and food preferences for an active and healthy life [21].*

From this definition one can identify four key dimensions of food security: availability of sufficient food; economic, physical, and social access to the resources needed to acquire food; stability of this availability and access; and utilization that includes nutrition, food safety and quality, clean water, and sanitation [22–25]. Consequently, it can be deducted food insecurity could result if one or more of the key dimensions fail to be fulfilled [12, 26].

So far there have been so many research works conducted in different parts of Ethiopia on food insecurity. However, many of the research works conducted [27–30] used a single indicator (mainly food balance model or calorie availability) to measure the food insecurity situation which could capture only a portion of the whole situation. Such an approach, however, could underestimate the prevalence of food insecurity and its consequences; make the diagnostics difficult; and lead to the design of "one-size-fits-all" interventions [31]. That is why our paper deviates from that traditional way of studying food insecurity by employing multiple indicators so that the situation could be understood comprehensively. In fact, the use of a suite of indicators is a call of the time which is advocated by authoritative organizations such as FAO, IFAD, and WFP and leading scholars such as Carletto et al. [23], Coates [31], to mention few.

Therefore, it is based on the existing high prevalence, recurrence of food insecurity, and the need to study food insecurity in a comprehensive way that this paper was conceived. The paper aims to analyze the food insecurity situation of the rural households in Boset district, which has been a hot spot for food assistance for a long period of time. Besides, identification of determinants of the food insecurity in the study area is the other aim of the study. Thereby, this study could build on similar findings on the topic and would enable to understand food insecurity in its multidimensionality.

Theoretical framework

There have been so many theories proposed to explain about food insecurity. In fact, the explanatory powers of those theories depend on the time and existing situation of a particular place. Moreover, having recognized the existence of multiple theories of food insecurity, it was contended that among the different theories of famine (*food insecurity in this case*) no single theory is dominant or capable of excluding the others [32]. Cognizant of this fact, we used three of the existing theories to explain the food insecurity situation of the study area.

The first theory considered is the political economy explanation which describes that "the lack of political conditions for an anti-famine contract revolve around anti-democratic tendencies that abrogate any existing democratic rights, thereby hindering timely and effective action to prevent famine, and can therefore be said to involve a famine crime" [32]. This approach attributes famine (*food insecurity*) occurrence, whatever the economic or natural shocks, to governments' incompetence and lack of commitment at best, or to a deliberate action or inaction at worst [33]. Similarly, Devereux [34] argues that "all famines are explained by a combination of 'technical' and 'political' factors, where political factors include bad government policies, failure of the international community to provide relief, and war." Thus, the political economy explanation suggests that whenever incumbent government authorities and even donors are not delivering what they ought to, food insecurity or famine (in the extreme case) could happen.

Second, the climatic and environmental theory deals with the fact that food insecurity happened with the possible increase of extreme events, in which natural hazards are magnified in intensity and frequency [32]. It was also illustrated that "a combination of arable land lost to population pressure, deforestation and overgrazing, together with the possibility of a long-term decline in rain-fall in dryland farming areas in Africa and Asia, will cause declines in crop production and exacerbate food insecurity" [34]. In a similar vein, it was stated that this approach considers drought (sometimes floods) and recently climate change factors in the explanation of disruption or reduction of food output, which may at the end result in food insecurity [33].

The third theory is concerned with food insecurity as an outcome of vulnerable livelihood. Accordingly, this theory gives explanation in the sense that food insecurity could result when households fail to secure access to the various forms of assets, or when the mediating processes (i.e., institutions, organizations, and social relations at work) are not serving what is expected and/or a combination of these factors when interacting with the existing context (history, trends, and vulnerability/shock) [5].

Methods

Description of the study area

Boset district extends between 8°24′–8°51′ north latitude and 39°16′–39°50′ east longitude. It is located in the northeast part of East Shewa Zone, Oromia National Regional State. It is bordered with Adama district in the west; by Amhara National Regional State in the north; by Fantale district in northeast; and by Arsi Zone in southeast. Data obtained from the population projection by Central Statistical Agency (CSA) [35] indicated that the total population of Boset district for the year 2017 was projected to be 189,795 out of which 42,793 (22.5%) are urban population and 147,002 (77.5%) are rural population.

Based on a report obtained from the district's Finance and Economic Development Office [36], Boset district is located in the midst of the Rift Valley, which extends from the north to south. The same document revealed that climatically most parts of the district (about 89%) belong to tropical (*kolla*) agro-climatic zone and the remaining small section (about 11%) is subtropical (*woina dega*) [36]. Similarly, the document [36] showed that the district is characterized by hot and dry weather with an average annual temperature which varies between 25–30 °C for the tropical (*kolla*) and 15–20 °C for the subtropical (*woina dega*). The rainfall is weakly bimodal with spring (a small rainy season) during the months of April and May, while summer (a long rainy season) during the months of July–September. The average annual rainfall ranges between 700 and 800 mm with the intensity and variability being high in the district. In terms of drainage system the district falls in the Awash River Basin, with no other major streams and lakes.

Research design and data collection

This study was conducted in Boset district, East Shewa Zone of Oromia National Regional State. It was undertaken as a cross-sectional survey using mixed methods research approach. The choice of mixed methods was dictated by the research problem under investigation and to benefit from the merits of using this research approach [37–40].

In terms of sources of data, both primary and secondary sources were utilized. The primary data were generated by employing household survey which was administered by 12 Development Agents (DAs[1]) who are familiar to the study area and conversant with the local language (Afan Oromo). After pretesting and fully developing the structured questionnaire, it was administered face-to-face. Key informant interviews (KIIs) were also held with heads of offices and focal persons from health, women's and children's affairs, water resources, irrigation, crop production, livestock production, natural resources management, disaster preparedness and prevention, World Vision Ethiopia (Boset Area Development Program), and community elders living in the sample *kebeles*.[2]

[1] Development Agents (DAs) are individuals trained in agriculture colleges for 2 years in areas of crop production, animal health, and natural resources management to promote the agricultural extension program.

[2] A *kebele* is the lowest community-level administrative organ consisting of a number of villages.

Similarly, focus group discussions (FGDs) with selected 6 men's and 4 women's groups were conducted separately with members comprising 6–10 individuals. The groups were formed on volunteer basis with the help of the DAs working in the respective sample *kebeles*. The criteria for inclusion in the group discussion were household heads lived in the *kebele* for more than 5 years and some knowledge on food insecurity issues. Lastly, personal observations coupled with informal discussions were also employed to generate primary qualitative data. In addition, we have utilized current and relevant journal articles which were mainly published within the last 10 years as sources of secondary information.

To have a full picture of the district, a total of 6 *kebeles* located at different places were selected purposely by the district-level experts after thorough discussion on the topic of the research. Besides, food insecurity status, access to irrigation facilities, and participation in the Productive Safety Net Program (PSNP) were also used as criteria for selecting the *kebeles*.

List of households living in each of the selected *kebeles* was taken as a sampling frame, and then respondents selected using systematic random sampling technique proportionate to the size of households living in each *kebele*. The systematic random sampling technique was employed because it is one of the probability sampling methods and is easy to manipulate during selection of the sample households [41, 42]. Using the formula illustrated by Israel [43], the sample size was calculated, which resulted in a total of 397 participants (48 female- and 349 male-headed households). In the determination of the sample size, a 95% confidence level and a p value of 0.05 for maximum variability were assumed.

Mathematically, the formula is presented as:

$$n = \frac{N}{1 + N(e)^2},$$

where n stands for the sample size, N signifies the total number of households in all the *kebeles*, e designates maximum variability which is 5% (0.05), and 1 stands for the probability of the event occurring.

Data analysis

Food insecurity analysis

Based on the nature of the variables measured, to analyze the data collected both descriptive and inferential statistics were employed. Accordingly, to measure the food insecurity status through different indicators, sources of food for households, and mechanisms of filling food gaps we have mainly used percentages, mean, and standard deviation. The final analysis of the quantitative data was performed using STATA, version 12, data management software after the coded responses to the questionnaires were entered into computer. In addition, results of the FGDs, KIIs, and the field observations were transcribed and analyzed according to themes emerged.

Different authors and organizations have suggested the necessity to comprehensively analyze the four dimensions of food (*in*)security [23, 44] and to use a suite of indicators to capture the complex realities of food insecurity [e.g., 23, 31, 45]. This is because each measure captures and neglects different phenomena intrinsic to the concept of food security, thereby subtly influencing prioritization among food security interventions [46].

Following those suggestions, the indicators we employed include: the identification of Months of Adequate Household Food Provisioning (MAHFP); Household Food Insecurity Access Scale (HFIAS); Household Dietary Diversity Score (HDDS); and assessing the access and use of water supply, sanitation, and hygiene (WASH) facilities.

According to Bilinsky and Swindale [47], when using the MAHFP, although the response options start with the month of January, the respondent is asked to think back over the previous 12 months, starting with the current month. This is done by adjusting the months according to when one conducts the survey so that the current month appears first. Hence, respondents could be asked to identify in which months (during the past 12 months) they did not have access to sufficient food to meet their household needs. The purpose of these questions is to identify the months in which there is limited access to food regardless of the source of the food (i.e., production, purchase, barter or food aid) [47].

Based on information obtained from Namana and Souli [48] and making some modifications households were classified into three categories of food insecurity: least food insecure which includes households that reported being able to satisfy their food requirement for 10–12 months; moderately food insecure which includes households that were able to satisfy their food needs for 7–9 months of the year; and most food insecure which includes households that cannot feed their household members for 6 and more months during the previous year.

The other indicator to food insecurity measurements based on subjective responses is HFIAS [23]. The HFIAS is based on the idea that there is a set of predictable reactions to the experience of food insecurity that can be summarized and quantified, allowing for measurement through household surveys [23, 49].

It was indicted that the HFIAS has a set of nine questions which represent universal aspects of the experience of food insecurity, capturing information on food shortage, food quantity and quality of diet to determine the status of a given household's access to food [23]. In addition, households and populations can be classified according to the severity of their food security status along a spectrum, by using data on the severity and frequency of their experiences over the previous 30 days [23].

Based on the description given by Coates et al. [49], the maximum score for a household is 27 (the household response to all nine frequency-of-occurrence questions was "often," coded with response code of 3) and the minimum score is 0. (The household responded "no" to all occurrence questions, frequency-of-occurrence questions were skipped by the interviewer and subsequently coded as 0 by the data analyst.) The higher the score, the more food insecurity (access) the household experienced. The lower the score, the less food insecurity (access) a household experienced.

The HDDS, on the other hand, is calculated by summing the number of unique food groups consumed during the last 24 h [23]. The value of this variable will range from 0 to 12, in which lowest HDDS value signifies higher food insecurity status and vice versa. Even though there is no international consensus on which food groups to include in the scores [50], the HDDS denotes 12 food groups in which the following are considered in this study: cereals; root and tubers; vegetables with tubers; vegetables which are leafy; fruits; meat, poultry, offal; eggs; fish; pulses/legumes/nuts; milk and milk products; oil/fats; and sugar/honey.

Based on the findings from a review of different empirical studies made by Ruel [51] and Faber et al. [52] the cutoff point for the HDDS in this study is made that HDDS ≤ 5 represents low dietary diversity, HDDS 6–7 medium dietary diversity, and HDDS ≥ 8 high dietary diversity. This categorization could signify most food insecure, medium food insecure, and food secure, respectively.

To obtain a complete picture of the factors that ultimately determine the food insecurity status of a given household, non-food factors were also considered. This is because the various non-food factors contribute to determining the level of the food insecurity outcomes [23]. Accordingly, access to basic services such as clean water and sanitation together with their use was considered in this study.

Hence, based on review of the empirical studies, the MAHFP is used to assess the availability dimension, HFIAS to examine the access dimension, HDDS to measure the quality [23, 51], and WASH to assess the utilization dimension. The stability dimension, on the other hand, could be captured by looking into the results of the other three components, as it cuts across all these dimensions [23, 45].

Specification of the logit model

To identify the determinants of household food (*in*) security a binary logistic regression model was used. Following the work of Tewodros and Fikadu [29] the functional form of the logistic model is presented as follows:

$$P_i = E\left(Y = {}^{1}\!/_{X_i}\right) = \frac{1}{1 + e^{-(\beta_o + \beta_j X_i)}}. \tag{1}$$

Substituting $(\beta_o + \beta_j X_i)$ by Z_i, the equation would become:

$$P_i = \frac{1}{1 + e^{-Z_i}} = \frac{e^{Z_i}}{1 + e^{Z_i}}, \tag{2}$$

where $P_i = E(Y = 1)$ is the probability that a household is food secure. z_i is a set of explanatory variables of the ith household. β_o and β_j are the parameters to be estimated.

If P_i is the probability that a household is food secure, as is given in Eq. 2, the probability of food insecurity is then expressed as:

$$1 - P_i = \frac{1}{1 + e^{z_i}}. \tag{3}$$

From this, the odds ratio in favor of food security thus could be:

$$\text{The Odds ratio} = \frac{P_i}{1 - P_i} = \frac{e^{Z_i}/1 + e^{Z_i}}{1/1 + e^{Z_i}} = e^{Z_i}. \tag{4}$$

Since logit model uses logarithmic transformation to assume linearity of the outcome variables on the explanatory variables, the specific logit model to predict the odds of household food security is given in Eq. 5.

$$\ln\left[\frac{P_i}{1 - P_i}\right] = Z_i = \beta_0 + \beta_j X_i, \tag{5}$$

where β_0 is the constant and β_i, where $i = 1, 2, \ldots j$, are the coefficients of the exogenous variables to be estimated. X_i is a vector of explanatory variables.

Description of the variables employed The dependent variable in this model is the result obtained from the HDDS value as measured by counting the food groups that households have consumed during the last 24 h of the study period. Households with HDDS ≥ 8 are food secure and represented by 1, and those with HDDS < 8 are food insecure and represented by 0. The independent variables constitute gender, age, educational status, household size, participation in non-farm activities, farmland size, access to irrigable land, total annual income, noticed frequent drought and flood occurrence, access to weather forecast,

Table 1 Variables hypothesized to affect food security in the study area

Explanatory variables	Description	Expected sign
Gender	Dummy takes the value of 1 if male and zero otherwise	+
Age	Continuous	+/−
Educational status	Dummy takes the value of 1 if formal education and zero otherwise	+
Household size	Continuous	+/−
Participation in non-farm activities	Dummy takes the value 1 if there is access and zero otherwise	+
Farmland size	Continuous	+
Access to irrigable land	Dummy takes the value 1 if there is access and zero otherwise	+
Total annual income	Continuous	+
Noticed frequent drought	Dummy takes the value 1 if they have noticed and zero otherwise	−
Noticed frequent flood occurrence	Dummy takes the value 1 if they have noticed and zero otherwise	−
Access to weather forecast	Dummy takes the value 1 if there is access and zero otherwise	+
Availability of *kebele* assistance	Dummy takes the value 1 if there is access and zero otherwise	+
Access to DA services	Dummy takes the value 1 if there is access and zero otherwise	+
Access to health extension services	Dummy takes the value 1 if there is access and zero otherwise	+
Access to credit services	Dummy takes the value 1 if there is access and zero otherwise	+
Distance from input/output markets	Continuous	−
Distance to road transport	Continuous	−
Distance to health services	Continuous	−
Availability of other (NGO) support organizations	Dummy takes the value 1 if there is access and zero otherwise	+

Table 2 Number of months respondents consume their own produce by *kebele*

Name of *kebele*	Number of months						Total	
	3–6 months		7–9 months		10–12 months			
	n	%	*n*	%	*n*	%	*n*	%
Buta Wagare	3	14.3	11	52.4	7	33.3	21	100.0
D/Wanga	9	30.0	13	43.3	8	26.7	30	100.0
Q/H/Mirqasa	50	31.3	100	62.5	10	6.2	160	100.0
Sara Areda	5	9.6	47	90.4	0	0.0	52	100.0
Sifa Batte	30	27.8	75	69.4	3	2.8	108	100.0
Tiri Birreti	8	30.8	13	50.0	5	19.2	26	100.0
Total	105	26.5	259	65.2	33	8.3	397	100.0

Pearson χ^2 (10) = 58.5444, Pr = 0.000

availability of *kebele* assistance, access to DA services, access to health extension services, access to credit services, distance from input/output markets, distance to road transport, distance to health services, and availability of other (NGO) support organizations. The variables hypothesized to affect the food security are presented in Table 1 together with their expected sign.

Results and discussion

Food insecurity as measured by MAHFP

The availability dimension of the food insecurity of the respondents as measured by the MAHFP indicates that only 8.3% of them were least food insecure, 65.2% moderately food insecure, and 26.5% most food insecure (see Table 2).

Results from Table 2 show across all the *kebeles* majority of respondents fall in the category of 7–9 months (moderately food insecure). The statistical test conducted also shows there is statistically significant difference ($p < 0.001$) in terms of the number of months respondents encounter food insecurity across the sample *kebeles*. For example, there was no respondent from Sara Areda *kebele* who fall in the least food insecure category as compared to 33.3% of the respondents in Buta Wagare *kebele*. Here it can be learned that interventions aimed to tackle food insecurity in the study area should consider the severity level of the problem.

In line with what was obtained in the survey, key informants and focus group discussants all indicated that food insecurity occurs on seasonal basis. In the words of the focus group discussants "always after Ethiopian Easter is celebrated" food insecurity occurs in most of the *kebeles* in the study area, which shows a cyclical pattern of inadequate availability and access to food. Focus group discussants in four of the sample *kebeles* (except Sifa Batte and Q/H/Mirqasa) expressed the months of June to September are those in which households experience the worst food shortage compared to other months. Related to the cyclical pattern of food insecurity, the finding may signify a need for mechanisms to be designed accordingly to avoid the depletion of resources while trying to compensate for the diminishing access to food.

As far as the consequences of the months of inadequate food provisioning is concerned, it was highlighted that seasonality and being food deficit from own production even in normal years can be implicated in making people less food secure [53]. It can also be deducted that the stability component of the respondents' food security could be highly compromised; this is because "stability in the availability of and access to food should be ensured regardless of sudden shocks like climatic crisis or cyclical events which involve seasonal food scarcity" [25].

To fill the gap that is created from consuming own production, opportunities of earning income from non-farm activities are so important. However, it is only 15.1% ($n=60$) of the respondents who reported to engage in different non-farm income-generating activities. Given such limited opportunity for generating income outside their farming, one can imagine what the food insecurity situation could look like, especially for those who are consuming their own produce for only 3–6 months.

Generally, households could have different sources of food for their consumption. And understanding the basic patterns of the sources and how they vary across locations, population groups, and over time will provide a particularly important starting point for understanding the general nature of the food security problem [54]. Accordingly, respondents have identified different sources of food as indicated in Table 3.

Results from Table 3 show all of the respondents obtained food from their own production. This is just because they were engaged in farming activities. Next is obtaining food by purchase (63.2%), through aid (40.3%), and through borrowing (21.9%). It can be learned that one source of getting food was not adequate to feed the whole family members throughout the year. Again, it can be seen that more than half of the respondents replied to have obtained food from purchase, and that issue of price fluctuation and market in general are important factors that merit special attention when dealing with food insecurity in the study area. This is because "given the heavy dependence of both the rural and urban poor on markets, inflation has potentially devastating effects on the food security of poor households" [55].

Table 3 Sources of food for the households (multiple responses)

Sources of getting food	Responses		Percent of cases
	n	Percent	
Get from their own production	397	44.4	100.0
Get from purchase	251	28.0	63.2
Get their food by borrowing	87	9.7	21.9
Get their food through aid	160	17.9	40.3
Total	895	100.0	225.4

Furthermore, it was demonstrated that particularly for the poor, who spend more than 50% of their income on food consumption, changes in the prices of major staple crops can have a dramatic impact [56, 57]. Likewise, scholars have again explained that an increase in food prices could have both direct impacts and reductions in real incomes for poor consumers who spend a large share of their income on food that will inhibit their access to available food [12, 58].

On the other hand, it can also be seen that getting food through borrowing has the least contribution (Table 3). This is because, according to focus group discussants, the ever increasing poverty of the rural residents and the monetization of everything have eroded the value of supporting each other among the community members.

Since majority of respondents were not found to have consumed only from their own produce throughout the year, and the unavailability of income diversification opportunities could indicate the existence of food gaps. Hence, findings about the mechanisms of filling food gaps from Table 4 show the rural households used different mechanisms of filling food gaps.

The finding shows multiple mechanisms that households used to fill the food gaps they are confronted with. However, non-farm activities, remittance, and borrowing contribute relatively minimum. Besides, most of the mechanisms used by respondents appear to be unreliable.

Looking closely at the mechanisms of filling food gaps some of them need to be treated with special precautions as their end result may not be the one which was intended. For instance, during the study period, it was observed that so many households including those

Table 4 Mechanisms of filling food gaps (multiple responses)

Food gap filling	Responses		Percent of cases
	n	Percent	
Fill food gap through aid	128	19.8	33.5
Fill food gap through borrowing	90	14.0	23.6
Engage in food-for-work	158	24.5	41.4
Engage in non-farm activities	60	9.3	15.7
Engage in casual labor	142	22.0	37.2
Fill food gap through remittance	67	10.4	17.5
Total	645	100.0	168.9

considered to be better came to the district agriculture office to complain about their exclusion from benefitting out of the PSNP. Even at times of informal discussions with residents in the district, people told that overcoming the food insecurity problem without the PSNP is unthinkable. Besides, staff members of the food security team in the district mentioned the difficulty in discriminating households who are going to graduate from the program, as there is no one who wants to graduate.

Related to what is physically observed and highlighted by the district staff members, one of the mechanisms that deserve careful handling is food aid. On this issue different scholars advised to handle food aid with caution. For example, it was reported food aid as one of the oldest forms of foreign aid and one of the most controversial [59]. Similarly, Barrett [60] mentioned "while there is effectively universal agreement as to the desirability of the goal of reducing acute and chronic food insecurity, there remains considerable dispute as to how effective food aid is in achieving the goal." It was emphasized the undesirable aspect, "negative dependency," which arises when meeting current needs comes at the cost of reducing recipients' capacity to meet their own basic needs in the future without external assistance [60]. Furthermore, the negative dependency typically arises when individuals, households, or communities alter their behavior in response to the provision of assistance that unwittingly creates disincentives to undertake desirable behavior (e.g., to grow a crop, or to allocate time to work) [60].

In line with the argument made in the preceding paragraph, it was explained that "food aid dependency undermines food security in Ethiopia at every level, from the household to the national government" [61]. This is because the government has little incentive to expend its scarce resources on food security programs as long as the international community remains willing to sink its food surpluses into Ethiopia [61]. Similarly, even very recently it was depicted that "despite the existence of the PSNP, there has not been a reduction in food aid to Ethiopia. In fact, the 2015 appeal issued by the Government of Ethiopia and humanitarian partners is one of the biggest ever at 1.4 bn USD" [62].

> Box 1. Reflection of a key informant from World Vision Ethiopia of Boset district. The key informant stated that engaging in daily labor (Merti and Africa Juice companies), charcoal making and firewood selling, petty trading, renting out land with cheap price, diversifying income sources through irrigation, and taking credit from private money lenders are the mechanisms through which the local people fill the food gap created.

As far as food gap filling mechanisms are concerned, it can be deducted that some of the mechanisms, such as food aid, charcoal and firewood selling, and renting out land could result in undesirable results that call for a critical review of the country's strategy. This is because community members (in the case of the study area) and even nations could fall addicted to such quick fixing but downgrading way of solving a problem.

Food insecurity as measured by HFIAS

Based on the nine generic questions (see Table 5) of the access-related conditions, the finding shows it is only 26.2% ($n=104$) who never worried having not enough food, whereas the remaining 73.8% ($n=293$) of surveyed households have experienced problems of both economic and physical access to food at varying levels of food insecurity.

Looking at the finding on the basis of the severity level, out of the total score of 2607, it can be observed that 30.8% of households encountered access problems "rarely," 41.9% "sometimes," and 27.3% "often" during the last 1 month of the study period.

Furthermore, the finding shows the mean score of HFIAS for the respondents is 12.9 with a standard deviation of 4.3, while the minimum and the maximum were 1 and 22, respectively. Based on the categorization made by FAO [63] on a study conducted in Mozambique to determine the cutoff point, a score of 0–11 was taken as "most food secure"; 12–16 medium food insecure; and a score above 17 most food insecure. Accordingly, it was found in the study that 39.8% ($n=158$) were most food secure; 38.5% ($n=153$) were medium food insecure; and 21.7% ($n=86$) were most food insecure.

Of course, the HFIAS is better interpreted when used to assess Household Food Insecurity Access Prevalence (HFIAP) [10]. Accordingly, the HFIAP indicator categorizes households into four levels of household

Table 5 Distribution of households by HFIAS condition

S/no.	HFIAS conditions	Severity status and number of households					
		Rarely		Sometimes		Often	
		n	%	*n*	%	*n*	%
1	Worried having not enough food	152	51.88	129	44.03	12	4.10
2	Not able to eat the food kinds he/she preferred	55	13.90	212	53.81	127	32.23
3	Able to eat only a limited variety of foods	6	1.52	167	42.39	221	56.09
4	Able to eat some foods that he/she did not want to eat	195	68.18	89	31.12	2	0.70
5	Able to eat a smaller meal than he/she felt needed	47	11.93	208	52.79	139	35.28
6	Able to eat fewer meals/day because there was not enough food	17	4.33	165	41.98	211	53.69
7	Absolutely no food to eat in the household	178	69.80	77	30.20	0	0.00
8	Slept at night hungry because there was not enough food	95	73.08	35	26.92	0	0.00
9	Went a whole day and night hungry b/c there was not enough food	58	85.29	10	14.79	0	0.00
	Total score	803	30.8	1092	41.9	712	27.3

Clues to the severity status

(1) Rarely (once or twice in the past 4 weeks)

(2) Sometimes (three to ten times in the past 4 weeks)

(3) Often (more than ten times in the past 4 weeks)

Table 6 Distribution of households by Household Food Insecurity Access Prevalence (HFIAP)

Question	Frequency		
	Rarely	Sometimes	Often
1a	152	129	12
2a	55	212	127
3a	6	167	221
4a	195	89	2
5a	47	208	139
6a	17	165	211
7a	178	77	0
8a	95	35	0
9a	58	10	0

Clue to the severity conditions:

(green)	Food secure /access/ = 9.4%
(grey)	Mildly food insecure /access/ = 27.2%
(yellow)	Moderately food insecure /access/ = 33.8%
(black)	Severely food insecure /access/ = 29.6%

food insecurity (access): food secure, and mild, moderately, and severely food insecure [49]. Thus, households are categorized as increasingly food insecure as they respond affirmatively to more severe conditions and/or experience those conditions more frequently. Results on the HFIAP are presented in Table 6.

Results from Table 6 show there were only few (9.4%) of the respondents who are food secure; i.e., such households experience none of the food insecure conditions, or just worry, but rarely. To the contrary, the result shows there are 3 times more respondents who were severely food insecure; i.e., households already cut back on meal size or the number of meals often, and/or experience some of the three most severe conditions.

Food insecurity as measured by HDDS

The results of the finding on the HDDS show respondents were found to have consumed an average of 6 food groups with a standard deviation of 1.53. Besides, the minimum HDDS value is 3 and the maximum HDDS value is 11. The detail of the HDDS value across the sample *kebeles* is presented in Table 7.

The statistical test made on the results of the overall HDDS shows there is a statistical significant difference across the sample *kebeles* about the HDDS at $p<0.001$ with a degree of freedom which equals 40. Hence, based

Table 7 HDDS for respondents across *kebeles*

HDDS	Names of *kebeles*												Total	
	Buta Wagare		D/Wanga		Q/H/Mirqasa		Sara Areda		Sifa Batte		Tiri Birreti			
	n	%	*n*	%	*n*	%	*n*	%	*n*	%	*n*	%	*n*	%
3	0	0.0	1	3.3	15	9.4	0	0.0	3	2.8	0	0.0	19	4.8
4	0	0.0	3	10.0	33	20.6	2	3.8	11	10.2	7	27.0	56	14.1
5	5	23.8	6	20.0	55	34.4	8	15.4	10	9.3	5	19.2	89	22.4
6	6	28.6	5	16.7	41	25.6	11	21.2	26	24.1	7	27.0	96	24.2
7	6	28.6	5	16.7	14	8.8	14	26.9	37	34.2	4	15.4	80	20.1
8	2	9.5	3	10.0	1	0.6	12	23.1	17	15.7	1	3.8	36	9.1
9	2	9.5	6	20.0	0	0.0	5	9.6	4	3.7	1	3.8	18	4.5
10	0	0.0	0	0.0	1	0.6	0	0.0	0	0.0	1	3.8	2	0.5
11	0	0.0	1	3.3	0	0.0	0	0.0	0	0.0	0	0.0	1	0.3
Total	21	100.0	30	100.0	160	100.0	52	100.0	108	100.0	26	100.0	397	100.0

Pearson χ^2 (40) = 150.0694, Pr = 0.000

Table 8 Whether respondents have access to safe drinking water by *kebele*

Name of *kebele*	Access to safe water				Total	
	No	%	Yes	%	*n*	%
Buta Wagare	18	85.7	3	14.3	21	100.0
D/Wanga	0	0.0	30	100.0	30	100.0
Q/H/Mirqasa	121	75.6	39	24.4	160	100.0
Sara Areda	42	80.8	10	19.2	52	100.0
Sifa Batte	43	39.8	65	60.2	108	100.0
Tiri Birreti	2	7.7	24	92.3	26	100.0
Total	226	56.9	171	43.1	397	100.0

Pearson χ^2 (5) = 120.2161, Pr = 0.000

on the categorization discussed earlier about 41.3% of the respondents were found to consume less dietary diversity, implying they are more food insecure due to lack of the means to acquire and consume a variety of foods. Those who have medium level of food insecurity account for 44.3%, and only 14.4% of the respondents have HDDS ≥ 8 that they were food secure and were able to acquire and consume a variety of foods. In fact, it should be noted that the HDDS value could be reduced if sugars and beverages are to be taken out, because they do not add to the nutritional quality of the diet [52].

The nexus between WASH and food insecurity

When explaining how much water supply and sanitation practices are crucial, it was stated that populations suffering from hunger are often the same as those that lack adequate water and sanitation [64]. On the other hand, it was argued that ensuring poor people's access to safe drinking water and adequate sanitation and encouraging personal, domestic, and community hygiene will improve the quality of life of millions of individuals [65]. Moreover, safe drinking water, sanitation, and hygiene (WASH) contribute significantly to the increased capacity of individuals to absorb and use the nutrients in their food [66].

Access to clean water

Access to safe drinking water has a direct contribution to the improvement of the food insecurity problem in different ways. Water is a key driver of agricultural production [67], and accessing safe drinking water has the benefits to reduce exposure to a variety of diseases that obstruct the intake and utilization of food and minimizing expenses related to health [68]. In light of these facts respondents were asked about their access to safe drinking water, and the responses are presented in Table 8.

Results from Table 8 show among the sample *kebeles* respondents of *D/Wanga* (100.0%), *Tiri Birreti* (92.3%), and *Sifa Batte* (60.2%) claimed to have better access to safe drinking water. Their access can be justified (for the first two *kebeles)* due to their proximity to Wolanchity

Table 9 Respondents' ownership of latrine by *kebele*

Name of *kebele*	Ownership of latrine				Total	
	No	%	Yes	%	*n*	%
Buta Wagare	4	19.1	17	80.9	21	100.0
D/Wanga	2	6.7	28	93.3	30	100.0
Q/H/Mirqasa	139	86.9	21	13.1	160	100.0
Sara Areda	11	21.2	41	78.8	52	100.0
Sifa Batte	23	21.3	85	78.7	108	100.0
Tiri Birreti	4	15.4	22	84.6	26	100.0
Total	183	46.1	214	53.9	397	100.0

Pearson χ^2 (5) = 181.6547, Pr = 0.000

town, and for the third *kebele* its proximity to Bofa (Gode Dhera) town. In the remaining sample *kebeles,* a significant number of the respondents were not lucky to have access to safe drinking water, thereby forcing to resort to other sources for drinking and other domestic purposes.

Findings of the six sample *kebeles* show that about 56.9% of the respondents did not have access to safe drinking water. During the FGDs participants informed that even those who have access did not feel the supply is reliable because the facilities were broken down frequently which forced them either to return back to the unsafe sources, or to spend lots of money to buy water for domestic purposes. When discussing how difficult it was getting safe drinking water, the discussants at *B/ Wagare kebele* pointed out that one jerry can that holds 20 L will cost them 10 birr after travelling 2 h on foot. Coupled with the high price, the queue might be so long that sometimes it was in the second day that they could collect the water. A key informant interview made with head of the district's water resources office also showed water shortage was really a serious problem due to deepening of the water table and frequent breakdown of the existing facilities.

As to the implication, it can be said that apart from the direct burden on households' health due to lack of safe drinking water, it has got so many negative implications. For instance, it was mentioned that water availability will be one of the limiting constraints for crop production and food security [69]. In a similar vein, scholars have indicated that farmers' inability to access or control water has an obvious direct impact on potential yields and income, and an indirect impact by reducing potential payoffs from investments in fertilizers, improved seed varieties, and learning technical skills [70].

In a more broader way, it was illustrated that "the lack of adequate water is linked to poverty—households facing water shortages are more likely to be poor or fall into poverty than households not facing such shortages" [71]. Thus, this could imply that households who lack access to water and those who are poor will be more vulnerable to food insecurity. And what is more frustrating is that "... water scarcity will become, more or less, a major threat to food security due to increasing food demand and competition for water resources among sectors" [4].

Among those households who did not get access to safe drinking water an attempt was made to distinguish their sources of drinking water. Accordingly, it was obtained that river/stream (95.6%) is their chief source of water. In fact, these unsafe sources could have detrimental consequences particularly on the health of individuals and on their food security situation in general. Among the negative effects, it was said lack of access to safe water could result in poor health and affect the physical well-being [68].

As far as the burden of collecting water for different purposes is concerned, FGDs held with both women's and men's groups admitted that collecting water was primarily the responsibility of women, followed by grown-up children. This could imply that whenever women are forced to spend more time to haul water, it will reduce their productivity [68] that could end in keeping women in a poverty trap.

Ownership and use of latrines

Among the WASH components one of them that were treated in this study is ownership and use of latrines. The findings from Table 9 show an interesting result that, except in Q/H/Mirqasa *kebele,* in all the remaining *kebeles* latrine was owned by the large majority of the respective households.

Despite the ownership of latrines by majority of respondents, however, there was a problem when it comes to the use. FGDs in all the sample *kebeles* revealed that shortage of water to clean the latrine was bringing bad smell and, as a result, compelled members of households to defecate in unsafe places. A key informant

Table 10 Respondents' mechanisms of waste disposal

Name of *kebele*	Areas of waste disposal						Total	
	Open field		In the garden		In a waste disposal pit			
	n	%	*n*	%	*n*	%	*n*	%
B/Wagare	5	23.8	3	14.3	13	61.9	21	100.0
D/Wanga	5	16.7	6	20.0	19	63.3	30	100.0
Q/H/Mirqasa	54	33.7	79	49.4	27	16.9	160	100.0
Sara Areda	13	25.0	23	44.2	16	30.8	52	100.0
Sifa Batte	38	35.2	19	17.6	51	47.2	108	100.0
Tiri Birreti	4	15.4	5	19.2	17	65.4	26	100.0
Total	119	30.0	135	34.0	143	36.0	397	100.0

Pearson χ^2 (10) = 69.9092, Pr = 0.000

interview held with the head of the district's health office also confirmed the weak usage of latrines due to shortage of water that could be used for cleaning. So the above percentage shows mainly the physical presence of latrines not necessarily their use.

On the other hand, respondents who did not own latrine were inquired to identify the area of defecation they use. Accordingly, about 70.0% ($n = 128$) expressed to have used bushes and/or forest areas, whereas the remaining 30.0% ($n = 55$) used open field for defecation. Because of such unsafe practice, the rural households could be denied of the benefits that one can obtain by having sanitation and hygiene facilities and good practices. Sanitation reduces or prevents human fecal pollution of the environment, thereby reducing or eliminating transmission of diseases from that source [65]. It was argued that high-tech solutions are not necessarily the best and some simple latrines can be very effective, while untreated sewage distributes pathogens in the environment and can be the source of disease [65].

Mechanisms of waste disposal

With recognition of the fact that proper waste disposal could help to avoid disease creating pathogens, respondents were asked to identify the mechanisms they use.

Findings from Table 10 show only 36.0% of the respondents dispose waste in a waste disposal pit. The remaining 64.0% of the respondents dispose waste in an unsafe way, either in an open field or in the garden. Such act of unsafe disposal of waste could play a role in transmission of diseases, which signifies the need to be cautious about the surrounding environment where people are residing. This is because "it is insufficient for an individual to receive an adequate quantity of food, if he or she is unable to make use of the food due to illnesses resulting from inadequate sanitation or poor sanitary practices" [23].

Determinants of food (*in*)security

As can be observed from Table 11, out of the 19 variables fitted in the binary logistic regression model, 9 significantly affected food insecurity status of the rural households in the study area (Table 11). The discussions made below are based on the findings from Table 11.

Educational status The educational status of the household heads was found to be important in determining their food security situation. With a significance level of less than 5% household heads with better educational status were more likely to be food secure with a 9.6% probability. In fact, the inverse relationship of educational status with food insecurity was also found in other empirical studies [27, 29].

Farmland size Having more cultivable land was strongly associated with 11.4% of being food secure at less than 1% significance level. This could mean that households with more cultivable land could produce more food, may purchase food for consumption from the income they get from their land, or even may diversify their crop to insure for crop failure. Hence, ways should be sought to lift off the pressure on farmland size.

Access to irrigable land A unit increase in access to irrigable land was associated with 9.3% probability of being food insecure with a 10% significance level. This could be due to the meddling of brokers who compel rural households to sell their products with lower prices to those who also cheat them on the scale when measuring the products. Besides, the products which were produced by the farmers require the use of modern inputs which were expensive and not easily available. Thus, the manipulation of the brokers and the cost of modern inputs keep them into a debt spiral that results in food insecurity and even poverty.

Table 11 Parameter estimates of the binary logit model with their marginal effects

Explanatory variables	Coefficients	p value	Marginal effects
Gender	1.146024	0.119	0.074412
Age	− 0.0012422	0.955	− 0.0014243
Educational status	0.931175	0.032**	0.0962907
Household size	− 0.0142931	0.883	0.0016966
Participation in non-farm activities	0.005792	0.990	− 0.001335
Farmland size	0.9697931	0.000***	0.1142719
Access to irrigable land	− 0.7562149	0.099*	− 0.0926803
Total annual income	0.0000991	0.000***	0.0000112
Noticed frequent drought	− 0.9546311	0.057*	− 0.1109556
Noticed flood occurrence	− 0.2651678	0.535	− 0.0221719
Access to weather forecast	0.8280069	0.413	0.0247848
Availability of *kebele* assistance	0.0351892	0.942	− 0.0084927
Access to DA services	0.7159894	0.185	0.0503655
Access to health extension services	0.2784892	0.556	0.035605
Access to credit services	− 0.6171401	0.145	− 0.0778475
Distance from input/output markets	− 1.676472	0.006***	− 0.1401994
Distance to road transport	− 0.842344	0.066*	− 0.1036628
Distance from health services	1.812996	0.003***	0.1759557
Availability of other support organizations	0.9402377	0.048**	0.0597707
LR χ^2 (19)	103.46		
Prob > χ^2	0.0000		
Pseudo-R^2	0.3167		

***,**,*Significant at 1%, 5%, and 10% probability levels, respectively

Total annual income The finding on the total annual income shows the existence of statistically strong evidence ($p < 1\%$) that an increase in annual income will increase the probability of being food secure, but the corresponding percentage of increment is much lower than 1%. It could be learned that creating more income-generating opportunities could be helpful by improving their purchasing power, let farmers use modern inputs that improve their production and productivity, and even make them rent in more farmland that could help them diversify their production activities.

Noticed frequent drought It was obtained from the finding that the more frequent the drought occurs, there is an 11.1% probability for households to become food insecure at 10% significance level. This is due to the fact that frequent drought could result in crop failure that impedes the availability of food and reduce income that households could have earned from their production.

Distance to input and output markets Distance to input and outputs markets was found to have a strong negative influence on the food security situation of the households; i.e., a unit increase in distance from input and output markets increases the probability of being food insecure with 14%. This finding is statistically significant at 1% level. It implies that market centers should be expanded so that distance could be shortened and people will easily get inputs which help them improve their production and productivity and also enable them sell their products to generate more income that will be used for consumption smoothing and diversify their income sources.

Distance to road transport The finding again shows that for a unit increase in the distance from road transport, there is a 10.4% probability for the households to be less food secure with a statistical significance level of less than 10%. This could be because when distance to road transport increases people may not be encouraged to diversify and produce marketable products.

Distance to health facilities The finding shows an unexpected result that an increase in distance from health facilities is associated with 17.6% probability of increase in being food security at p value of less than 1%.

Availability of other supporting organizations It was found that the availability of supporting organizations could improve the food security status of households with a probability of 6%. This finding is statistically significant at less than 5% level. With a great care not to create dependence syndrome, introducing support by partner organizations could assist in smoothing consumption and even may enable households support themselves.

Conclusion

Food insecurity is more worrisome now than ever before due to the unprecedented variability of the climate and the poverty trap that rural people are in. This paper aimed to examine the food insecurity status and identify their determinants for the rural households in Boset district, East Shewa Zone. The results revealed that households which account 26.5%, 21.7%, and 41.3% are *highly* food insecure through MAHFP, HFIAS, and HDDS, respectively. In addition, the WASH results show 56.9%, 46.1%, and 64% of the households did not have access to safe drinking water, did not own latrine, and dispose waste in an unsafe way, respectively. As far as determinants of food security are concerned, it was found that educational status, farmland size, total annual income, distance to health services, and the availability of supporting organizations were positively associated with being food secure. To the contrary, access to irrigable land, occurrence of frequent drought, distance to input and output markets, and distance from road transport were negatively associated with being food secure.

It can be observed that frequent occurrence of drought is found to have statistical significance in making households food insecure. This finding is in line with what the climatic and environmental theories propose in explaining food insecurity. On the other hand, distance from inputs and outputs markets, and from road transport could show the governments' incompetence and lack of commitment in curbing food insecurity as was argued in the political economy explanation. Lastly, the inability to diversify households' income and the frequent drought occurrence could imply the applicability of the theory that contends food insecurity as an outcome of vulnerable livelihood. Our study suggests that due to the seasonal nature of food insecurity, future research may need to be conducted based on longitudinal data so that consistency of findings could be confirmed and interventions could be made efficient. Besides, decision makers in the area should insist in the future about the use of multiple indicators to clearly understand the nature of the prevailing food insecurity and respond accordingly.

Abbreviations

CSA: Central Statistical Agency; DAs: Development Agents; FAO: Food and Agriculture Organization of the United Nations; FGDs: focus group discussions; HDDS: Household Dietary Diversity Score; HFIAS: Household Food Insecurity Access Scale; HFIAP: Household Food Insecurity Access Prevalence; IFPRI: International Food Policy Research Institute; KIIs: key informant interviews; MAHFP: Months of Adequate Household Food Provisioning; MoARD: Ministry of Agriculture and Rural Development; NGOs: non-governmental organizations; PSNP: Productive Safety Net Program; WASH: water supply, sanitation, and hygiene.

Authors' contributions

GTM designed the study, collected the data, performed the analysis, and developed the manuscript. DT and NS contributed to the research design and analysis, reviewed and made editorial comments on the draft manuscript. All authors read and approved the final manuscript.

Acknowledgements

We are grateful for the financial support provided by Addis Ababa University to cover part of the field expenses. We also wish to extend our thanks for all individuals involved in the study from Oromia National Regional State to the *kebele* levels.

Competing interests

The authors declare they have no competing interests.

Funding

Expenses for part of the data collection were covered by Addis Ababa University.

References

1. Peacock KW. Global issues: food security. New York: Facts on File Inc.; 2012 **(An imprint of Infobase Learning)**.
2. Webb P, Coates J, Frongillo EA, Rogers BL, Swindale A, Bilinsky P. Measuring household food insecurity: why it's so important and yet so difficult to do. J Nutr. 2006;136:1404S–1408S. http://jn.nutrition.org/content/136/5/1404S.full.pdf+html. Accessed 6 June 2017.
3. Godfray HCJ, Beddington JR, Crute IR, Haddad L, Lawrence D, Muir JF, Pretty J, Robinson S, Thomas SM, Toulmin C. Food security: the challenge of feeding 9 billion people. Science. 2010;327:812. https://doi.org/10.1126/science.1185383.
4. Chartres C, Sood A. The water for food paradox. Aquat Procedia. 2013;1:3–19. https://doi.org/10.1016/j.aqpro.2013.07.002.
5. Degefa T. Rural livelihoods, poverty and food insecurity in Ethiopia: a case study at Erenssa and Garbi communities in Oromiya Zone, Amhara National Regional State. Ph.D. thesis, Norwegian University of Science and Technology, NTNU, Trondheim; 2005.
6. Messay M. Resettlement and food security nexus in Ethiopia: a case study from Nonno district. Ph.D. thesis, Addis Ababa University. LAP LAMBERT Academic Publishing; 2012.
7. Meskerem A, Degefa T. Household food security status and its determinants in Girar Jarso *Woreda*, North Shewa Zone of Oromia region, Ethiopia. J Sustain Dev Afr. 2015;17(7):118–37.
8. Belay K, Dawit A. Agricultural research and extension linkages: challenges and intervention options. Ethiop J Agric Sci. 2017;27(1):55–76.
9. Berhanu A. The food security role of agriculture in Ethiopia. Agricultural and Development Economics Division of FAO. Electron J Agric Dev Econ. 2004;1(1):138–153. https://ageconsearch.umn.edu/bitstream/12012/1/01010138.pdf. Accessed 13 June 2018.
10. Guyu FD. Household vulnerability to green famine: component based analysis of indicators in Belo-jiganfoy District (case study area), Benishangul-gumuz region, Ethiopia. Appl Sci Rep. 2015;9(3):139–56. https://doi.org/10.15192/PSCP.ASR.2015.9.3.139156.
11. Oromia Bureau of Agriculture. Oromia region 2007 E.C. (2014/2015 Fis-

cal year) proposed safety net plan, Finfinne/AA/. Unpublished report; 2014.
12. Jones AD, Ngure FM, Pelto G, Young SL. What are we assessing when we measure food security? A compendium and review of current metrics. Adv Nutr. 2013;4:481–505. https://doi.org/10.3945/an.113.004119.
13. International Food Policy Research Institute (IFPRI). Reaching sustainable food security for all by 2020: getting the priorities and responsibilities right. Washington, DC; 2002. http://www.ifpri.org/publication/reaching-sustainable-food-security-all-2020-0. Accessed on 13 June 2018.
14. Schanbacher WD. The politics of food: the global conflict between food security and food sovereignty. Santa Barbara: ABC-CLIO, LLC; 2010.
15. Behnnasi M, Yaya S. Food crisis mitigation: the need for an enhanced global food governance. In: Behnassi M, Draggan S, Yaya S, editors. Global food insecurity: rethinking agricultural and rural development paradigm and policy. Berlin: Springer; 2011. p. 93–125. https://doi.org/10.1007/978-94-007-0890-7.
16. Maxwell S. Food security: a post-modern perspective. Food Policy. 1996;21(2):155–70. https://doi.org/10.1016/0306-9192(95)00074-7.
17. Hoddinott J. Choosing outcome indicators of household food security. In: Hoddinot J, editors. Food security in practice: methods for rural development projects. Washington, DC: International Food Policy Research Institute; 2001. pp. 39–45. http://www.ifpri.org/sites/default/files/publications/fspractice_01.pdf. Accessed 13 Aug 2014.
18. Carr ER. Postmodern conceptualizations, modernist applications: rethinking the role of society in food security. Food Policy. 2006;31:14–29. https://doi.org/10.1016/j.foodpol.2005.06.003.
19. Food and Agriculture Organization of the United Nations (FAO). Coping with a changing climate: considerations for adaptation and mitigation in agriculture. Environment and natural resources management series no. 15. Rome, Italy; 2009.
20. Conceicao P, Levine S, Lipton M, Warren-Rodríguez A. Toward a food secure future: ensuring food security for sustainable human development in Sub-Saharan Africa. Food Policy. 2016;60:1–9. https://doi.org/10.1016/j.foodpol.2016.02.003.
21. FAO. Climate change and food security: a framework document. Rome; 2008b. http://www.fao.org/forestry/15538-079b31d45081fe9c3dbc6ff34de4807e4.pdf. Accessed 2 Dec 2017.
22. Schmidhuber J, Tubiello FN. Global food security under climate change. PNAS. 2007;104(50):19703–8. https://doi.org/10.1073/pnas.0701976104.
23. Carletto C, Zezza A, Banerjee R. Towards better measurement of household food security: harmonizing indicators and the role of household surveys. Glob Food Secur. 2013;2:30–40. https://doi.org/10.1016/j.gfs.2012.11.006.
24. Poppy GM, Jepson PC, Pickett JA, Birkett MA. Achieving food and environmental security: new approaches to close the gap. Philos Trans R Soc. 2014;B369:20120272. https://doi.org/10.1098/rstb.2012.0272.
25. FAO. The state of food and agriculture: climate change, agriculture and food security. Rome; 2016. http://www.fao.org/3/a-i6030e.pdf. Accessed 18 Oct 2016.
26. Du Toit A, Ziervogel G. Vulnerability and food insecurity: background concepts for informing the development of a national FIVIMS for South Africa; 2004. http://www.agis.agric.za/agisweb/FIVIMS_ZA. Accessed 15 Aug 2014.
27. Abebaw S, Ayalneh B. Dimensions of food insecurity and livelihood strategies among rural households in Dire Dawa, Eastern Ethiopia. Trop Sci. 2007;47(2):73–80. https://doi.org/10.1002/ts.199.
28. Guyu FD. Ethno-culture disparity in food insecurity status: the case of Bullen District, Benishangul-Gumuz regional state, Ethiopia. Afr J Food Sci. 2014;8(2):54–63. https://doi.org/10.5897/AJFS2013.1096.
29. Tewodros T, Fikadu T. Determinants of households food security and coping strategies for food shortfall in Mareko District, Guraghe Zone southern Ethiopia. J Food Secur. 2014;2(3):92–9. https://doi.org/10.12691/jfs-2-3-4.
30. Guyu FD, Muluneh W. Determinants of seasonal food insecurity in the 'green famine' belt of Ethiopia: the case of households in Belo-jiganfoy District, Benishangul-gumuz region. Afr J Food Sci. 2016;10(11):278–91. https://doi.org/10.5897/AJFS2016.1434.
31. Coates J. Build it back better: deconstructing food security for improved measurement and action. Glob Food Secur. 2013;2:188–94. https://doi.org/10.1016/j.gfs.2013.05.002.
32. Wisner B, Blaikie P, Cannon T, Davis I. AT RISK: natural hazards, people's vulnerability and disasters. 2nd ed. London: Routledge; 2004.
33. Ali H. Vulnerability to drought risk and famine: local responses and external interventions among the Afar of Ethiopia, a study on the Aghini pastoral community. Ph.D. dissertation, Department of Social Geography, Faculty of Biology, Chemistry and Earth Sciences University of Bayreuth, Germany; 2008.
34. Devereux S. Transfers and safety nets. In: Devereux S, Maxwell S, editors. Food security in Sub-Saharan Africa. Pietermaritzburg: University of Natal Press; 2001. p. 267–93.
35. Central Statistical Agency (CSA). Population projection of Ethiopia for all regions at *Wereda* level from 2014 to 2017. Federal Demographic Republic of Ethiopia, Addis Ababa; 2013. www.csa.gov.et/…/ehioinfo-internal?…population-projection-of-ethiopia-for-all-regions. Accessed 21 Oct 2017.
36. Boset District Finance and Economic Development Office (2012).Physical and socioeconomic profile of Boset district. A document submitted to Oromia Finance and Economic Development Bureau. **(Unpublished report)**.
37. Johnson RB, Onwuegbuzi AJ. Mixed methods research: a research paradigm whose time has come. Educ Res. 2004;33(7):14–26. http://mintlinz.pbworks.com/w/file/fetch/83256376/Johnson%20Mixed%20methods%202004.pdf. Accessed 11 Feb 2017.
38. Creswell JW. Research design: qualitative, quantitative, and mixed methods approach. 3rd ed. Thousand Oaks: SAGE Publications Inc; 2009.
39. Creswell JW. Educational research: planning, conducting, and evaluating quantitative and qualitative research. 4th ed. Boston: Pearson Education Inc; 2012.
40. Creswell JW, Clark VLP. Designing and conducting mixed methods research. 2nd ed. Thousand Oaks: SAGE Publications Inc; 2011.
41. Babbie E. The basics of social research. 4th ed. Stamford: Thomson Wadsworth; 2008.
42. Bryman A. Social research methods. 4th ed. Oxford: Oxford University Press; 2012.
43. Israel GD. Determining sample size. Institute of Food and Agricultural Sciences (IFAS), University of Florida; 2013. https://www.psycholosphere.com/Determining%20sample%20size%20by%20Glen%20Israel.pdf. Accessed on 2 May 2018.
44. FAO, IFAD, WFP. strengthening the enabling environment for food security and nutrition. Rome: FAO; 2014. p. 2014.
45. Leroy JL, Ruel M, Frongillo EA, Harris J, Ballard TJ. Measuring the food access dimension of food security: a critical review and mapping of indicators. Food Nutr Bull. 2015;36(2):167–95. https://doi.org/10.1177/0379572115587274.
46. Barrett CB. Measuring food insecurity. Science. 2010;327:825–8. https://doi.org/10.1126/science.1182768.
47. Bilinsky P, Swindale A. Months of Adequate Household Food Provisioning (MAHFP) for measurement of household food access: Indicator guide (v. 4). Washington, DC: FHI 360/FANTA; 2010. https://www.fantaproject.org/sites/default/files/resources/MAHFP_June_2010_ENGLISH_v4.pdf. Accessed 11 Apr 2017.
48. Nanama S, Sóuli K. Brief: comparison of two methods for measuring household food security and vulnerability—evidence from the Zondoma food security initiative, Burkina Faso. Afr Food Secur Rev. 2007. Washington DC: Africare/Headquarters. http://www.africare.org/at_work/offd/index.html. Accessed 17 Nov 2017.
49. Coates J, Swindale A, Bilinsky P. Household food insecurity access scale (HFIAS) for measurement of household food access: Indicator guide (v. 3). Washington, DC: Food and Nutrition Technical Assistance Project, Academy for Educational Development; 2007.
50. FAO. Guidelines for measuring household and individual dietary diversity. Prepared by Gina Kennedy, Terri Ballard and Marie Claude Dop, Nutrition and Consumer Protection Division; 2010. http://www.fao.org/docrep/014/i1983e/i1983e00.pdf. Accessed 13 Jan 2018.
51. Ruel MT. Operationalizing dietary diversity: a review of measurement issues and research priorities. J Nutr. 2003;133:3911S–26S.
52. Faber M, Schwabe C, Drimie S. Dietary diversity in relation to other household food security indicators. Int J Food Saf Nutr Public Health. 2009;2(1):1–15.

53. Ellis, F. Human vulnerability and food insecurity: policy implications. Forum Food Secur South Afr. 2003. https://www.odi.org/sites/odi.org.uk/files/odi-assets/publications-opinion-files/5605.pdf. Accessed 18 Feb 2017.
54. Frank R, Nancy M, Bruce C, Laura B, Eric K. Food security indicators and framework for use in the monitoring and evaluation of food aid programs. Arlington: Food Security and Nutrition Monitoring Project (IMPACT), ISTI, Inc., for the U.S. Agency for International Development; 1999. http://pdf.usaid.gov/pdf_docs/Pnacg170.pdf. Accessed 13 Jan 2018.
55. Ministry of Agriculture and Rural Development (MoARD). Food security program 2010–2014: productive safety net; 2009. http://extwprlegs1.fao.org/docs/pdf/eth144896.pdf. Accessed 18 Apr 2017.
56. Kristkova ZS, Van Dijk M, Van Meijl H. Projections of long-term food security with R&D driven technical change—a CGE analysis. NJAS Wagening J Life Sci. 2016;77:39–51. https://doi.org/10.1016/j.njas.2016.03.001.
57. Wheeler T, von Braun J. Climate change impacts on global food security. Science. 2013;341:508. https://doi.org/10.1126/science.1239402.
58. Ringler C, Zhu T, Cai X, Koo J, Wang J. Climate change impacts on food security in Sub-Saharan Africa: insights from comprehensive climate change scenarios. IFPRI Discussion Paper 01042; 2010. http://ebrary.ifpri.org/cdm/ref/collection/p15738coll2/id/6983. Accessed 13 Jan 2018.
59. FOA. The state of food and agriculture: food aid for food security? 2006. ftp://ftp.fao.org/docrep/fao/009/a0800e/a0800e.pdf. Accessed 2 Dec 2017.
60. Barrett CB. Food aid's intended and unintended consequences. ESA working paper no. 06-05; 2006. http://www.fao.org/3/a-ag301t.pdf. Accessed 2 Dec 2017.
61. Devereux S. Food insecurity in Ethiopia: a discussion paper for DFID; 2000. https://www.ids.ac.uk/files/FoodSecEthiopia4.pdf. Accessed 27 Feb 2016.
62. Sandstrom S, Juhola S. Continue to blame it on the rain? Conceptualization of drought and failure of food systems in the Greater Horn of Africa. Environ Hazards. 2017;16(1):71–91. https://doi.org/10.1080/17477891.2016.1229656.
63. FAO. Report on use of the household food insecurity access scale and household dietary diversity score in two survey rounds in Manica and Sofala Provinces, Mozambique, 2006–2007; 2008a. http://www.fao.org/fileadmin/user_upload/eufao-fsi4dm/doc-training/moz_diet.pdf. Accessed 13 Jan 2018.
64. Thompson B, Cohen MJ, Meerman J. World food insecurity and malnutrition: scope, trends, causes and consequences. In: Thompson B, Cohen MJ, editors. The impact of climate change and bioenergy on nutrition. Berlin: Springer; 2012. p. 21–41. https://doi.org/10.1007/978-94-007-0110-6.
65. Pruss-Ustun A, Bos R, Gore F, Bartram J. Safer water, better health: costs, benefits and sustainability of interventions to protect and promote health. Geneva: World Health Organization; 2008. http://apps.who.int/iris/bitstream/10665/43840/1/9789241596435_eng.pdf. Accessed 11 Jan 2017.
66. Rautanen SL, While P. Using every drop—experiences of good local water governance and multiple-use water services for food security in far-western Nepal. Aquat Procedia. 2013;1(2013):120–9. https://doi.org/10.1016/j.aqpro.2013.07.010.
67. Hanjra MA, Qureshi ME. Global water crisis and future food security in an era of climate change. Food Policy. 2010;2010(35):365–77. https://doi.org/10.1016/j.foodpol.2010.05.006.
68. Degefa T, Tesfaye T. Linkages between water supply and sanitation and food security: a case study in four villages of East Hararghe zone, Oromia region; 2008. http://www.rippleethiopia.org/library.php/files/file/20080908-wp6-water-and-food-security. Accessed 4 Jan 2017.
69. Kang Y, Khan S, Ma X. Climate change impacts on crop yield, crop water productivity and food security—a review. Prog Nat Sci. 2009;19:1665–74. https://doi.org/10.1016/j.pnsc.2009.08.001.
70. Giordano M, de Fraiture C, Weight E, van der Bliek J. (eds) Water for wealth and food security: supporting farmer-driven investments in agricultural water management. Synthesis report of the Ag Water solutions project. Colombo: International Water Management Institute (IWMI); 2012. https://doi.org/10.5337/2012.207.
71. Faurès J-M, Santini G. (eds) Water and the rural poor: interventions for improving livelihoods in sub-Saharan Africa. Rome: FAO; 2008. http://www.fao.org/3/a-i0132e.pdf. Accessed 2 Dec 2017.

Assessment of production and marketing constraints and value chain of sweet potato in the municipalities of Dangbo and Bonou

Vincent Ezin[1*], Florent Quenum[1], Roméo Houénagnon Bodjrenou[1], Chabrolle M. I. Kpanougo[1], Emeric M. G. Kochoni[2], Bienvenu Ifagbémi Chabi[3] and Adam Ahanchede[1]

Abstract

Background: Sweet potato [*Ipomoea batatas* (L.) Lam.] plays an important role in ensuring food security and incomes for local communities. It is particularly grown in southern region of Benin, and its production over the last few years has declined due to disease and severe pest infestation. Therefore, it was imperative to carry out this study to identify the main constraints related to the recent decline in sweet production and more specifically analyzing the production and marketing constraints of sweet potato in the townships of Dangbo and Bonou.

Results: The results of the survey revealed that many varieties of sweet potato were grown, mainly "Vobodouaho" (local name in Goun language), the most preferred variety by producers. Weedicides and insecticides were used to control weeds and insect pests, respectively. There was no efficient control methods against black rot, *Alternaria blight, cylas* spp., and caterpillar among the pathogens and insects that damaged sweet potato production in the study areas. The main constraints reducing sweet potato production in the townships of Dangbo and Bonou are: lack of adequate methods for controlling pest and diseases, lack of financial resources due to no access to credit, absence of modern equipment, scarcity and high cost of labor and absence of best methods for long-term storage. The average yield of sweet production per hectare ranged from 6.5 to 15.1 tons fresh matter/ha. The market value of sweet potato is still low at farmers' level when compared to that of other stakeholders along the value chain.

Conclusion: This study calls for the attention of public and private sectors and policy makers for more investment in sweet potato research. Avenues for further research on sweet potato have also been suggested.

Keywords: Actors in sweet potato, Horticulture, Limitations, Productivity, Revenue

Background

Benin is an essentially agricultural country where 70% of the population lives in rural areas [1]. This population basically practices agriculture of subsistence. Among the main staple food crops cultivated are tubers and roots. In order to encourage the development of these crops and improve their contribution to the food security, Benin government, through its policy of diversification of agricultural production and fight against poverty, has initiated the Program of Development of Tubers and Roots (PDRT) in collaboration with partners for development such as International Fund for Agricultural Development (IFAD) and West African Bank of Development (BOAD) [1]. The program aims at rationalizing sustainably production, transformation and marketing of tubers and roots through reinforcement of technological capacities. Among tubers and roots are cassava, taro, yam, potato and sweet potato. The findings of the present research strengthen the Ministry of Agriculture in its program of intensify tubers and roots and ultimately stimulate calls for attention for more investment in sweet potato research in order to help farmers increase their productivity and revenues. Sweet potato can adapt to different ecological conditions and requires

*Correspondence: ishola.vincent@yahoo.com; awevincent@gmail.com
[1] Department of Crop Production, Faculty of Agricultural Science, University of Abomey-Calavi, Abomey-Calavi, Benin
Full list of author information is available at the end of the article

few inputs for its cultivation [2]. It is an important food and vegetable crop in most developing countries where it was ranked fifth economically after rice, wheat, maize and cassava [3]. Sweet potato is grown mainly for the fresh market. The fresh market sweet potato tubers are sold in the surrounding markets around the study areas, which need to be improved upon for high price and subsequent high income. The sweet potato value chain of different innovative platform approaches is required to facilitate work in the farms and postharvest operations and at the same time improve the quality of sweet potato products including improve nutritional value and food and nutritional security; and strengthen the capacity of all stakeholders in order to alleviate poverty along the value chain.

From nutritional point of view, all the parts of sweet potato are rich in food fibers, the leaves and the stems contain high content in minerals, particularly iron and carotene, vitamin B2, vitamin C and vitamin E [4]. The tubers of sweet potato are rich in starch, sugar, vitamin C, β-carotene, iron and several other minerals [5–7]. According to [6], sweet potato tubers proved to be richer in proteins than cassava and the yam tubers. All these features indicate that sweet potato is an important crop capable of playing a significant role in food security. In spite of these assets, sweet potato still remains a marginalized crop in the development programs of the country. Moreover, from 2012 to 2015, the area of production of sweet potato compared to other tubers and roots declined from 2.21 to 2.01% and a decrease in production from 69,829.43 to 55,404.97 tons was also recorded; henceforth, a slight reduction in production area, i.e., 0.20% and a drastic yield reduction of 20.65% [8]. It has been basically due to a certain number of factors such as pathogens, weevils (*Cylas* spp.), sweet potato virus disease, low market value tubers harvested by farmers, lack of certified planting materials resistant to abiotic and biotic elements and modern storage facilities for long-term conservation. Jackai et al. [9] reported that there are losses of millions of dollars annually only due to infestation of sweet potato by weevils and these insect pests are more prevalent in the tropical regions. Based on this decline in sweet potato production and taking into account its high potentiality, it could be important to intensify research efforts to increase its production for the benefit of the producers and consumers and sort out the problem of food insecurity. To this end, this survey was conducted in the valley of Benin more precisely in the townships of Dangbo and Bonou to analyze the techniques of production systems and marketing of the sweet potato by identifying the constraints that prevent the development of this sector.

Methods

Study site and choice of study area

The municipalities of Dangbo and Bonou are located in the department of Ouémé. Dangbo Township is bounded in the north by Adjohoun Township, in the south by Aguégués Township, in the east by Akpro-Missérété Township and in the west by Sô-Ava Township. As for Bonou, it is bounded in the north by the Ouinhi Township, in the south by Adjohoun, in the east by Sakété and Adja-Ouèrè Townships and in the west by Zê and Zogbodomé Township. Dangbo and Bonou cover a total land area of 149 and 250 km^2, respectively.

With the poor distribution of rainfall observed nowadays, the seasons underwent noticeable change (PDC2 Bonou, 2013–2017). In the last 20 years, the total average annual rainfall is 1237 mm (ASECNA, 2012) with a temperature that varies between 25 and 30 °C. The number of rainy days turns around 135 days per year and the period of vegetative growth varies between 229 and 251 days.

The valley of Ouémé consists of four municipalities, namely Dangbo, Bonou, Adjohoun and Aguégué. Our survey was conducted in Dangbo and Bonou Municipalities. Each municipality is divided into several villages among which ten were investigated (Fig. 1).

These villages were chosen for two main reasons:

- Favorable zones for sweet potato production. Given that the needs in water of the sweet potato are estimated at 600 mm of water per cycle, with high quantity of water at the early stages of crop development and adequate temperatures between 22 and 33 °C, and it is also required between 120 and 210 days for crop development [10]. We can say that the rainfall conditions and temperature in the survey zone are propitious to the requirements of the sweet potato production; and
- Areas of high sweet potato production.

Sampling and data collection

In order to achieve the objectives of this work, the survey was carried out in two (02) phases: an exploratory phase and an in-depth survey phase.

The exploratory phase consisted in establishing contact with people in charge of CARDER (Regional Action Center for Rural Development), the different producers of the two townships. The contact was established through the person in charge of rural development and the adviser in crop production representing the ministry of agriculture in the locality for the list of farmers growing sweet potato.

The exploratory survey enabled us to obtain some primary information about sweet potato through

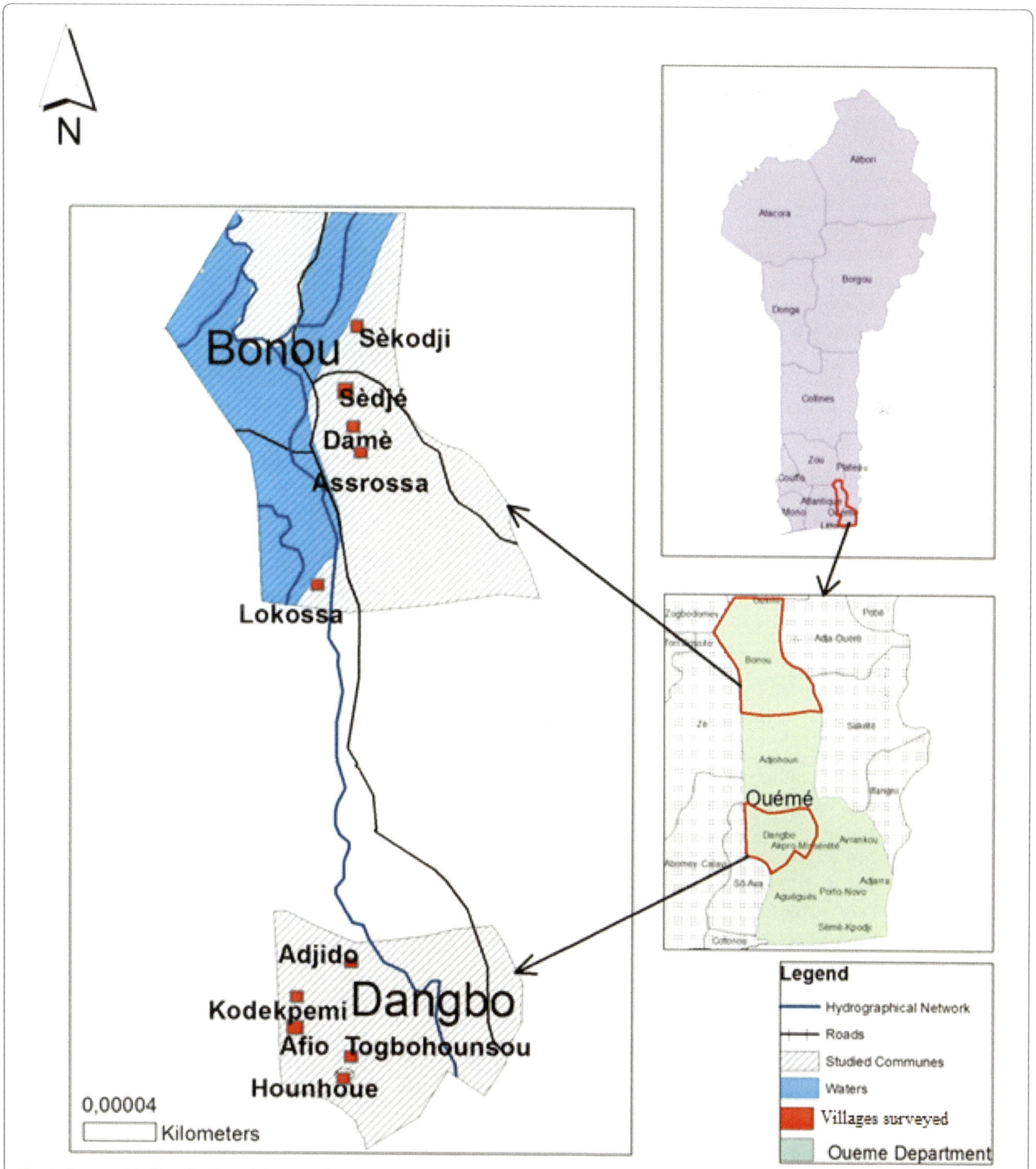

Fig. 1 Geographical localization of the ten villages investigated

interview with producers (target group and individual discussion) selected randomly and observations made in the field. This phase was also of great help for setting proper questionnaire for the in-depth phase.

In the in-depth survey phase, a total of 150 producers were randomly surveyed but due to the heavy infestation of caterpillars and *cylas* spp. and some pathogens in the last few years, complete information were obtained only from 100 producers while incomplete information were obtained from those who abandoned sweet potato production as at the time of the study. Thus, 100 producers with complete information were used in this study. Therefore, the producers were surveyed through interviews, participating and/or direct observations and focus group. Qualitative and quantitative data were collected. The parameters included sex, age, size of the household, level of education, cropped area, number and types of varieties, mode of land acquisition, and types of labor, soil fertility management, development phase and yield of the sweet potato. The producers were ranked based on the size of areas cropped as follows: cropped area lesser than 1 ha as small producers; between 1 and 2 ha as average scale farmers and cropped area greater or equal to 3 ha as large scale producers. As far as the constraints are concerned, they were ranked based on farmers' percentage.

Data analysis

The collected data in survey sheets were examined and codified and then entered into Microsoft Excel. The descriptive statistics with the aid of Minitab software was computed and averages, frequencies, minimum and maximum of some parameters such as sex, age, size of the household, the level of education and the availability of the land were determined.

Results

Profile of farmers

The producers of sweet potato investigated in the township of Bonou and Dangbo were all native (40% were from sèto and 60% from Whémé). Majority of sweet potato producers in study area were men (56%) with ages between 40 and 60 years while those of less than 40 years and older than 60 years were 12% (Table 1). This important number of men in sweet potato production could be justified by the toughness of the works on the vertisols. About 80% of respondent households were within the range of working age (< 40 to 60 years old), whereas only 12% of them were elder (> 60 years old). Despite the high number of men, an increase in women in some tasks such as cutting of vines, planting and tuber gathering for storage were observed.

The education level of respondents surveyed is presented in Fig. 2. From the analysis of this figure, most of producers did not attend schools and were primary school dropouts.

Table 2 shows the classification of respondents according to the cultivated areas. From the analysis of the results, 44% of respondents grew sweet potato on farmland of 1 to 2 ha, 36% in lesser than 1 ha of land and 20% of respondents in 3 ha of land.

Figure 3 shows the distribution of the respondents according to the mode of access to land. The analysis of the results of Fig. 3 reveals that the mode of access to land the most observed was the inheritance (50%) followed by purchase (30%) and renting (20%). It should be noted that women were less involved in farm activities due to the fact that they are not entitled to land heritage in Africa context, mainly because of their significant roles in the household affairs.

Table 3 shows the different types of labor used by producers. The analysis of the results of this table reveals that the most observed was labor (75%) followed by family workforce (25%). Mutual aid was also observed in complement of family workforce. Family labor was only used during weekend and holidays. Male producers made use of the two types of labor whereas female producers

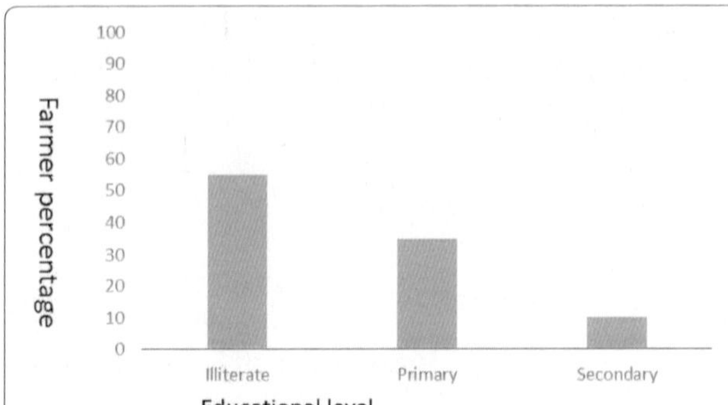

Fig. 2 Educational level of producers investigated

Table 1 Socio-demographic characteristics of respondents according to their age, sex, ethnic group and townships

Township	Age						Ethnic group	
	< 40		40–60		> 60		Sèto	Whémé
	M	F	M	F	M	F		
Bonou	10	0	32	10	8	0		60
Dangbo	2	2	24	8	4	0	40	
Total	12	2	56	18	12	0	40	60
Frequencies (%)	12	2	56	18	12	0	40	60

Table 2 Typology of producers according to the areas cultivated

Areas cultivated	Sex	Townships				Total	
		Bonou		Dangbo			
		Number	Fre %	Number	Fre %	Number	Fre %
< 1 ha	Male	16	16	8	8	36	36
	Female	4	4	8	8		
1–2 ha	Male	20	20	2	2	44	44
	Female	4	4	18	18		
> 3 ha	Female	14	14	4	4	20	20
	Female	2	2	0	0		
Total		60	60	40	40	100	100

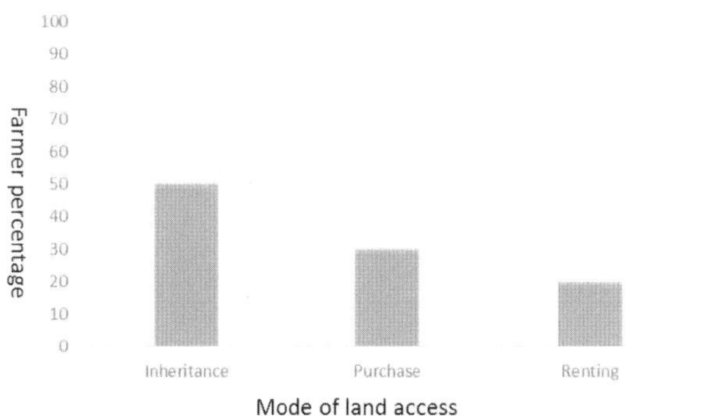

Fig. 3 Distribution of producers according to the mode of land access

Table 3 Different types of labor used by the producers

Labor	Number	Frequency (%)
Labor	75	75
Family aid	25	25
Total	100	100

often dependent on their husbands for help and occasionally used labor to supplement the family assistance. It should also be noted that in addition to the high cost paid to manpower, producers also provided foods to the workers during the period they carried out the work in the field.

Capital funds invested in the production of sweet potato by farmers include their own funds and loans. Their own funds were from other agricultural-related activities. Producers took loans at interest from different agricultural banks such as CLCAM (Local Bank for mutual agricultural loan) and ASF (Association of Financial Services), to be able to face labor payment and other input activities. The loans were repaid at interest rate of 22%. The main constraints to loan access are several formalities to complete at the banks, the long distance that separates these banks from producers, the non-return of the loans by some farmers in past and the high interest rate.

Sweet potato production and storage: constraints and adaptive strategies

The propagation by cuttings was performed 2 weeks after the application of herbicides. It should be noted that problem of input supply arises in the study area. The problem of weedicide and insecticide supply in the study area could be justified by the long distance between the area of cultivation and CARDER headquarter especially because of the river of Ouémé separating the two sites. Some producers buy an important quantity (liters) of pesticides that is more than what is needed for their farms and resell at high price to other producers. The cost of a liter of the inputs ranged between 4000 and 5000 FCFA (US \$8 and US \$10) when bought from CARDER and 5000–6000 FCFA (US \$10–12) when bought from local sellers. The majority of farmers used sweet potato planting material from their previous crop or from their neighbors.

The land preparation consists in cutting weeds with the aid of sickles on the ridges of previous season. According to the majority of producers, this allows the decomposition of weed residues which are sources of nutrients to the growing sweet potato plants for better productivity. Moreover, it also helps to manage the harmful pests by destroying their habitat and also by preventing the resumption of the adventitious tubers. To facilitate the work, the producers applied SHARP herbicide at a rate of 480 g/L, 2 weeks before land clearing.

Irrigation (Fig. 4) was applied during the dry season to facilitate the plowing and water supply to plants. About (80%) of producers interviewed applied irrigation during dry season for out-season production.

Fig. 4 Land irrigation

Table 4 Traits of varieties of sweet potato grown in Bonou and Dangbo

Morphotypes	Color skin	Water content	Ability for storage
Adjlè	White	Moderately low	Good
Atakan	White	Moderately low	Good
Ahotonon	Red	Moderately low	Poor
Amiwèdé	White	Low	Good
Zohoun	Red	Moderately low	Good
Vobodouaho	White	High	Poor
Adogoun	White	Moderately low	Poor
Minché	White	Moderately low	Poor

The water content is based on farmer perspective

All producers investigated (100%) plowed their lands manually and planted on ridges. The height of the ridges was 30 cm; the distance between the ridges was 50–80 cm. The length of each ridge ranged from 3 to 6 m.

Results on cropping systems indicated that most of farmers did not practice mixed cropping. According to the producers, the association of sweet potato to other crops would affect the productivity and the expected yield of sweet potato due to the competition for nutrients, sun, air etc. They did not also practice crop rotation. Weeding was carried out at least twice, this depends on the proliferation of weeds, but some producers did a single weeding because of the high cost of labor. The majority of respondents did not use chemical fertilizers. Producers did use some methods to maintain soil fertility such as the burying of weeds and crop residues in the soils.

The planting material used was obtained from at least 5 months old plants. The lengths of the planting material vines varied between 30 and 40 cm and were immediately planted at a depth of 20 cm with a density of 33,333–46,667 plants per hectare. When the ridges to be made for planting were not ready, the vines were kept in moist soils under the shades in order to maintain their viability. The majority of respondents conserved the planting materials in the fields for next planting. It should be noted that none of the producers interviewed used certified planting materials because breeders and seed agencies did not make this available to them.

Table 4 presents the eight varieties of sweet potato used by farmers and their different traits. The variety "Vobodouaho" was grown by all producers interviewed. The choice of this variety is based on the following: preferred by consumers, short cycle of 4 months, high yields and big tubers when compared to other varieties. "Vobodouaho" means you can be in debt without worrying. According to the producers, this variety by its high productivity enables them to pay their debts.

According to the producers surveyed, the harvests were performed at harvest maturity but there was also early harvest which was done at physiological maturity due to flooding. The potato reaches its physiological maturity when the stems and leaves began yellowing and leaves started dropping. The harvest usually starts in April and ends in August. When it rains and the ridges are half flooded, the producers immediately harvest them. The men carried out the harvesting and the women picked up with a basin and then transport them to the place of packaging.

The producers surveyed have several storage methods. A first method consists on gathering the tubers in heaps and covering them with plant debris (Fig. 5). The storage period varies between 1 and 2 months if the ridges were not flooded before harvest. A second method consisted in packaging the sweet potatoes in the bags and covering them with vegetable debris (Fig. 6). None of the producers interviewed had sophisticated facilities for long period of storage.

All the producers surveyed estimated that for two ridges (of 6 m long each) an average of 55 kg was harvested but in case of pest attack at least six ridges of sweet potato were required before obtaining the 55 kg which reduced the average yield to about 6.5 tons/ha. Therefore, the average yield of sweet production per hectare without disease and pest infestation and at optimal conditions was 15.125 tons.

The main constraints were diseases and insects viz black rot, *Alternaria* and caterpillars (Lepidoptera larva), which cause serious damage to the tubers (Fig. 7). This insect is called "Hlo" in local language. The causal agent of black rot was identified as *Ceratocystis fimbriata* while that of *Alternaria species* was *Alternaria solani* and

Fig. 5 Heaping of sweet potatoes on the farm

Fig. 6 Storage in bags of sweet potatoes

insect was identified as *Acraea acerata* commonly known as "wanvi" in local language by producers. Another insect species identified was *Cylas puncticollis*, a weevil, known as **vâ** in local language, and it does cause galleries (hollows) on tubers (Fig. 8). It causes the root losses resulting in crop failure. It also leads to poor quality of tubers that are no longer accepted in the market. *Ceratocystis fimbriata* caused the yellowing of young leaves and the blackening of the underground parts of the stem while it caused circular holes on the tubers. *Alternaria solani* showed brown rings on the leaves and black lesions on the petiole and stem. The matured leaves were the most attacked. *Acraea acerata* attacked the leaves by perforating them. In the dry season, this caterpillar caused tremendous damage and resulted in poor yields (Fig. 7). To control this insect, the producers used active ingredient of super action insecticides and super cydim containing cypermethrin. It should be noted these insecticides are normally used for other market gardeners, which are not specific to sweet potato pests. None of these pesticides control *Cylas puncticollis* as reported by the farmers. Further studies must then be initiated to find the most suitable pesticides and other control measures for the control of sweet potato pests and diseases.

Other constraints identified by farmers surveyed were high temperatures, high moisture content of tubers, flooding of potato fields and late harvesting. Producers also listed weeds as serious constraints in the Ouémé valley where the lands are rich in nutrients. Some producers used herbicide SHARP 480 SL as control measures.

Table 5 presents the various constraints in order of importance as perceived by the producers and the adaptation strategies developed.

The adaptation strategies developed by the producers aim at reducing the effects of these constraints for a good productivity of sweet potato.

Marketing of sweet potato products and constraints

Producers marketed their sweet potato produces in several local markets, namely Dangbo, Ouando (Porto-Novo), Tatonnonkon, Dasso, Assrossa, Bonou and Dantokpa (Cotonou). But none of the producers did export their products to the subregion and international markets. The marketing options include direct sales to consumers and selling to wholesalers or retailers. During the periods where sweet potato tubers are much more available, the supply becomes greater than the demand. This causes the producers to sell their products at a low price. This is due to the lack of storage facilities and the financial needs of producers. The sweet potato tubers were sold either in a basin (Fig. 9) or packed in bags and transported by canoes (Fig. 10) to the markets. The basket was between 3500 and 4000 FCFA and the bag was 6000 FCFA. There is rapid growth in demand for sweet potato in urban areas because it is fried every afternoon in every corner of the cities and it is also boiled but in rural areas it is boiled and pounded together with yam and eat as food with sauce. This has created huge opportunities to sell and market sweet potato in urban and rural areas. The main constraints in marketing sweet potato tubers is due to the infestation of some tubers by insect pests which lower its quality and nutritional value and therefore make the consumers not to be interested in it because oftentimes some farmers fail to remove the damaged tubers from the healthier ones. But most often

Fig. 7 Decayed tubers due to the attack of caterpillar

Fig. 8 Some sweet potato tubers attacked by *Cylas puncticollis*

it is sold at low price due to the low marketable root yields, which lead to low income.

Value chain of sweet potatoes

The actors involved in the value chain are: producers, wholesalers, semi-wholesalers, retailers, transporters and consumers (Fig. 11). The producers were the sweet potato farmers. They were involved in field production and they do receive input supply from CARDER and informal agencies who bought their inputs from Nigeria. After harvest, the sales were carried out in the fields. The wholesalers were big traders of the locality; they bought directly from the producers and the basket of sweet potato is sold at 3500 FCFA. The semi-wholesalers bought the wholesale from wholesalers; they were traders and they bought in large quantities at a price that could be estimated at 5000 FCFA/basket. The retailers bought agricultural products at relatively high prices (7000 FCFA/basket) from semi-wholesalers compared to the purchase price practiced by semi-wholesalers from the wholesalers and the wholesalers from producers. They were market traders. The consumers bought from retailers for their own food needs. They bought at a relatively higher price. The semi-wholesalers sell 4 to 5 tubers per heap (200–400 FCFA) to the consumers. The producers preferred to sell their products to wholesalers instead of going to the market in order to avoid paying transport fees. The transporters were responsible for transporting products from the fields to the markets.

Discussion

The leaders of households interviewed were mostly males as the case in other African countries. Our results are similar with those of [11] who reported that over 60% of the farmers surveyed were men in their study in Kenya. 55% of the respondents were illiterate and only 35% completed primary education (Fig. 2). This is below the mean national literacy level of adults to adopt some agricultural technologies. So, the illiteracy of farmers could be the raison behind some poor agricultural practices observed in the field, and the low yield recorded year after years. They could not read the instructions on the label on the inputs used unless they are associated with somebody who can help. But these producers were experienced enough to apply some good agricultural practices. These results are not consistent with those of [12] who reported that most (54.2%) of the interviewed households completed primary education in southern Ethiopia, same results were obtained by [13].

The production of sweet potato in the studied areas is still traditional. Crop production practices are carried out manually using rudimentary tools such as hoe, cutlass. The main difficulty was related to the preparation of the soil by making ridges, a difficult task on clay soils of the areas of study due to the rudimentary tools used. The 88% of respondents within the range of working age possess the physical strength required to face the challenge of sweet potato production with the use of rudimentary tools. Under these conditions, the use of modern tools for the production of sweet potato should be explored in order to reduce the difficulties associated with sweet potato production. In another words, the use of oxen and tractors would be of great help in plowing, weeding and harvesting, which will reduce root damages especially during harvesting. In relation to planting, there is always problem of plant material scarcity, i.e., lack of vines or seed tubers. Producers often do not

Table 5 Constraints in order of importance and coping strategies

Constraints	Adaptive strategies	Rank	% farmers
Climate hazards (flood and submergence)	Temporary abandonment of fields	1	39
Absence of modern's equipment for production	Use of available means and materials of production	2	20
Lack of adequate methods for pests management	Use of chemical products intended to market gardening to reduce pest damage	3	10
Scarcity and expensiveness of labor	Use of family workforce, labor and mutual aid	4	8
Toughness of activities	High use of labor	5	7.5
Absence of best storage methods for a long time	Immediate sale of tubers after harvest	6	6
Inexistence of road system for tubers transport to the market	Using pirogue to transport sweet potatoes	7	5.5
Slump in sales during harvest of tubers	Early harvest of tubers and choice of short-cycle varieties (Vobodouaho et minché)	8	3
Lack of training and aid to the members of their associations and from agricultural extension agents	Use of traditional practice for the production	9	1

have planting material for the next growing season. This problem is exacerbated by the flood from the overflowing of the river which invades their field coupled with the lack of supervision and training. Therefore, a training of producers is necessary in order to provide all the technical assistance for the best cultivation of the crop [1], particularly on the techniques of planting and multiplication of cuttings and their conservation for next growing season. Sweet potato was mainly planted on ridges and the density of seedlings was about 33,333–46,667 plants per hectare. These values were relatively higher than the ones recommended by [14], which is 30,000 plants per hectare for better productivity. The average yield of sweet potato ranges from 6.5 tons to 15.125 tons/ha. The high yields were obtained with adequate control measures and were of 275 basins or 15.125 tons per ha which was above the average yield in other African countries such as Kenya and Ethiopia. [15] in their review on sweet potato production constraints reported that the average yield of sweet potato in Ethiopia was 8 tons/ha but below 18 tons/ha in the Asian countries. This indicates that the production constraints need to be properly addressed in order to improve the productivity of sweet potato in Benin so as to better the yields obtained by Asia countries.

Fig. 9 Basin of sweet potato tubers

Furthermore, the cultivation of sweet potato crop is facing with other difficulties such as weeding. This is a serious problem due to the nature of the soil (clay) and the lack of manpower in the areas of study. The results of our surveys showed that some producers, considering the high cost of labor, reduced the number of weeding during the sweet potato development cycle. The decline in yield observed in some respondents' fields was due to the lack of weeding. Our results are consistent with those of [16] who reported yield losses of 22% in India and 57% in Africa from non-weeded farms of sweet potato. Another constraint is related to the attack of pathogens and insect pests such as black rot, *Alternaria* blight, *cylas* spp. (weevils) and caterpillars which cause enormous damage and reduced the market value of the produce. This could be due to the single-crop farming adopted year after year in the areas. Tefera et al. [17] have also reported sweet potato virus disease as a major constraint of sweet potato production in Ethiopia and they also, reported that *Alternaria* blight was a constraint even though not serious. As a control measures, [18] in Nigeria reported that a mixture of 50 mL of neem seed oil extract and 30 mL of diazinon in either 2 or 10 L of water and cuttings dipped for 30 min (unconventional

Fig. 10 Transport of sweet potatoes

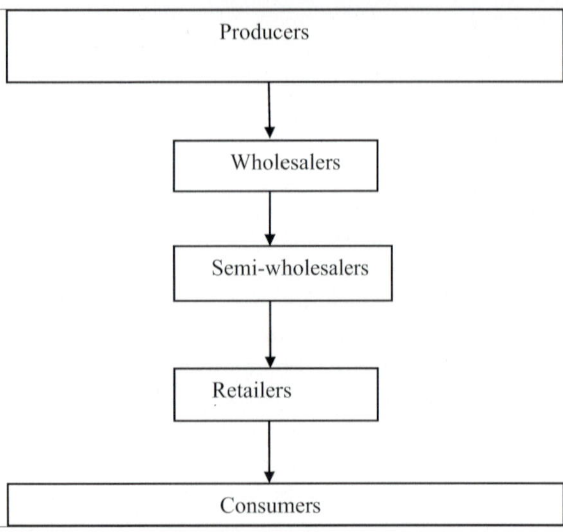

Fig. 11 Sweet potato value chain

weevil control method) controlled *Cylas puncticollis* by 35.5%. Producers used the sweet potato planting materials from the previous crop because of the non-availability of certified planting materials of sweet potato tolerant to pathogens and pests in the country. This indicates the reason why there are disease and pest infestations in the field and low yields. Fuglie [19] carried out a survey work on constraints across many countries and similarly found that lack of clean and enough planting material was a major constraint in most of African countries. The respondents' training could help limit the number of successive crops on the same plot and adopt crop rotation or crop associations to avoid the buildup of diseases and insects of sweet potato in the farms. According to [13], these practices have many advantages in optimizing production and maintaining soil fertility. In the same vain, [20] also reported the need to improve and build the capacity of farmers. They further stated that the lack of resistant varieties as well as disease- and pest-free sweet potato planting materials were the major hindrance to pest and disease control. Henceforth, breeders need to identify resistant varieties as well as promote production of clean seed to improve sweet potato production. This is consistent with the study of [21] on resistance gene expression on sweet potato virus resistance. The farmers did not apply mineral fertilizers to sweet potato plants due to the fact that they considered their fields fertile. They stated that there is frequent inundation in every rainy season due to the overflow of the rivers and drought in dry season which help in maintaining the fertility of their soils. Producers are also confronted with water problems because of the poor distribution of rainfall in recent years due to climate change, which hampers the development of seedlings and do affect negatively the productivity. After the harvest, producers were compelled to sell their produces at low price because of the storage problem, i.e., the difficulties of conservation due to lack of adequate facilities and the recalcitrant nature of sweet potato tubers. The different methods of short storage observed by respondents are the collection of tubers by women during harvest and put them together in heaps and covered them with plant debris as showed in Fig. 5. The storage period varies between 1 and 2 months if the ridges are not flooded before harvest. A second method consists in conditioning the sweet potatoes in the bags and protecting them with vegetable debris (Fig. 6). The storage period using this second method is just 7 days for the sweet potatoes to be sold to avoid rotting. Sweet potato producers really need better strategies that will allow them to better conserve their crops and sell high-quality sweet potatoes under the best storage practices, especially when demand is high in the market and for high income. For sustainable field preservation in order to avoid insect attack, [8] suggested the cutting of all the potato biomass at the end of the rainy season, the remaking of the ridges so as to damage the eggs of the weevils and covering the exposed tubers and the regrowth should be regularly cut. The tubers are thus kept in the field until the beginning of the rains. This practice also makes it possible to better conserve the tubers and to sell them during periods of high demand [8]. Hue and Low [22] also reported that many works have testified about the significant effectiveness of *Bacillus thuringiensis* in controlling diverse pests of sweet potato including *cylas* spp.

Conclusion

Sweet potato is a crop with high nutritional value and is classified as the most energy-rich tuber. The findings of this work reveal that the lands are fertile to grow sweet potato in the area but it is yet to be grown in a large scale as the farm size varies from < 1 to > 3 ha. The main constraints are flooding, diseases and insect infestations which reduce the productivity of sweet potato in the area of study. Farmers don't have access to certified planting materials resistant to pathogens and insect pests. Therefore, appropriate measures need to be taken for their proper control in order to increase the return of farmers. This is followed by poor tubers due diseases and insects, low market price, scarcity and high cost of labor, and the lack of better conservation methods for a long time, low yield. Additionally, farmers don't benefit from their labor when considered the profit made by other actors along the value chain and this need to be tackled by policy makers so as lessen the suffering of the producers. The policy makers must also invest in inputs which they could subsidy for the benefit of farmer, and also invest in modern tools for long-term storage and in research at national level for the researchers to develop certified and resistant sweet cultivars to pathogens and insect pests which could be available to farmers year round. This will definitely lead to sweet potato tubers free of diseases and pest infestation (high marketable sweet potato) which will also result in improving the high market value of sweet potato along the value chain.

Authors' contributions

VE designed the study, RHB and CMIK collected and analyzed data and drafted the manuscript. VE, QF, BFC, EMGK and AA supervised data analysis and revised the manuscript. All authors read and approved the final manuscript.

Authors' information

RHB and CMIK are Research Assistants at University of Abomey-Calavi. VE and QF (PhD) are lecturers in Plant breeding and genetics at University of Abomey-Calavi, Department of Crop Science. BFC is a lecturer in food science at University of Abomey-Calavi. AA is a Professor at Department of Crop Science, University of Abomey-Calavi. EMGK is a PhD candidate at Institut National de la Recherche Scientifique – Eau Terre Enironnement (INRS-ETE), Canada.

Author details

[1] Department of Crop Production, Faculty of Agricultural Science, University of Abomey-Calavi, Abomey-Calavi, Benin. [2] Institut National de la Recherche Scientifique – Eau Terre Enironnement (INRS-ETE), Quebec, Canada. [3] Department of Food Science, Faculty of Agricultural Science, University of Abomey-Calavi, Abomey-Calavi, Benin.

Acknowledgements

We really thank sweet potato producers for their time spent giving us information related to production and marketing of sweet potato.

Competing interests

The authors declare that they have no competing interests.

Funding

Not applicable.

References

1. Adégbola P. Analyse de la filière patate douce au Bénin, PDRT/MAEP. Rapport final 2003; 162p.
2. Egbe OM, Afuape SO, Idoko JA. Performance of improved sweet potato (Ipomea batatas L.) varieties in Makurdi, Southern Guinea Savanna of Nigeria. Am J Exp Agric. 2012;2(4):573–86.
3. Tottappily G, Loebenstein G. The sweet potato. New York: Springer; 2009.
4. Ishida H, Suzuno H, Sugiyama N, Innami S, Tadokoro T, Maekawa A. Nutritive evaluation on chemical components of leaves, stalks and stems of sweet potatoes (*Ipomoea batatas* poir). Food Chem. 2000;68(3):359–67.
5. Laurie SM, van Jaarsveld PJ, Faber M, Philpott MF, Labuschagne MT. Trans-b-carotene, selected mineral content and potential nutritional contribution of 12 sweet potato varieties. J Food Compos Anal. 2012;27:151–9.
6. Oloo BO, Shitandi A, Mahungu S, Malinga JB, Ogata BR. Effects of lactic acid fermentation on the retention of beta carotene content in orange fleshed sweet potatoes. Int J Food Stud. 2014;3:13–33.
7. Sanoussi AF, Dansi A, Ahissou H, Adebowale A, Sanni LO, Orobiyi A, Dansi M, Azokpota P, Sanni A. Possibilities of sweet potato [Ipomoea batatas (L.) Lam] value chain upgrading as revealed by physico-chemical composition of ten elites landraces of Benin. Afr J Biotechnol. 2016;15(13):481–9.
8. Countrystat Benin. Food and agriculture data network 2017.
9. Jackai LE, Sosinski B., Jackson DM et al. Occurrence and intra-specific variation of sweetpotato weevil (Brentidae: Coleoptera) in relation to its potential spread in southern United States of America and the Caribbean. In: ISHS acta horticulture 703: II international symposium on sweetpotato and cassava: innovative technologies for commercialization, p. 197–204, International Society for Horticultural Science, Leuven, Belgium; 2006.
10. Agueguia A, Fontem DA, Mbahe RE, Mboua JC, Mouen M, Ngo Som J, Poné Kamdem D, Tchuanyo M, Zok S. La patate douce: un potentiel trop peu connu; 2000. http://www.fao.org/wairdocs/x5695f/x5695f05.htm. Accessed 12 Feb 2017.
11. Muthoni JJ, Shimelis H, Melis R. Potato production in kenya: farming systems and production constraints. J Agric Sci. 2013;5(5):182–97.
12. Gebru H, Mohammed A, Dechassa N, Belew D. Assessment of production practices of smallholder potato (Solanum tuberosum L.) farmers in Wolaita zone, southern Ethiopia. Agric Food Secur. 2017;6:31. https://doi.org/10.1186/s40066-017-0106-8.
13. Doss CR. Understanding farm level technology adoption: lessons learned from CIMMYT's micro surveys in Eastern Africa. CIMMYT Economics Working Paper. Mexico, DF: CIMMYT. 2003; p. iv, 20.
14. Aho N, Kossou DK. Précis d'agriculture tropicale. Bases et Eléments d'applications, les éditions du Flamboyant 1997, Cotonou, p. 464.
15. Shonga E, Gemu M, Tadesse T, Urage E. Review of entomological research on sweet potato in Ethiopia. Discourse J Agric Food Sci. 2013;1:83–92.
16. Degras L. Le Technicien d'Agriculture Tropicale, La Patate Douce. Paris: Editions Maisonneuve et Larose; 1998.
17. Tefera TT, Handoro F, Gemu M. Prevalence, incidence and distribution of Sweet potato virus: it's effect on the yield of sweet potato in Southern region of Ethiopia. Int J Sci Res. 2013;2:591–5.
18. Ehisianya CN, Ukeh DA, Isah MD, Lale NES, Umeozor OC. Field efficacy of Neem seed oil and diazinon in the management of sweetpotato weevil, Cylas puncticollis (Boh.) in south eastern Nigeria. J Plant Stud. 2013;2:135–44.
19. Fuglie KO. Priorities for sweet potato research in developing countries: results of a survey. HortScience. 2007;42:1200–6.
20. Degu G, Markos D, Bekele A, Kassie M. Community survey and on-farm trials for conservation agriculture to enhance adoption and its impact. Int J Sci Eng Res. 2013;4:1225–35.
21. McGregor CE, Miano DW, LaBonte DR, Hoy M, Clark CA, Rosa GJM. Differential gene expression of resistant and susceptible sweet potato plants after infection with the causal agents of Sweet potato virus disease. J Am Soc Hortic Sci. 2009;134:658–66.

Determinants of food security: evidence from Ethiopian Rural Household Survey (ERHS)

Kedir Hussein Abegaz*

Abstract

Background: Though agriculture is obstructed by a number of both endogenous and exogenous problems, it is the dominant generator of means of livelihoods for the majority of people in the least developing countries like Ethiopia. The aim of this study is to look at the status of food security and to identify its determinants in the rural Ethiopia.

Methodology: The pooled data for this study were obtained from the sixth and seventh round of the Ethiopia Rural Household Survey (ERHS). Bar chart and frequency distribution tables were used to illustrate the given data. Binary multivariable logistic regression was employed to identify the determinants of food security.

Results and conclusion: Majority of the households were found to be food insecure, though the figure of food insecurity decreased in 2009 when it is compared with 2004. Food security was significantly determined by rain shock, lack of off-farm income, and region of the households. To assure food security, the farmers should have to consider every rain seasons in the farming activity and the availability of off-farm income-generating activities should have to be enhanced. Finally, there is a need for an international policy regarding the adoption of mitigation strategies to control climate change, the main cause of agriculture and rainfall shock.

Keywords: Agriculture, ERHS, Ethiopia, Food security, Pooled cross section

Background

For several decades in the globe, agriculture has been the main source of livelihoods for the developing world subsisting for a significant portion of their nations. It has been the key sector providing employment opportunities for nearly seventy percent of the rural population and contributing the largest share to their national gross domestic product (GDP). However, its productivity has been challenged by a massive amount of exogenous and endogenous shocks. These shocks arise from climate changes as well as man-made calamities of civil strife and prolonged war [1, 2]. In consequence, it has resulted in lower agricultural outputs putting rural people at greater risks of food insecurities.

In the developing world, the agricultural sector is characterized by miserable performance often carried out by traditional means and farming technologies. Poor infrastructural developments ensure a poor level of productivity. Productivity can be affected at different stages of the agricultural activities starting from preparations of farming land, through growing, harvesting, storing, and distributions of the final outputs. This unique nature has made the sector to be more vulnerable and susceptible to wide ranges of risks, natural and artificial, which in turn increases the risks of food insecurity and malnutrition.

Conversely, population growth has been increasing since long times widening the gap between food supply and food demand. In the twenty-first century, this gap has reached intolerable levels, where securing food supply to the most rural destitute is the most challenging part of social welfare objectives and global development endeavors. Food and Agriculture Organization of the United Nations (FAO)'s successive World Food Summits since 1996 to date and the global development objectives of the millennium attest the above fact with their long

*Correspondence: kedir6300@gmail.com
Biostatistics and Health Informatics, Public Health Department, Goba Referral Hospital, Madda Walabu University, Bale Goba, Ethiopia

and grip vision of "the world free from hunger and malnutrition" [3].

The agricultural shock impacts on food security have been identified as a major area of concern owing to climatic variations in many parts of the world. The predominance of rain-fed agriculture in much of the developing world especially in sub-Saharan African (SSA) has resulted in food systems that are highly reliant and sensitive to rainfall variability. Agricultural shocks, driven by climate changes, affect agricultural production and hence food security in a multitude of ways.

Food insecurity status is increasing from time to time. According to recent estimates of FAO [4], over 870 million people are chronically malnourished and food insecure around the world. Even though considerable efforts were made to reduce food insecurity, the number of people suffering from malnutrition and hunger remains unacceptably high. In the developing countries, the vast majority, over 850 million people, are estimated to be undernourished. Similarity, Tobin et al. [5] have pointed out that the number of food-insecure people in SSA is showing unabated signs, increasing from 170 million in 1990 to over 200 million in 2004.

The developing regions of the world are the most vulnerable groups to food insecurity in the world, as it has been well documented in the literature. Of the developing world, SSA appears to be the hardest area to food security risks. Several factors contributing to this insecurity in the region were identified, and Tobin et al. [5] listed poor agricultural productivity as the major factor. Agricultural productivity is constrained by poor technology, poor infrastructure, natural and man-made shocks, poor marketing, etc. The region is also lagging behind others in terms of using agricultural inputs. As a result, productivity has stagnated for several decades.

In spite of the rapid and impressive progress in tackling poverty in recent years, it has been reported in subsequent food security assessment studies that on average nearly 35% of Ethiopians have been suffering from chronic food insecurity and undernourishment every year mainly caused by agricultural shocks and climate changes [6, 7].

In response to how agricultural shocks affect the level of food security at household level, the sound policy formulation with a view to reducing the level of food insecurity risks and increasing resilience in an economy requires a detailed understanding. This study is important in the sense that it gives helpful insight to the true nature of agricultural shock in the country and its determinants on the level of food insecurity along with the channels through which agricultural shocks affect food security. Food security was one of the MDGs that was expected to be achieved through "Eradication of extreme poverty and hunger." Therefore, the result of this study and its resultant recommendation is very important for policymakers at all levels.

The objective of the study is to look at the status of food security and to identify determinants of food security in the rural Ethiopia.

Food security concepts and definitions

Food security concept is believed to have originated three decades ago in the mid-1970s in the first world food conference and was narrow in its coverage and definition. This concept initially paid attention to the national and international level and was defined from the perspective of the food supply with special attention to stable food price and food availability [8].

A situation where all people, at all times, have physical and economic access to sufficient, safe, and nutritious food that meets their dietary needs and food preferences for an active and healthy life is known as food security [9]. With boosting awareness and dynamism of vulnerability to risks of food shortages, a more comprehensive and multidimensional approach to the concept of food security became emergent. Accordingly, the food security concept encompasses the following four components: food access, food availability, food utilization, and sustainability [10, 11].

Food access is ensured when households and all individuals within them have adequate resources to obtain appropriate foods for a nutritious diet. Additionally, food availability is said to be achieved when sufficient quantities of food are consistently available to all individuals within a country [10, 12, 13].

Food utilization refers to the consumption of food through adequate diet, clean water, sanitation, and health care to reach a state of nutritional well-being where all physiological needs are met. Food sustainability, on the other hand, refers to a situation where the above three components of food security are fulfilled at any time [10]. In other words, to be food secure, a population, household, or individual must have access to adequate food at all times. This component of access to food security implies that people should not be in any situation of risks owing to sudden shocks of economic or climatic crisis or cyclical events. The concept of stability can, therefore, refer to both the availability and access dimensions of food security. Food insecurity, in contrast, is viewed as the denial of the above rights either at household, individual, or community levels [3, 5].

The food system is another broader concept which encompasses food security. It refers to all human food chain activities of producing, processing, distributing, and consuming food to a range of social and environmental contexts [14]. It is quite evident that any of the above human food chain activities are affected by climate changes directly or indirectly. On the other hand, human activities of producing, processing, packaging, distributing, retailing, and consuming food are partly responsible for changing the world's climate through emissions of greenhouse gasses. These activities contribute also to global climate change through changing fresh water supplies, air quality, nutrient cycling, biodiversity, and soil.

All the definitions above imply that food security is a wide concept and requires taking into account a wide array of causes and measurements. This paves the way for approaching the issue from a different perspective and provides evidence for consideration of the wider definition given by FAO (1996) encompassing all elements of availability, accessibility, utilization, and sustainability.

Agricultural shock is another important term to be defined in this study. It refers to any unexpected, intense, and distressing experience in agricultural activities, which has a sudden and often devastating effect on agricultural yield. The collapse in agricultural yields can be attributed to several underlying causes. These factors include untimely rainfalls, lack of water, too much rain, pest infestation, floods, and animal disease outbreaks, and frost [7]. The occurrences of any of the above factors profoundly affect the quantity and quality of agricultural yields leading to immediate risks of an increased crop failure and loss of livestock.

In general, it is important to note that the impacts of agricultural shocks on food security must be viewed within the larger framework of changing global climatic dynamisms and observable changes in multiple socioeconomic and environmental variables. This paper seeks to illuminate the impacts of such shocks on food security as viewed from availability and accessibility perspective.

The link between food security and agricultural shocks

Recently, one-sixth of humanity is undernourished more than ever before. Changing climatic conditions are projected to affect food security through their impact on local food systems. Climate change will generate significant and intensified weather events such as floods, tornados, and hurricanes; increased drought; loss of coastal areas and water shortages; and changes in the incidence of disease [6, 15].

Causes of food insecurity in sub-Saharan Africa

Limited rural development, weak infrastructural development of power, road, and market access aggravate food insecurity in SSA. Weak government policy that adversely afflicted the agricultural sector is another factor contributing to food insecurity in the region. Poor health condition also poses its impact on food insecurity. The region is also characterized by the prevalence of contagious and fatal but preventable diseases of HIV/AIDS, malaria, tuberculosis (TB), and other diseases. This reduces rural labor participation in agriculture and off-farming activities contributing to food insecurity. Moreover, rising global commodity prices and climate change will likely further exacerbate food insecurity in the region. Agricultural productivity in SSA, as measured by grain yield, is only about 40% of the rest of the world's developing countries, and the gap has widened over the years [5, 13].

Food security in Ethiopia

Food insecurity emerged as a key problem and development challenge in Ethiopia in the early 1970s and became pervasive in the subsequent decades although the issue is a century-old concern for FAO and other donating institution in the country. More significantly, since the 1980s the happening of severe drought and large-scale starvation ignited the need for food security and food aid initiatives in the country. Conversely, the concept has become more complex due to a move in the level of analysis from global and national to household and individual levels [16].

In Ethiopia, the special programs for food security (SPFS) have begun in 1995 in two regions: Amhara and Tigray. In the year 1996, the food security strategy was launched and it was revised in 2002, to include elements of water harvesting, environmental rehabilitation, and the introduction of high-value crops, livestock, and agroforestry development. The project was continued by Italian support in 1998 supporting 4062 participating households and 24,500 beneficiaries. The project was further expanded to other regions of the country in 2001 with a view to improving the nutritional status and food security of the population. New coalition for food security was established in 2003 with the aim of supporting chronically food-insecure households to reach a level of food security necessary to survive and thrive. In 2004, the government of Ethiopia designed and implemented new collusion program of Food Security Program (FSP) and expanded its endeavor to fight against food insecurity, malnutrition, and hunger by allocating more resources [16].

In mid-2005, the National Program for Food Security (NPFS) was launched focusing on three broad components: productive safety nets (PSN), household asset building (HAB), and voluntary resettlements. Like other developing countries, Ethiopia applied a wide array of strategies to lessen incidences of food insecurity risks to ensure food security at household, local, and regional levels. In addition, in 2006, improving food security was recognized within the framework of Sustainable Development Poverty Reduction Paper (SDPRP) as a central concern of government [17]. This program, on food security, in the country was guided by two fundamental principles: firstly, by the principle of reliance, where rural food-insecure farmers are made reliant on food aid to help them use their own resources to overcome food insecurity. Secondly, by breaking away the perpetual food aid dependence, so that they become food self-sufficient.

In many parts of Ethiopia, as a cause of food security combinations of natural and man-made factors have resulted in serious and growing food insecurity problem. The immediate causes of food insecurity include frequently recurring droughts and erratic rainfall patterns. Ecosystems degradation, rapid population growth, poor rural infrastructure, and legacies of the past policy constraints are also considered as basic causes of food insecurity and widespread poverty in the country [18–20].

Operational definitions

Food insecurity occurs in two forms as chronic and transitory. Chronic food insecurity happens when a household is unable to meet the minimum amount of food needed for healthy life over a long period (three or more months). But if this food insecurity is less than 3 months, it will be transitory food insecurity [2, 20].

Food security is ascertained in the households if the given household is free from the two kinds of food insecurity: chronic and transitory.

Extension visit is simply a situation where the household's land is visited by an extension agent during the last main season.

Off-farm income is any kind of income that households had got beyond farm activity including the government food aid.

Rain shock refers to the unexpected event regarding rain such as over rain, less rain, the rain out of the rain period, and rain at a time of harvesting and collection of crops.

Methodology

Data source, study design, and Analysis

A database describing each of a number of individuals, households, etc., across a sequence of time periods is known as pooled data. Thus, this blends characteristics of both cross-sectional and time series data. Pooled data are vital to the analyst because they contain the information necessary to deal with both the intertemporal dynamics and the individuality of the entities being investigated [21].

A pooled cross-sectional study was conducted by using the last two datasets of the Ethiopian Rural Household Survey (ERHS), 6th and 7th round datasets. After pooled the two datasets and cleaned the missing information in all variables, 2722 rural households have been included in the study. The households' information under the study is presented through the use of frequency distribution tables and bar chart. The bivariate binary logistic regression was employed to show the crude association between the food security and the exposure variables. Lastly, the multivariable binary logistic regression has been used to identify the determinants of food security, by using the adjusted association between food security and the exposure variables. STATA-14 was the tool to make all the analyses.

Results and discussion

From the total 2722 households included in the study, the food-secure households from pooled data were only 1040 (38.21%) households. But when we explore it at different single data, the food security level was boosted from 35.64% in 2004 to 40.28% in 2009 (Fig. 1).

The sample households were identified from the four regions of Ethiopia based on proportional to size allocation method. As a result, the minimum, 261 (9.59%) households were from Tigray, and the majority of households 591 (37.48%) were from Oromia. In the pooled result, rain and crop shocks were among the main constraints of agricultural productivity, 2385 (87.62%) being affected by rain shock and 2302 (84.57%) by crop shock, respectively. The study identified that more than half of the population were not visited by extension workers; only 901 (33.1%) had the chance to be visited. Nearly half (49.71%) of the population in the rural Ethiopia did not have off-farm income during the study periods, and only 10 (0.37%) of husbands died (Table 1).

From the result of Table 2, among 1682 food-insecure households 1517 (90.19%) have experienced rain shock and 868 (83.46%) food-secure households also experienced seasonal rain shock. In addition, 1440 (85.61%) food-insecure and 862 (82.88%) food-secure households experienced crop shock. Therefore, this shows that rain shock and crop shock affected both food-secure and food-insecure households. In general, food-insecure households had the following characteristics; 90.19 and 85.61% of them experienced rain shocks and crop shock, respectively, only 32.05% of them are visited by extension agents, and 54.16% of insecure households have off-farm income. The concentration of food-secure households across regions is 3.08% in Tigray, 40.29 in Amhara, 37.21 in Oromia, and 19.42% in SNNPR households were in food security status. Relatively speaking, households in Tigray region were the least food-secure households when we compare with the other three regions (Table 2).

In the crude analysis, the association of food security with the exposure variables has been identified. Accordingly, from the exposure variables, rain shock, crop shock, region, off-farm income, and extension visit have

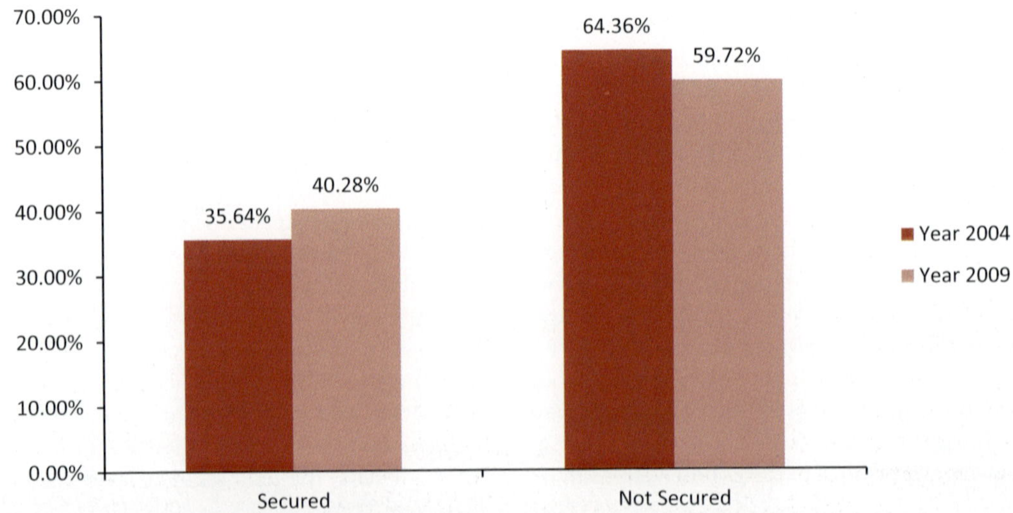

Fig. 1 Food security status of the rural Ethiopia in 2004 and 2009

Table 1 Descriptive details of food security and the determinants in rural Ethiopia

Variables with category	Frequency	Percentage
Food security		
Secured	1040	38.21
Not secured	1682	61.79
Region		
Tigray	261	9.59
Amhara	760	27.92
Oromia	900	33.06
SNNPR	801	29.43
Rain shock		
Yes	2385	87.62
No	337	12.38
Crop shock		
Yes	2302	84.57
No	420	15.43
Husband death		
Yes	10	0.37
No	2712	99.63
Year of the survey		
2004	1215	44.64
2009	1507	55.36
Extension visit		
Yes	901	33.10
No	1821	66.90
Off-farm income		
Yes	1369	50.29
No	1353	49.71

Table 2 Association between food security and determinant variables

Variables with category	Food security		Crude OR (CI)	*p* value
	Not secured	Secured		
Region				
Tigray	229	32	–	
Amhara	341	419	8.79 (5.91–13.07)	0.000
Oromia	513	387	5.39 (3.64–7.99)	0.000
SNNPR	599	202	2.41 (1.61–3.61)	0.000
Rain shock				
No	165	172	–	0.000
Yes	1517	868	0.59 (0.44–0.69)	
Crop shock				
No	242	178	–	0.056
Yes	1440	862	0.81 (0.66–1.01)	
Husband death				
No	1675	1037	–	0.595
Yes	7	3	0.69 (0.18–2.68)	
Extension visit				
No	1143	678	–	0.137
Yes	593	362	1.13 (0.96–1.33)	
Off-farm income				
No	771	582	–	0.000
Yes	911	458	0.67 (0.57–0.78)	

an association with the dependent variable, food security. However, husband death has no association with food security (Table 2).

In the adjusted analysis, the households that have experienced rain shock are 73% less likely to be food secured when it is compared to the households which were not experiencing the rain shock by keeping other variables constant. The association was statistically significant with a p value of 0.000. Households that have got off-farm income are 62% more likely to be food secure than those without off-farm income by controlling the other variables, and it has a significant association with food security, with p value 0.000. The sample households in Amhara region are 7.97 times, in Oromia region 5.15 times, and in SNNP region 2.14 times more likely to be food secure than the households in Tigray region, respectively (Table 3).

I have greatly worked on data cleaning before the study begins. Also, I have used the stepwise logistic regression for the model selection. Having this, it was checked for confounding, interaction, and the changes in coefficients (β). On the selection procedure, variables having a p value of less than 0.15 [22] were included in the final model then there was no any confounding or interaction effect in the final logistic model.

Finally, to check the correctness of the final formulated model, Hosmer–Lemeshow test for overall goodness of fit was used through a level of significant value $\alpha = 0.005$, and the value became 0.1649 that is insignificant, which means the final fitted model was correct. To ensure whether the fitted model was predicted well, I used the receiver operating characteristic (ROC) curve and its value of the area under the curve was 0.6261. Therefore, the determinant variables were capable of predicting well food security.

Conclusion and recommendation

Though the majority of the households were found to be food insecure, the figure of food insecurity decreased in 2009 when it is compared with 2004. Food security was significantly determined by rain shock, lack of off-farm income, and region of the households. Thus, these households were suffering from food insecurity by rain shock, lack of off-farm income, and their region of residence.

Based on the result of this study, the following conclusions can be drawn:

First, rain shock is found to be the main determinant for households' food security. The problem of rain shock is a result of global climate change. As a result, the following points are forwarded as a recommendation; there is a need for an international policy regarding the adoption of mitigation strategies to control climate change, the main cause of agricultural and rain shocks, and this strategy should be integrated into the development planning.

Second, off-farm income has a significant contribution in moving households to food security. As a result, off-farm employment opportunities should have to be created and expanded.

Third, region of the households was also the main determinant of food security. In response to this, the agricultural experiences have to be modified in all of the regions, to make households food secured.

Table 3 The adjusted analysis of pooled food security of the rural Ethiopia

Variables with category	Adjusted OR (CI)	*p* value
Region		
Tigray	–	–
Amhara	7.97 (5.34–11.89)	0.000
Oromia	5.15 (3.46–7.65)	0.000
SNNPR	2.14 (1.42–3.21)	0.000
Rain shock		
No	–	–
Yes	0.73 (0.58–0.93)	0.011
Off-farm income		
No	–	–
Yes	1.62 (0.53–0.73)	0.000

Acknowledgements
None.

Competing interests
The author declares that he has no competing interests.

Funding
No organization funded this study.

References

1. UNDP. Africa Human Development Report in towards food Secure future UN. United Nation's Development Program: Piaza, New York. 2012. http://www.undp.org/content/dam/malawi/docs/general/Africa_HDR_EN_2012.pdf.
2. Welteji D, Mohammed K, Hussein K. The contribution of productive safety net program for food security of the rural households in the case of Bale Zone, Southeast Ethiopia. Agric Food Secur. 2017;6(1):53. https://doi.org/10.1186/s40066-017-0126-4.
3. FAO. The state of food insecurity in the world, F.a.A. Organization, editor. Rome, Italy. 2008a. http://www.fao.org/tempref/docrep/fao/011/i0291e/i0291e00.pdf.
4. FAO. The state of food insecurity in the world. Rome, Italy. 2012. http://www.fao.org/docrep/016/i3027e/i3027e.pdf.
5. Tobin JC. Hunger effort and food security. New York: Nova Science Publishers, Inc.; 2009.
6. FAO. Crop and food security assessment mission in Ethiopia: integrating crop and food supply in the Emergency of Food Security Assessment. Rome: Food and Agricultural Organization; 2009.
7. DFID. Summary of DFID's Work in Ethiopia. D.F.I. Development, DFID Ethiopia. British Embassy, Addis Ababa; 2011.
8. Young H, Jaspers S, Brown R, Frize J, Khogali H. Food-security assessments in emergencies: a livelihoods approach. Overseas development institute (ODI). Humanitarian practice network (HPN). 2001. https://www.files.ethz.ch/isn/96840/networkpaper036.pdf.
9. World Food Summit. The world free from hunger. Rome; 1996.
10. FAO. Food security. F.a.A. Organization FAO, Paris; 2006.
11. Gregory PJ, Ingram JS. Climate change and the current 'food crisis'. CAB Rev Perspect Agric Vet Sci Nutr Nat Resour. 2008;3(99):1.

12. Schmidhuber J, Tubiello FN. Global food security under climate change. Proc Natl Acad Sci. 2007;104(50):19703–8. https://doi.org/10.1073/pnas.0701976104.
13. Ingram JS, Gregory PJ, Izac AM. The role of agronomic research in climate change and food security policy. Agr Ecosyst Environ. 2008;126(1):4–12. https://doi.org/10.1016/j.agee.2008.01.009.
14. Liverman D, Kapadia K. Food systems and the global environment: an overview. In: Ingram J, Ericksen P, Liverman D, editors. Food security and global environmental change. London, Washington, DC: Earthscan; 2010. p. 3–24.
15. Selvaraju R. World food security: the challenges of climate change and food energy climate change and food security in South East Asia. New York: Springer Science and Business Media; 2011.
16. Sabates-Wheeler R, Tefera M, Bekele G. assessing enablers and constrainers of graduation: evidence from the Food Security Programme, Ethiopia. http://opendocs.ids.ac.uk/opendocs/handle/123456789/2189.
17. Haan N, Nisar M, Darcy J. A review of emergency food security assessment practice in Ethiopia in a study commissioned by and prepared for the World Food Programme, HPG, editor. ODI, London; 2006. http://www.odi.org.uk/sites/odi.org.uk/files/odi-assets/publications-opinion-files/864.pdf.
18. FAO. Crop and food security assessment mission in Ethiopia. Rome: Food and Agricultural Organization of the United Nation; 2012b.
19. FAO. Crop and food security assesment mission in Ethiopia. Rome: FAO/WFP; 2004b.
20. Galhena DH, Freed R, Maredia KM. Home gardens: a promising approach to enhance household food security and wellbeing. Agric Food Secur. 2013;2(1):8. https://doi.org/10.1186/2048-7010-2-8.
21. Sjursen IH. Determinants of child mortality in Angola: An econometric analysis. Master's thesis, The University of Bergen. http://hdl.handle.net/1956/5704.
22. Dielman TE. Pooled cross-sectional and time series data: a survey of current statistical methodology. Am Stat. 1983;37(2):111–22. https://doi.org/10.1080/00031305.1983.10482722.

Ensuring food security: an analysis of the industrial smoking fishery sector of Ghana

Berchie Asiedu[1*], Pierre Failler[2] and Yolaine Beygens[3]

Abstract

Background: The smoked fishery sector of Ghana plays an important role in terms of food security, employment creation, income generation and foreign exchange earnings aiding sustainable development of the Ghanaian economy. Fish smoking is the leading technique of fish processing in Ghana (70–80% of local fish consumption is smoked). As fish consumers are becoming more aware of fish safety issues, it is vital to analyse developments in the smoked fishery sector of Ghana in order to ensure food safety, nutritional needs, fish food security and employment. The purpose of this paper is to analyse the industrial fish smoking sector, especially, exports to the European market (Ghana's biggest customer in terms of earnings). Also, an analysis of the various characteristics and interests of fish smoking will be conducted. Managing the fishery resources through value addition and trade will help Ghana to achieve sustainable development targets.

Results: Smoked fish provides food, nutrients and livelihoods to a number of Ghanaians, especially the poor. Aside from traditional fish smoking production, about 67 t of industrial smoked fish [in live weight (Euro stat Conversion Factor for smoked fish (into live weight) is 3.31)] is produced annually by smoking establishments, most of it going to Europe. The market for Ghanaian industrial smoked fish products is the African diaspora living in European countries and is worth 64 t (in live weight) valued at 80,000 EUR in 2016. The major species smoked are from inland sources (mainly *Chrysichthys* sp., Tilapia sp., *Lates* sp., *Synodontis* sp., *Hydrocynus* sp., Cyprinus *carpio* and *Clarias* sp.) and marine sources (*Sphyraena* sp., *Caranx* sp., Sardinella, herrings and *Penaeus* sp.). Industrial smoking establishments are supplied by approved fishing establishments selected only based on EU food safety standards, by Ghanaian Competent Authority (CA). The approved smoking establishments are supplied by just a few fishing vessels supplying marine species, and mainly rely on small-scale fisheries for their raw material. Both the plant and the Competent Authority regularly monitor all stages of the production chain, the landing sites, plant structure and processing process.

Conclusions: Industrial smoked fish products have high value added. It is important to support this processing industry. Generally, the smoking industry in Ghana is characterised by up to 50% capacity utilisation. Fish waste is generally sold to livestock and aquaculture feed farms. Quality requirements of smoked products not only involve polycyclic aromatic hydrocarbons (PAHs) levels but also *Listeria*, histamine (for some species) and *Clostridium*. Further attention and support should be given to the small and medium enterprises and processing associations trying to reach the EU PAHs threshold (2 ppm/kg) by improved methods, making traditional smoking safer and directly benefiting the Ghanaian population. Moreover, the capacity of the Ministries of Fisheries and Aquaculture Development, Trade and Industry, Health, Food Research Institute, Ghana Export and Promotion Authority, Ghana Standards

*Correspondence: berchieasiedu@yahoo.com
[1] Department of Fisheries and Water Resources, University of Energy and Natural Resources, P. O. Box 214, Sunyani, Ghana
Full list of author information is available at the end of the article

Authority and the Food and Drugs Authority should be built to effectively monitor processing establishments in order to produce safe smoked products for European diasporas and Ghanaians.

Keywords: Fish, Ghana, Processing, Fish smoking, Food security, Quality, Small-scale fisheries

Background

Fish plays a key role in the supply of animal protein, micronutrients and essential fatty acids needs of Ghanaians. The annual per capita consumption of fish in Ghana is estimated at about 25 kg which is higher than the world's average of 20 kg [1]. Globally, fish contribute around 18% of the total animal protein intake [2]. Fish is the preferred and cheapest source of animal protein in Ghana; about 75% of the total annual catch in the country is consumed locally. Fish consumption represents 60% of animal protein intake by Ghanaians [3] which is higher than the world's average. Ghanaians prefer consuming fish in smoked form because of its flavour and delicacy (70–80% of local fish consumption is smoked). Smoked fish products offer the lowest values of saturated fatty acids [4].

Aside it being consumed locally, some fish and fishery products are also commodities for the export business (regionally with the neighbouring countries, internationally with the European Union mainly, UK, France, Spain, and the Netherlands). Ghana is among major countries in West Africa that consume large quantities of traditional smoked fish and also export to the international markets in Europe and USA for the Diaporas [5].

Fish processing in Ghana can be broadly classified into traditional and modern. Traditional fish processing methods include smoking, drying, salting, fermenting and combinations of these while modern fish processing methods include canning and freezing [6]. Traditional fish processed products such as smoked fish, dried and salted fish are mostly patronised by the average Ghanaian due to their affordability and good taste. In Ghana, fish smoking is the most practised preservation method. Practically, all species of fish available in the country can be smoked. Fish smoking in Ghana is traditionally carried out by women in coastal towns and villages, groups/associations, and inland fishing communities particularly along the shores of Lake Volta. This is common in most sub-Saharan African countries [7].

Traditional smoking establishments use basic facilities and equipment including smoking ovens or kilns, stainless knives, fuel-wood and plastic bags. Water used to wash fish before smoking is retrieved from public standpipes or fetched from the nearest well, river, or marine shore [8]. Most traditional smoked fish processors in Ghana operate in dusty and unsanitary environments without a good refuse disposal system [9, 10]. Conducted in the open or under a shelter, there is little quality control of the establishment structure and process. This puts millions of smoked fish consumers (mostly local consumers) at health risk. Furthermore, traditional smoking processors have serious health drawbacks due to the smoking technique, such as respiratory and eye illnesses, and also develop cancer susceptibility [11] due to the smoke. Smoked products also contain carcinogenic and mutagenic substances, known as polycyclic aromatic hydrocarbons (PAHs) [11–13]. These compounds are found also in food as a result of food processing techniques like drying, roasting and grilling [14]. The International Association of Food Inspectors (IAFI) published a position paper in 2017 at the World Seafood Congress 2017, evaluating the PAHs content to 32 ppb/kg in West Africa traditional smoked fishery products [25]. Therefore, these traditional products do not comply with EU food safety standards, particularly PAHs maximum residues limits [12], such as for benzo[a]pyrene (2 ppb/kg) and for the involved four PAHs (12 ppb/kg). As a result, some industrial smoking emerged in the country fulfilling this food safety needs, protecting consumer health, in order to satisfy the demand of the African population living in European countries. Fish species smoked by traditionally smoked fish processors include catfish, tilapia, sardinella, horse mackerel, tuna, and salmon.

Canned fish, usually from industrial processing, are patronised by the affluent Ghanaians. It is in this regard that some Ghanaian businesses have indeed entered into industrial fish smoking, supplying smoked fish to overseas markets and to a much lower extent in local distribution supermarkets. In 2017, only four industrial smoking plants were approved by the official Ghanaian Competent Authority to export to Europe, their main customer [15]. According to the EuroStat website,[1] the average production of these industrial smoking establishments over the last decade (2007–2016) is approximately 20 t of smoked products per year (worth 67 t in live weight[2]), bringing to the Ghanaian economy approximately 80,000 EUR per year (within the same time period). This helps to cushion the economy in sectors like infrastructure development,

[1] EuroStat website: http://ec.europa.eu/eurostat/data/database.

[2] Conversion factor of EuroStat between dry weight and live weight is 3.31.

Table 1 Summary of primary data collection methods used

Method employed	Number of people	Targeted group
Interview	20	Fishermen, smoking workers and establishments, quality control managers, Government officials, NGOs, traders, Ghana Standards Authority
Validation workshop	63	Smoking workers and establishments, quality control managers, Government officials, NGOs, traders, Standards Authority, project managers, private sector development practitioners
Market survey	5	African diasporas, stores selling hot-smoked fish products

education, health, fisheries, agriculture, rural development and foreign reserves.

Purpose of study

Fish smoking plays an important role in the socio-economic development of Ghana. The overall objective of this paper is to analyse developments in the smoked fishery sector of Ghana. Specifically, trade pattern, the smoked fish chain flow, food safety importance in the smoking sector, the performance of the processing sector and value added by smoking establishments, and different commitments will be assessed. Understanding the industrial smoked fish value chain and its performance will help in identifying opportunities in the value chain that can be exploited for commercial purposes as well as helping in the preservation of Ghana's fishery resources for future generations.

Research context

This paper is structured in this way: in the section following the introduction is the methodology of the study. This is followed by an assessment of the industrial fish smoking sector including trade pattern, smoked fish chain flow, food safety importance in the smoking sector, the performance of the processing sector as well as its value addition, and level of commitments regarding quality. The paper ends with a conclusion section, highlighting recommendations.

Methods

Primary data

Interview and stakeholders' workshop

Interviews were held with stakeholders, namely fishermen, industrial fish processors, fish exporters, distributors, government institutions, policy- makers, Non-Governmental Organisations operating along the fish processing chain as well as with supporting institutions and management bodies. Tailored interviews with 20 selected stakeholders, along with field visits, were carried out in Accra and the Volta Region (Ghana), actors being involved directly in the fish smoking industry (Directors, Quality Managers, smoking workers, fishermen) fisheries management and control (see Table 1). The interview focused on the following areas: the fish smoking flow, trade pattern, the current function of the industrial smoked fish chain, its trends, and bottle-necks, the performance of the processing sector and value added by smoking establishments, capacity index, as well as various requirements. Face-to-face interviews were followed by email and telephone exchanges for further clarification. Additionally, a 1-day workshop was held with 63 participants (see Table 1) involved in the fish production and processing sector, provided opportunities for stakeholders to discuss and agree with the main findings [5].

Market survey

A brief qualitative survey was carried out in five stores selling hot-smoked fish products in Brussels (Africa area), where diasporas bought their food (see Table 1). The aim was to get general information on species smoked, their packaging, prices, and country of origin.

Secondary data

Document analysis

Document analysis was used both to prepare the study and to complement the primary sources (i.e. interviews and workshop). A number of publications and databases produced by individuals and organisations were retrieved and analysed as secondary data sources (i.e. Eurostat, Ghana Standards Authority, Food and Drugs Authority, Ministry of Fisheries and Aquaculture Development, Fisheries Commission, West African Regional Fisheries Project, Ghana Statistical Service, Ministry of Trade and Industry, Custom Excise and Preventive Service, Food and Agricultural Organisation).

Table 2 Exported smoked fish from Ghana to EU (dry weight)

	2007*	2008*	2009*	2010*	2011*	2012*	2013*	2014*	2015*	2016*	Average
Qty (t)	16	25	22.5	23.6	22.8	15.6	12.7	20.8	29.4	13.9	20.2

*Estimated figures; *Source*: Eurostat (http://epp.eurostat.ec.europa.eu/portal/page/portal/statistics/search_database)

Table 3 Exported smoked fish from Ghana to EU (EUR)

	2007*	2008*	2009*	2010*	2011*	2012*	2013*	2014*	2015*	2016*	Average
EUR (000)	73.862	77.658	80.653	74.040	93.599	61.744	27.447	64.381	129.023	57.838	79.024

*Estimated figures; *Source*: Eurostat (http://epp.eurostat.ec.europa.eu/portal/page/portal/statistics/search_database)

Table 4 Production of smoked fish in some countries

Tonnes	2013	2014	2015	2016	Average
Thailand	3773.9	4298.8	2578.7	3647.0	*3574.6*
Vietnam	1987.6	2250.1	1390.1	1924.3	*1888.0*
China	1786.3	2048.7	1188.6	1722.7	*1686.6*
Ghana	*12.7*	*20.8*	*29.4*	*13.3*	*19.1*
Philippines	3.5	2.0		16.6	*7.4*
Senegal		4.2		1.0	*1.7*
The Gambia	3.5	2.0	0.3	17.8	*5.8*
Kenya			0.3	1.2	*0.8*

Italic values indicate Ghana's production figures

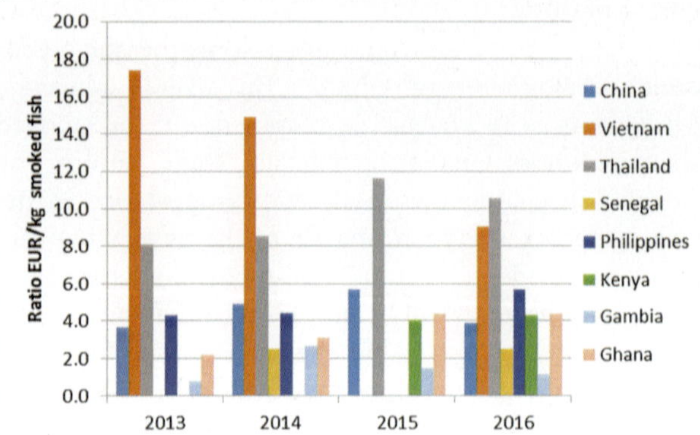

Fig. 1 Graphical representation of smoked fish price per kg exported in Europe for producing countries

Results and Discussion

Analysis of developments in the industrial fish smoking sector and the way forward

Trade pattern

According to Eurostat, over the last decade, an average of 20 t of Ghanaian smoked fish products have been exported annually (*Tilapia* sp. and other smoked fish[3]) to Europe (Table 2), and worth in average nearly 80,000 EUR within the same period (Table 3).

For smoked fish species, Eurostat places Ghana in the first position among the African producing countries in terms of quantity. For 20 years now, Ghana is one of the African major players in the international fish trade, and in 2015, fish exports were worth approximately US$ 300 million of which US$ 160,000 emerged from approved smoking establishments [16–18]. However, main producing countries of industrial smoked fish are found in most of the Asiatic continent (Table 4), that are flooding the European market.

Figure 1 represents the price per kg of the industrial smoked fish exported to Europe between 2013 and 2016 for some producing countries of Africa and Asia.[4] In a few cases, the shortage or absence of production in 2015 could be partly explained by the duty to comply with the new EU Regulation No 835 on PAHs of smoking establishments [12]. In Ghana, the ratio varies from 2.2 EUR/kg in 2013 to 4.3 EUR/kg in 2016, increasing steadily. The other producing African countries (The Gambia, Senegal, Kenya) follow the same price pattern with a low average for The Gambia (1.5 EUR/kg), a median average in Senegal (2.5 EUR/kg) and the highest average with Ghana and Kenya (4.3 EUR/kg). A few of the Asiatic countries have the same trend as China (4.5 EUR/kg) with Philippines producing at 4.8 EUR/kg. However, the largest Asiatic producers of smoked fish exported to Europe are Thailand and Vietnam, the average price per kg reach 9.7 EUR/kg and 13.8 EUR/kg, respectively. Such a difference in price is primarily to be related to the various fish species (e.g. tilapias and catfish in Africa, and carps and *Pangasius* in Asia). The Asiatic countries produced high-valued smoked fish compared to African countries.

[3] The species accounted by Eurostat were smoked *Tilapia*, Catfish, *Pangasius* sp., *Silurus* sp., *Clarias* sp., *Carps* sp., Nile Perch, including fillets (excluding offals) and other smoked fish including fillets (excluding salmons, mackerels, halibuts, herrings, eels and trouts).

[4] In both continents, smoked and dried fish are often produced by the same establishment and may have been mixed up in reports at the EU border.

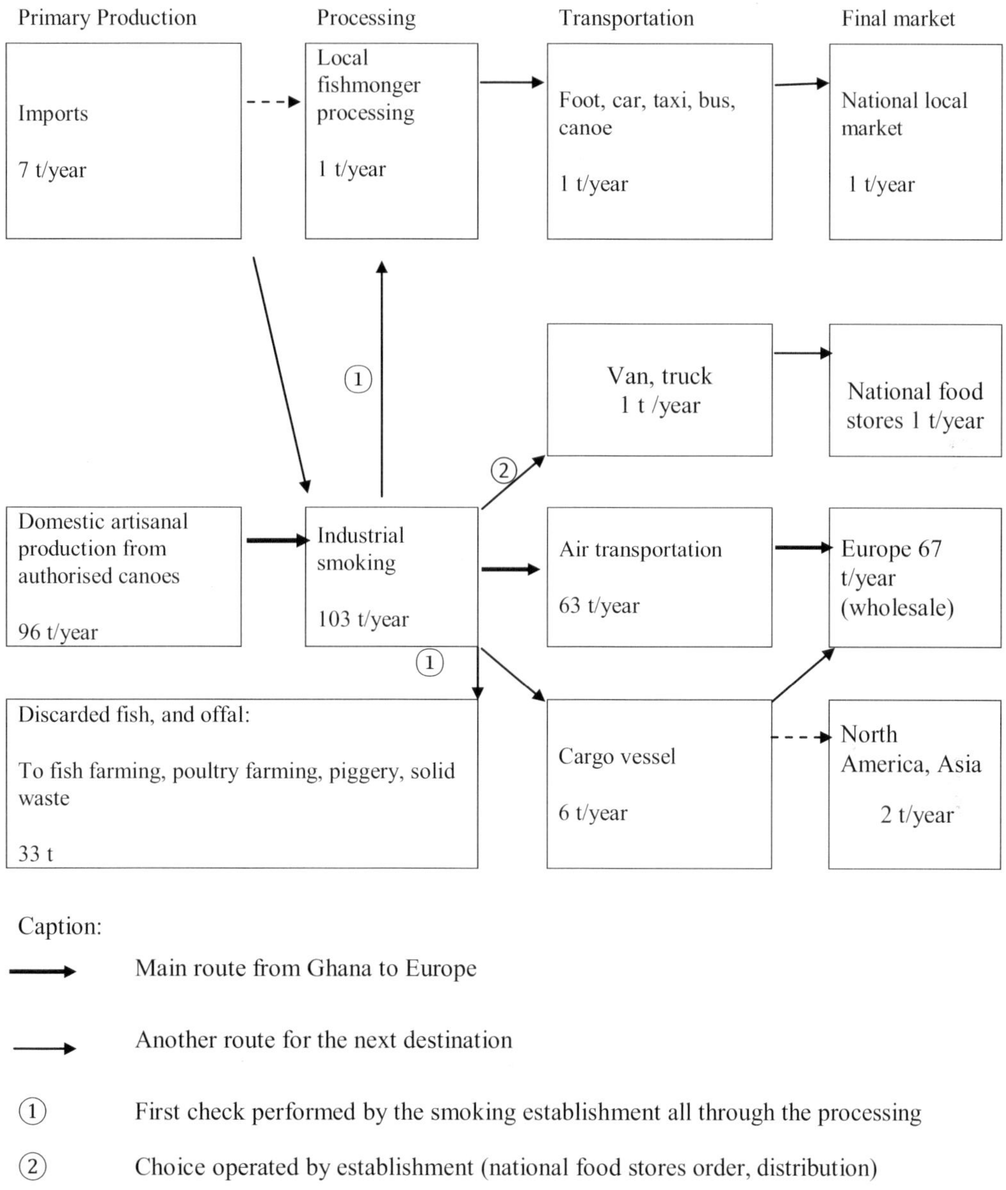

Fig. 2 Smoked fish processing value chain (live weight) [Euro stat Conversion Factor for smoked fish (into live weight) is 3.31 [33]] and product flows estimation

Smoked fish processing chain

A quick overview of the Ghanaian industrial fish smoking industry for average production (20.2 t smoked products) between 2007 and 2016 is presented in Fig. 2. Figure 2, expressed entirely in live weight (wet weight), reflects the industrial smoked fish value chain flow through (1: raw material; 2: processing; 3: transportation and 4: final market).

For the raw material, processing plants buy fresh fish from authorised small-scale canoes and to a lesser extent, from approved inshore vessels, to export to international market. In lean fishing season, processing establishment may also acquire fish from imports (i.e. frozen *sardinella)* to process them.

Before the processing can start, the processing establishment carries on own-checks (organoleptic and temperature checks) to verify the fish quality. Additionally, processing plant must have authorisation from Competent Authority inspectors, checking the quality, before they can actually start. Some fish supplied may be discarded for not being fresh enough to reach the exporting market and are directed to artisanal processing for local markets or alternatively, some other waste disposal practices for fish waste (fish farming, poultry farming, piggery, or solid waste). Increase in spoilage bacteria number is triggered by factors such bad storage conditions, cross-contamination, and ambient temperature that may have affected the raw material, thereby lowering the quality of the fish, and processors revenue [19]. During processing and just before distribution, own-checks are performed by the smoking plant, as well as those of the Competent Authority inspectors. Inspectors also check and sample the final product.

Transport to the market is the third stage. Industrial smoking companies are mainly geared towards Europe, but also to a minor extent, North America, and Asia. While a small fraction of the industrial smoked fish production is ordered by Ghanaian supermarkets, smoking products are mainly transported by air freight to Europe and reach the other destination market by cargo vessels. Traditional smoking processors use local means of transport (taxis, bus, truck, and canoe) to be sold on national or regional markets.

In respect of final international markets, approved smoking establishments contract agreements with wholesalers, then the latter will sell the fish to retailers, and the retailer to end consumer. In the study conducted by [20], in most cases, importers of smoked fish were selling directly to retailers or even have a grocery themselves.

Food safety in Ghanaian industrial fish smoking

International markets require that all imported food must abide by their food safety standards.[5] The approved country of origin must have an authorised Competent Authority that could be a public or private institution, responsible to undertake official controls over the food production chain and be able to grant a reliable Health Certificate to outgoing food products. The Competent Authority (in the case of Ghana is the Ghana Standards Authority) must also approve the processing plant, guaranteeing structure and ensures that processes are in conformity with exporting country food regulations and monitor the establishment. The Competent Authority and the establishment are also in charge of verifying the food quality of processed products through analysis. Further details of the central role of the Competent Authority in the food safety control of the industrial exporting fish industry are explained in various articles [21, 22]. Ghana's Competent Authority has the capacity and technical know-how to undertake checks on smoked fish to meet the international standards. The Ghana Standards Authority is a large statutory body in charge of the national quality infrastructure embracing metrology, standards and conformity assessment (certification, inspections and testing). The GSA is expected to become a model of excellence in standardisation in Africa [21]. This will help Ghana and other African countries to export to the EU, thereby deriving maximum benefits from fish trade.

In Ghana's industrial fish smoking case, the raw material and smoked products are tested by organoleptic test and laboratory analysis (polycyclic aromatic hydrocarbons or (PAHs), *Listeria* and *Clostridium,* histamine [12, 23, 24].

Since the PAH content is too high (around 32 ppb/kg) in traditional smoked fish in West Africa, it can no longer be exported to Europe [12, 25]. In the same way, exporting smoked products to neighbouring countries by industrial smoking establishments (producing a smoked fish below 2 ppb/kg of benzo-a-pyrene and 12 ppb/kg of PAH4) does not meet the market requirements in terms of taste: other African countries would still prefer the traditional smoked fish. Fish consumers taste is difficult to change which can affect demand and supply, irrespective of the quality and nutritional value of the fish species.

To produce smoked products that meet European requirements, an industrial smoking plant needs expensive equipment and materials such as smoking device, dryers, chilled storage room, stainless steel material as to be operated in proper premises. Water in the plant originates from a source, regularly controlled by public authorities.

Selected canoes for the small-scale fisheries

The contribution of the small-scale fisheries to the industrial smoked fish value chain cannot be underestimated. Smoked fish value chain starts from small-scale fisheries since smoking establishments buy their fish from selected local fishermen (authorised suppliers). Annually, about 96 t of fresh fish is supplied by the small-scale fisheries sector for industrial fish smoking. Each authorised small-scale fisherman has to be selected by establishments on the basis of their reputation as fishermen, their use of some good hygienic practices when fishing, landing the fish and handling. The fishermen must have their municipal health card in order, giving proof of their good health. Those fishermen have to possess a canoe in a good state of repair and maintenance (authorised canoe). The fishermen have to be registered by the Fisheries Commission,

[5] https://ec.europa.eu/food/sites/food/files/safety/docs/ia_trade_import-cond-fish_en.pdf (Accessed 12 March 2018).

in this case, their activities can be regulated and monitored. This can be a mechanism of ensuring good fishing practices as well as checking illegal, unreported and unregulated (IUU) fishing practices that affect the sustainability of the fishery resources, the ecosystem and biodiversity [26]. It is estimated that IUU cost about 2.3 billion USD in revenue annually to West Africa countries including Ghana [27]. This also poses threat to food security, health of stocks [1, 28] and has socio-economic consequences (poverty, crime, unemployment, insecurity).

Fishermen are trained by establishments in additional Good Hygiene Practices when fishing and landing the fish (on-site) complying with all regulations. Special attention is given to proper icing methodology with clean flake ice. The process of fishermen training includes a visit to the smoking plant to familiarise themselves with the process and what the fish establishments are looking for (species, freshness and smoking process). Authorised fishermen are given ice-boxes by smoking establishments and are taught how to clean them with approved detergent and disinfectant. Most of the time, establishments supply authorised fishermen with some clean ice from the plant. Icing preserves the raw fish quality [29].

The choice of landing location where the authorised canoe is berthed is also important. It should be in line with European Regulations [21]. It is fundamental that the landing location is well kept and clean by the fishing community/Beach Fisheries Committee Management and away from any source of contamination (dumping area, animals, and unhygienic practices on the landing site). Latrines should be provided. Electricity supply should be available throughout the day when smoking establishments have refrigerator and/or freezing equipment to keep fish in good condition.

In general, establishments are dependent on the requests from their client. Fish that are most commonly asked for are:

- Inland origin: *Chrysichthys* sp., Tilapia sp., *Lates* sp., *Synodontis* sp., *Hydrocynus* sp., *Carpio* sp. and *Clarias* sp.
- Marine origin: *Sphyraena* sp., *Caranx* sp., Sardinella, Herrings, *Penaeus* sp.

Concerning inspections of authorised canoes and landing sites, CA inspectors have set up a landing site monitoring schedule on six major landing sites (marine and inland), at the frequency of three times a year. Smoking establishments are in charge of transporting the fish from the landing site to the processing site in a refrigerated van. It represents a cost for smoking establishments which are located some distance from the landing site.

Table 5 Value addition in industrial fish smoking industry

	Quantity (kg)	Unit price (EUR)	Value (EUR)
Value of production	103,000	2	206,000
Value of processing	21,450	12	257,400
Value added	(206,000 + 257,400) × .40 = 185,360 EUR		

Performance of the processing industry and value added by smoking establishments

The industrial approved fish smoking industry is characterised by a high level of additional value between fresh fish and smoked products.

The price of raw fish from fishermen in Ghana is 2 EUR/kg, and the price of finished products for export is 12 EUR/kg (Table 5). This is a high value added that is generated since other intermediate consumptions do not amount to more than 3 EUR/kg. Industrial equipment needed to process the fish is around 300,000 EUR. A survey in a major commercial street selling smoked fish in Brussels conducted in March 2014 (African area) gave an average price of smoked fish sold by retailers as 30 EUR/kg.[6] The price of the fish is multiplied by 15 from the landing site to the end retailer in Europe. Thus, the smoked processing can be used as an avenue to create employment and generate income, especially for the youth. Over the years, the Government of Ghana has embarked on a number of initiatives (example: Youth Enterprise Support Programme, Youth in Agriculture and the Nation Builders Corps, Planting for Food and Jobs, etc.) to tackle unemployment without seriously considering fish smoking as one of the best options.

Smoked fish production capacity The current capacity of the smoking industry is presented in Table 6. The industrial smoking establishments have larger equipment than necessary (up to 50% capacity utilisation). The year 2013 has not been a good year in terms of exports. The smoking industry is relying on high cost equipment. They work on the order, intensively and they rely on casual workers. Their storage is filled in, but they do not keep fish too long in case there is a shortage of electricity. Smoked fish is cooled down and then packaged in labelled cardboard boxes with brown paper.

Level of commitments regarding quality

Producers keep up with any changes CA inspectors request in terms of quality. The last change they had to overcome is to trace the fish down to the landing site.

[6] Five shops selling smoked products of different fish species were visited.

Table 6 Capacity index of smoking industry

Parameters	Performance
Capacity utilisation	Up to 50%
Storage	120 m^3
Logistics and marketing	Can easily be contacted by phone, email and a few, websites Are easily found on Internet as partners of some projects
Labour productivity	Up to 1.3 t/person/year
Technology and processes	High technology equipment (bought in Europe, North America) in most cases composed of: Dryer Smoker Chiller
Waste material utilisation	Selling the waste to various farms (piggery, poultry and fish) to mix their food
Products quality	Depending on species, and on the order of the client: either whole or sliced, gutted sometimes light brown to dark colour
Packaging and labelling	Packaging in brown soft paper, and in boxes, labelled with the approved number

Women are the most active personnel at the smoking plant. They earn income to support their families. Smoking plants usually deal with offal and have a good waste disposal practice as are they have an agreement with livestock and feed farms. This is an effective way to support the aquaculture development in the country which is contributing to fish production (around 11% per annum) [30].

Quality in the smoked processing industry cannot be compromised. Quality control helps to reduce the occurrence of food borne diseases throughout the fish chain, from farm/capture to table [11]. Importers are consistent with the quality required by European standards but end buyers do not necessarily know. Whereas importers are interested in a good price, they also look at experienced suppliers that produce good smoked fish that meet the demand. Few importers are concerned about the environment. End buyers, mostly African Diaspora population (and some Asian population) are longing for the smoked product, coming from their region or country. Aside from this, consumers are becoming more aware that fish and fishery products can be a healthy alternative to meat from farm animals [2].

Employment

Smoking industry including the small-scale fisheries sector in the production chain provides employment to selected artisanal fishermen as well as to processing specialists (educated Quality Managers, Operation Manager) and workers. This employment helps to earn a living to an average of 4.4 household-members, including contribution to education, rent, medical fees, and food, and at the same time, to all the community. Similarly, it is offering jobs to the rural (49.1%) and the urban (50.9%) areas [31]. In 2014, the largest fish smoking establishment gave employment to 90 people, divided into 20 authorised fishermen, 40 processing workers, 20 people upstream and 10 people downstream [5].

While permanent workers have job security and can plan the life on the long-term, casual workers are working on orders and hence are able to defray short and medium term services.

A key role is held by women in the fish chain, as they represent 40% of stakeholders. While the gender ratio, in favour of women, is about 80/20 in handling and grading, distribution, and processing subsectors and 60/40 in quality control and marketing subsector, men were predominant in the production subsector at 70/30.

Trained personal by smoking establishments increase their knowledge, enabling them to raise awareness of their colleagues, beach community and association members, as well as friends. It gives them a higher rank in their community.

Fish trade

Fish is an important component of Ghana's economy and trade since Ghana is both a fish exporting and importing nation. Fish has become one of the important non-traditional export (NTE) products. Ghana is a net importer of fish, and the country is about 43% fish self sufficiency [30]. Annually, about 7 t of fish is imported for processing by industrial smoking firms. In 2015, Ghana's fish exports were worth approximately US$ 300 million of which US$ 160,000 emerged from approved smoking establishments. The Asiatic countries (mainly China, Philippines Thailand and Vietnam) are taking the lead in smoked fish export to the EU thereby deriving maximum benefits from the fish trade business. Almost all West African fishing grounds are either fully exploited or overexploited [32]. If the Asiatic countries managed their fishing grounds properly, then they will have more stocks to harvest, smoked and export. This will adversely affect the smoked fish trade and food security situation in West Africa in, particularly Ghana. Already, Africa has the lowest fish per capita consumption in the world (around 10 kg/year) [1] and food in-secure. The only comparative advantage that Ghana and other West African countries have is cheap labour. The Government of Ghana should encourage foreign investors to invest in the smoked fish business to realise the needed benefits.

Economic performance

According to the survey of the approved smoked fish industry in Ghana undertaken in 2014, few essential characteristics emerge [5].

Ghanaian fish smoking industry falls into the small and medium enterprises, in general family owned, and cover a very small part of the market, though extremely important for the African diaspora living in European countries,[7] longing for African-made smoked fish products.

The fish smoking industry is characterised by a high level of additional value between fresh fish in Ghana (average price 1 EUR/kg) and smoked product selling net price to European importers (approximately 3.5 EUR/kg). Therefore, each kilogram of smoked fish product generated about a value added of 1.5 euro as the difference between the revenue per kilo and the intermediate consumption,[8] respectively, 3.5 and 2 EUR).

Smoking establishments operated high technology equipment bought from the USA or Europe. These were larger than necessary, and consequently, the industry was approximately worth 50% capacity of utilisation. These industrial equipment, costly to buy and maintain, comprised at least a dryer, a smoker (or a combination of both), a chill storage as well as an in-house laboratory to undertake some microbiology analysis (own-checks). The chill storage (120 m^3 per establishment) was only filled in several subsequent orders. Although enterprises possess gasoline generator for equipment processing, they fear the shortage of electricity because of the gasoline cost. Between 2012 and 2016, Ghana experienced severe nationwide energy crisis that affected most fish smoking establishments and other sectors of the economy. Smoking enterprises usually work on the order and intensively, recruiting casual workers to complete the main staff. The labour productivity of the smoking establishment was evaluated at 1.3 tonne/person/year.

The operating costs of smoking establishments are not only salaries and wages of workers but mainly linked with food safety requirements compliance, such as specific smoking equipment purchase and maintenance (plus transport and installation costs), laboratory analysis cost and Competent Authority-related expenditures. Consequently, smoking enterprises often meet financial difficulties to sustain their fish smoking business. As a result, they diversify their offer in dried fish, salted dry fish, salted smoked fish and smoked agricultural products.

A brief telephone survey was undertaken in Brussels in 2018 at an intermediate wholesaler regarding smoked fish origin, species and price, providing retailers for African Diaspora. The bulk of the arrival came from re-export of another European country, or directly from developing countries. Smoked fish species were catfish, mackerel, lizard fish, barracuda rings, herrings, and bonga shad. The price of smoked fish ranged from 27 EUR/4 kg for catfish, 36 kg/5 kg for barracuda rings with frozen smoked mackerel at 18 EUR/4 kg, 36.9 EUR/5 kg (lizard fish), excluding Value Added Taxes.

Diasporas normally buy smoked fish from the retailers direct, themselves purchasing at the intermediate wholesalers. In the brief market study in 2014 carried out at five retailers stores in Brussels, the average price of smoked fish was 12 EUR/kg, but the African origin was not common (5 on 27 samples: 18.5%). No fish came from Ghana on the day sampling was undertaken. The bulk of the products are sourced from Asia (mainly Thailand China and Vietnam).

Environmental performance

Environmental characteristics have to be taken into account in setting the establishment, so as to ensure food safety. The supply of clean water at premises, equipment and materials and also fish has to be issued from a controlled distribution system. At the smoking plant, producers are conducting good waste disposal practices (e.g., small fish farming, livestock farms as a feed supplement). This is an effective way to support aquaculture development in the country, which is contributing to fish production (around 11% per annum) [30].

Conclusion and recommendations

Smoked products from inland and marine sources (67 t/year) continue to provide food security and nutritional needs of many Ghanaians. The industrial smoke products are geared towards the export market, which must comply with high standards. It would be worthwhile to support this processing industry. Usually, in terms of performance, the smoking industry in Ghana is characterised by up to 50% capacity utilisation using high technology such as dryers, chillers, and smokers. Quality requirements of smoked products are not only PAHs level but also *Listeria*, histamine (for some species) and *Clostridium*. CA inspectors must check both raw material and processed products and also sample them for

[7] http://belgium.iom.int/activities (Accessed the 18 March 2018).

[8] The intermediate consumption is mostly composed of transport fee.

official controls. To meet the European Commission requirement on the very low level of carcinogenic PAHs, companies have to invest in expensive equipment based on the separation of the drying and smoking phase and/or on indirect smoking. Notwithstanding the expense of the equipment, establishments also have to bear with costly tests.

Discarded fish species by industrial establishments are directed to the artisanal processing sector for the local markets. Fish waste is used to support fish farming, poultry, piggery and solid waste, thereby ensuring better use of the fishery resources.

The means of transportation is through local transport (taxis, bus, truck, and canoe), cargo vessels with air transport being the highest. Thus, the final product is geared towards the export market, namely Asia, North America and Europe.

Recommendations

The following is recommended to improve the industrial smoked fishery sector of Ghana:

- The Government of Ghana through the responsible ministries and agencies, namely Fisheries, Trade and Industry and Export Promotion Authority, should work strongly with fish processors so that they can meet the requirements of the international market for smoked fish. This will help gain trade benefits which will go a long way to sustain the Ghanaian economy.
- The capacity of the relevant monitoring institutions, namely Ministries of Fisheries and Aquaculture Development, Trade and Industry, Ghana Standards Authority and the Food and Drugs Authority, should be improved urgently to effectively monitor processing establishments in order to produce safe smoked products to meet food security and nutritional needs of Ghanaians and diasporans.

Abbreviations

EU: European Union; EUR: Euro; PAHs: polycyclic aromatic hydrocarbons; t: tonnes.

Authors' contributions

BA was responsible for conceptualising, sourcing of the literature, data analysis, and writing of the manuscript. PF participated in sourcing of the literature and writing of the manuscript. YB participated in sourcing of the literature, data collection at Brussels commercial streets, and writing of the manuscript. All authors read and approved the final manuscript.

Author details

[1] Department of Fisheries and Water Resources, University of Energy and Natural Resources, P. O. Box 214, Sunyani, Ghana. [2] Economics and Finance Group, Portsmouth Business School, University of Portsmouth, Richmond Building, Portland Street, Portsmouth PO1 3DE, UK. [3] Fish Quality Independent Consultant, Dieweg 30A, 1180 Brussels, Belgium.

Acknowledgements

We thank all processors, shop owners, government agencies, and institutions who assisted us with data. The contributions of the anonymous reviewers are acknowledged.

Competing interests

The authors declare that they have no competing interests.

Funding

This work was supported by the UNIDO/MOTI Trade Capacity Building Programme for Ghana US/GH/06/005.

References

1. FAO. The State of World Fisheries and Aquaculture 2016. Contributing to food security and nutrition for all. Rome: FAO; 2016.
2. FAO. The future of food and agriculture—trends and challenges. Rome: FAO; 2017.
3. Sarpong DB, Quaatey NK, Harvey SK. The Economic and social contribution of fisheries to gross domestic product and rural development in Ghana. FAO Sustainable Fisheries Livelihoods Programme (SFLP) GCP/INT/735/UK. 2005.
4. Akintola SL, Brown A, Abdullahi B, Osowo OD, Bello BO. Effects of hot smoking and sun drying processes on nutritional composition of giant tiger shrimp (*Penaeus monodon*, Fabricius, 1798). Pol J Food Nutr Sci. 2013;63(4):227–37.
5. Failler P, Asiedu B, Beyens Y. Value chain analysis of the fishery sector in Ghana with focus on quality, environmental, social, sustainable, food safety, organic requirements and its compliance in infrastructure. United Nations Industrial Development Organization Trade Capacity Building for Ghana Project. 2014. https://doi.org/10.13140/rg.2.1.2421.5201.
6. Nunoo FKE, Asiedu B, Kombat EO, Samey B. Sardinella and other small pelagic value and supply chain of the fishery sector, Ghana. The USAID/Ghana Sustainable Fisheries Management Project (SFMP). Narragansett, RI: Coastal Resources Center, Graduate School of Oceanography, University of Rhode Island and Netherlands Development Organisation. GH2014_ACT044_SNV. 2015. http://www.crc.uri.edu/download/GH2014_ACT044_SNV_FIN508.pdf.
7. Ahmed A, Dodo A, Bouba AM, Clement S, Dzudie T. Influence of traditional drying and smoke-drying on the quality of three fish species (*Tilapia nilotica, Silurus glanis* and *Arius parkii*) from Lagdo Lake, Cameroon. J Anim Vet Adv. 2011;10(3):301–6.
8. Axtell B, Fellows PJ. Setting up and running a small meat or fish processing enterprise. Opportunities in food processing series Wageningen: ACP-EU Technical Centre for Agricultural and Rural Cooperation (CTA). 2003. ISBN 92-9081-275-3.
9. Doe J. An analysis of quality control of smoked fish business in the Sunyani Municipality. B.Sc. thesis, University of Energy & Natural Resources, Sunyani, Ghana. 2017.
10. Kogbe M. Microbial and heavy metal contamination of skipjack and yellowfin tuna from beaches in the Greater Accra Region. M.Sc. thesis, Department of Theoretical and Applied Biology, Kwame Nkrumah University of Science and Technology. 2015.
11. FAO. Food safety policies and regulatory frameworks. Rome: FAO; 2009. p. 19.
12. EC. Commission Regulation (EU) No. 835/2011 of 19 August 2011 amending Regulation (EC) No. 1881/2006 as regards maximum levels for polycyclic aromatic hydrocarbons in foodstuffs. 2011.
13. Simko P. Determination of polycyclic aromatic hydrocarbons in smoked meat products and smoke flavouring food additives. J Chromatogr B. 2002;770(1–2):3–18.
14. Palm LMN, Carboo D, Yeboah PO, Quasie WJ, Gorleku MA, Darko A. Characterization of polycyclic aromatic hydrocarbons (PAHs) present in smoked fish from Ghana. Adv J Food Sci Technol. 2011;3(5):332–8.
15. FVO. Approved establishments in Ghana, Fishery products. Validity 00061. 17 July 2017. 2017.
16. Asiedu B, Failler P, Beyens Y. The performance of tuna processing fishery sector to sustainable fish trade and food security in Ghana. J Energy Nat

Resource Manage. 2015;2(1):8–14.
17. EC. Commission Decision (EC) No 98/421/EC laying down the conditions of fishery products imports from Ghana. p. 5. 1998. http://eur-lex.europa.eu/legal-content/EN/TXT/PDF/?uri=CELEX:31998D0421&from=FR.
18. Ministry of Fisheries and Aquaculture Development (MoFAD). 2015 Annual report. MoFAD, Accra. p. 78. 2016.
19. Valtysdottir KL, Margeirsson B, Margeirsson B, Lauzon HL, Martinsdottir E. Guidelines for precooling of fresh fish during processing and choice of packaging with respect to temperature control in cold chains. Technical report 40–10, Matís, Reykjavík, Iceland. 2010. http://www.matis.is/media/matis/utgafa/40-10-Guidelines-for-precooling-and-packaging.pdf.
20. Ward A. A study of the trade in smoked-dried fish from West Africa to the United Kingdom. FAO Fisheries Circular No. 981, Rome, FAO, p. 17. 2003.
21. Beyens Y, Failler P, Asiedu B. Institutional challenges and constraints for Ghana in exporting fishery products to the European Union. Food Rev Int. 2018;34(3):265–89.
22. Neeliah S, Goburdhun D. National food control systems: a review. Food Rev Int. 2007;23(1):35–51.
23. EC. Commission Regulation (EC) No. 1881/2006 of 19 December 2006 setting maximum levels for certain contaminants in foodstuffs. 2006.
24. EC. Commission Regulation (EU) No. 1019/2013 of 23 October 2013 amending Annex I to Regulation (EC) No. 2073/2005 as regards histamine in fishery products. 2013.
25. IAFI. Statement from the International Association of Fish Inspectors on the safety of smoked fishery products. Position paper, relating particularly to polycyclic aromatic hydrocarbons. World Seafood Congress, Reykiavik, Iceland. 10–13 Sept 2017. 2017.
26. Pauly D, Christensen V, Guenette S, Pitcher T, Sumaila UR, Walters C, Watson R, Zeller D. Toward sustainability in world fisheries. Nature. 2002;418:689–95.
27. Doumbouya A, Camara OT, Mamie J, Intchama JF, Jarra A, Ceesay S, Guèye A, Ndiaye D, Beibou E, Padilla A, Belhabib D. Assessing the effectiveness of monitoring control and surveillance of illegal fishing: the case of West Africa. Front Mar Sci. 2017;4:50. https://doi.org/10.3389/fmars.2017.00050.
28. Daniels A, Gutierrez M, Fanjul G, Guerena A, Matheson I, Watkins K. Western Africa's missing fish. The impacts of unreported and unregulated fishing and under-reporting catches by foreign fleets. London: Overseas Development Institute; 2016.
29. Graham J, Johnston WA, Nicholson FJ. Ice in fisheries. FAO technical paper no. 331. Rome, FAO. 1993.
30. Ministry of Fisheries and Aquaculture Development (MoFAD). 2016 Annual report. Accra: MoFAD; 2017, p. 78.
31. GSS. Population and housing census 2010. Summary report of final results. Accra: Ghana Statistical Service; 2012.
32. FAO. Contribution of fisheries to national economies in West and Central Africa—policies to increase the wealth generated by small-scale fisheries, new directions in fisheries e a series of policy briefs on development issues. Rome: FAO; 2006.
33. Eurostat. Eurostat, conversion factors new, Annex 8 - CF per CN8. 2014.

Impact of climate variability on household food availability in Tigray, Ethiopia

Shishay Kahsay Weldearegay* and Dawit Gebremichael Tedla

Abstract

Background: Currently, climate variability is a hot issue across the globe, especially in countries where rain-fed agriculture is a means of livelihood. Climate variability increases the risks of hunger in the region as it affects all four components of food security: *food availability, food accessibility, food utilization and food stability.* Rainfall shortage or excess hampers food production in the region, causing food insecurity and escalating famine. Extreme weather is affecting people indirectly through the sequential rather direct depletion of their assets. Therefore, looking on the impact of climate on food security is crucial to build a sustainable production system.

Results: Randomized data from 150 households were collected and analyzed using SPSS software. Household Food Balance Model analysis indicates that 15.7% of the households met the recommended daily calorie intake of 2100 kcal per adult equivalent/day, while 84.3% of the households fell below the daily recommended calorie intake. The average calorie intake of the households covers only 56.2%. Female-headed households are found to be more food insecure than their male counterpart. Those food insecure households use different coping mechanism to avert the shortage like daily laborer, reducing meal, borrowing, selling productive asset and switching meals. Moreover, the annual rainfall of a 20-year record averaged 725 ml (with a standard deviation of 6.7), with the minimum and maximum records at 354.5 and 1037 ml, respectively. Similarly, the average precipitation concentration index (PCI) stood at about 24.4 ml, recording a minimum and maximum PCI of 15.6 ml and 41.7, respectively. Finally, the coefficient of variability of rainfall variability is 0.16, which indicates the variation in inter-annual rainfall distribution is medium.

Conclusions: The result shows a high prevalence of food insecurity in the rural community with high prevalence in female-headed household. Therefore, strong intervention is required by the government and should exhaustively work on promoting irrigation, providing credit to farmers and subsidizing farm inputs to improve food security in rural Tigray.

Keywords: Food security, Climate variability, Coping strategies, Rural Tigray

Background

Climate-induced extremities like drought, intermittent rainfall and flood compromise the ability of households to meet their food requirements [1]. Food availability is diminished by change in agricultural productivity, and the land devoted for crop production, however, decreasing in rainfall quantity leads also to deterioration of food availability and health [2]. According to the Food Available Decline (FAD) Theory/Approach, failure in physical availability of food causes food insecurity, which is emanated mainly from the impact of natural hazards through increasing burden on the available natural resources [3]. Besides, [4] as cited in [3].

> *In case of the bad weather people are not directly victim of famine but their assets depletes sequentially. If drought occurred they cannot produce enough food then producer becomes net buyers by selling their assets especially livestock at distress value as livestock are thin and they become die due*

*Correspondence: sssk2002@gmail.com
Food Security, Shire Campus, College of Agriculture, Aksum University, P.O. Box 314, Shire Endaselassie, Tigray, Ethiopia

to shortage of feed. At last, victim people are fleeing to market due to failure of self sufficiency, break of formal system support, no employment opportunities then famine will happen.

Food availability decline is an attribute of environmental tribulations, population growth, political instabilities, poorly conceived developmental policies and fluctuation in food prices. It is also caused by natural disasters, and other socioeconomic factors, which emaciate household food security situation coupled with weak household asset accumulation [5–7]. However, a community where their livelihood is subsistence farming, rainfall is the dominant determinant of food production. This is due to Wetter years are generally associated with higher food production; conversely, dry years are linked to lower production [8–11]. Basically, in mid-latitude, high-latitude and higher-income countries climate change has positive impact on agricultural production or crop yield, and on the other hand, lower-latitude and lower-income countries experience a negative effect on agricultural production [12]. Developing countries are, however, most vulnerable compared to developed countries. There are many conditions, which enhance the vulnerabilities for developing countries, such as low level of technological progress, lack of resources to mitigate the adverse effect of climate change on agriculture. Additional conditions include a greater dependence of their large population on agriculture for their livelihood [13]. This would increase the severity of disparities in cereal yields between developed and developing countries [14].

Generally, climate variability is not a recent situation to Ethiopia because of the numerous and frequent droughts occurrences over the years: 1889–1892, 1972–1974, 1984–1985, 2002–2003 and 2015–16, which are due to the climate variability. Climate variability increases the risks of hunger in the region as it affects all four components of food security: *food availability, food accessibility, food utilization and food stability.* Rainfall shortage or excess hampers food production, causing food insecurity and escalation of famine in the region. Bad weather also affects people indirectly through the sequential rather than direct depletion of their assets. When drought occurs, the people cannot produce enough food to meet their needs, whereas livestock suffer from the shortage of pasture. Rural farmers, therefore, revert to coping strategies, such as livestock sale to generate income and purchase food. Livestock serves as a buffer in times of hardship, with farmers disinvesting in them to buy food [15]. Therefore, governmental and non-governmental organizations should device policies in favor of improving climate risk management, strengthening resilience community-based development, building advanced weather forecasting, strong risk reduction structure and enhancing social protection to hamper the prevalence of food insecurity, particularly in vulnerable segment of the community [16].

In Ethiopia, several empirical studies have been conducted to identify the impact of social and economic factors on household food security. For example, studies [17–20], conducted in Oromia, Amhara, Tigray and South Nation Nationalities and Peoples, are few among the studies.

The Tigray region, where the study area is located, is a drought- and famine-prone area. The people mainly derive their livelihood from subsistence agriculture, which is characterized by mixed farming system on fragmented land, over utilized land and affected by erratic rainfall [17]. Although the *woreda* is food insecure and persistently benefiting from safety net program [21], there is difference in the degree of food insecurity among households. So, considering this issue is also imperative as previous study conducted [20] in the area has a methodological gap where the researcher employed safety net graduation benchmark as food security/insecurity measurement. This method involves measuring of wealth in terms of money. Such methodology does not show clear picture of household food security. The extended civil war carried out in the study area during the socialist period and Ethio–Eritrea border conflict led many women to become heads of households. Given the reasons above, this study aims to measure household food security status and to see the disparity of food insecurity between female- and men-headed households. This study will further identify the coping strategies employed in the study area determine the temperature and rainfall trends and their implication to household food security.

Research

Study location/description of study site

The study was conducted in Aksum National Regional State of Tigray, located 1024 km north of Addis Ababa. It has both urban and rural *Woredas.* The study focuses in the rural *Woreda* of Laelay Maichew. It is one of the eleven *Woredas* of central zone. It is bounded by Merelbeke in the North, Adwa in the East, Werileke and Naeder Adiet in the South and Tahtay Maichew in the West. It is located at an elevation of 1200–2050 m above sea level with an average annual rainfall and temperature of between 650 mm and 23.5 °C, respectively [21]. According to the *Woreda* Plan and Finance [22] of the rural *Woreda*, has a total population of 84,682, of which 50.2% of female. The Woreda has a total household of 19,246 (27% are female-headed), with an average family size of five people. The total area of the *woreda* is about 5,383,300 hectares; of which 1,455,200 ha is arable land, the rest is covered with forest, grazing land and mountain. It is home to rugged and gentle slope arable lands. It has two climatic zones occupied by 15 peasant associations.

Like other parts of Ethiopia livelihood of the *woreda* is dominated by subsistence agriculture. In addition to the subsistence agriculture, they also generate some extra income from petty trading, working as daily laborer, mining and other sources of income. According to *Woreda* Bureau of Agriculture [21], average size of cultivated land owned by household with five family sizes is about 0.75 ha. Teff, barely, wheat, horse bean, sorghum, finger millet and chickpea are the major crops grown in the area. Due to the small land holding per household, lack of improved agricultural technologies and inadequate extension services, as well as recurrent drought, the area is exposed repeatedly to food shortage. A map indicating the study location is presented in Fig. 1.

Research techniques

A mixed of quantitative and qualitative method approach is used, because according to [23] the use of a combination of qualitative and quantitative research techniques is better than the use of one of them to address and assess impacts and food security status.

Sampling techniques

Both probability and non-probability sampling systems were used. Study area was purposively selected, because the *woreda* is either seasonally or chronically food insecure and rarity of studies conducted in food security. At last but not least, we grew up in the area and have valuable information about the life experience of the community, topography of the *woreda* and can entertain the regional language, which helps us to communicate easily. Rural households are the unit of analysis. The *woreda*

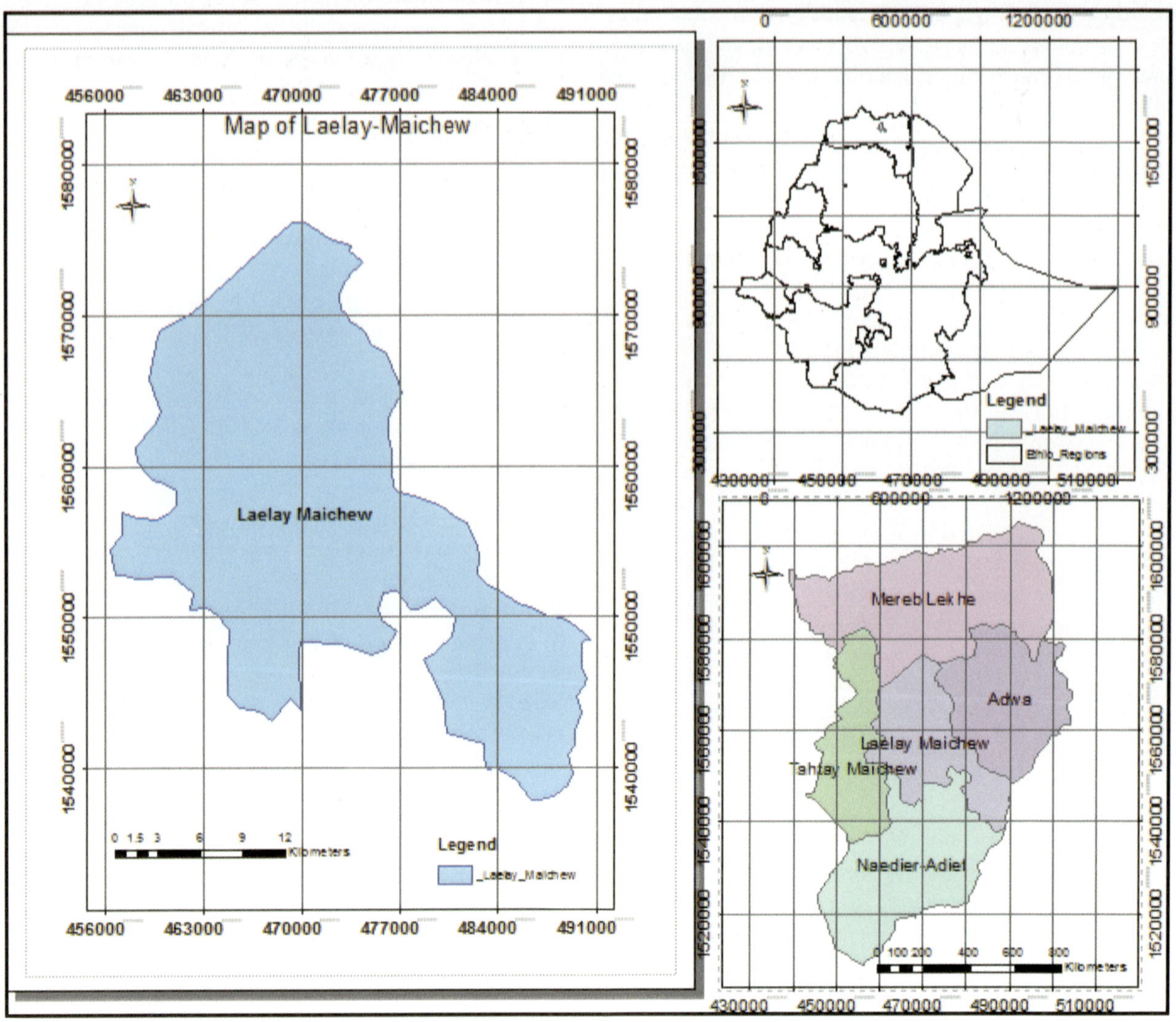

Fig. 1 Study area location. *Source*: Adapted from Kahsay and Mulugeta (2013)

has 15,785 households. A multistage random sampling method was employed to select respective agroecologies, kebeles and respondents. The *woreda* was clustered into two agroecological zone *kolla* and *woinadega*. One kebele from *kolla* and two kebeles from *woinadega* representative kebeles were selected. Accordingly, Miha, Lesalso and Dura were selected using simple random sampling. This sample is representative, because the households are living in similar agroecological setups, have similar livelihoods and are believed to be homogeneous. Since the study considered both female-headed and male-headed households. It is also stratified into female-headed and male-headed households to take representative samples from each household. Respondents were taken from a list of alphabetical order of latest tax registration document from each kebele. The first household was selected by simple random sampling; then, subsequent households were selected using systematic random sampling after n interval of the first selected household.

Sample size is determined by a formula developed by [24] for the determination of sample size.

$$n = 385/((1 + (385/N);$$

where n = sample size and N = population size. Accordingly, 150 respondents were selected (Table 1).

Data sources and collection procedures

Data were derived from primary and secondary sources. Primary data were collected by household survey questionnaires and interviews.

Household survey

One hundred and fifty questionnaires were distributed to collect data regarding the households.

Focus group discussion

Focus group discussants were selected purposively, and two focus group discussions were held in each kebele with ten to twelve participants in each focus group discussion, and a total of six focus group discussants from different economic status, different age groups, different religions, different educational levels, different sex and different positions were participated. To minimize dominance and biasness, equal chances were provided to each participant to express their feelings.

Observation

Observing the environment and life of the community.

Secondary data were collected from published books, journals, articles, maps and proceedings from various institutions, organizations and websites, *woreda* bureau of agriculture, *Woreda* plan and finance and Regional Meteorology Service Agency (RMSA).

Data analysis

Measuring food security

Food security has four basic dimensions, namely availability, access, utilization and stability. However, for this study food availability was measured using Household Food Balance Model (HFBM). Since the focus of the title is on food availability. Accordingly, total food production of each household was divided by the number of families' adult equivalent to get the per capita food availability of household in kg. This kg was converted into their respective kilo calorie using the FAO and EHNRI food composition tables (Additional file 1). Finally, it was calculated as daily calorie requirement/adult equivalent/day. According to the Ethiopian government, 2100 kcal/adult equivalent/day or 225 kg of cereals per person per annum set as minimum acceptable weighted average requirement [25, 26]. Based on this minimum daily calorie intake requirement, food secure and food insecure households were identified.

In Eq. (1), HFBM is calculated using the equation below adapted from [27, 28];

$$\begin{aligned} \text{NGA} =& [\text{GP} + \text{GB} + \text{FA} + \text{GG} + \text{FW} + \text{MP} + \text{DP}] \\ &- [\text{HL} + \text{GP} + \text{GU} + \text{GS} + \text{GSE}] \end{aligned} \tag{1}$$

NGA: Net grain availability (quintal/household/year), GP: Total grain production (quintal/household/year), GB: Total grain bought (quintal/household/year), FA: Quantity of food aid obtained (quintal/household/year), GG: Total grain obtained (quintal/household/year), FW: Food for work (quintal/household/year), MP: Total meat products (kilogram/household/year), DP: Total dairy products

Table 1 Selected kebele and sampled households. *Source*: *Woreda* Plan and Finance Office (WPF 2012)

PA	Year of study	Total households of the PA			Sampled households from Each PA		
		M	F	Total	M	F	Total
Miha	2013	492	182	674	24	9	33
Lesalso	2013	997	369	1366	44	16	60
Dura	2013	899	333	1231	39	18	57
Total	–	2388	884	3272	107	43	150

Source: *Woreda* Plan and Finance Office (WPF 2012)

(kilogram/household/year), HL: Postharvest losses (quintal/household/year), GU: Quantity of grain reserved for seed (quintal/household/year), GS: Amount of grain sold (quintal/household/year), GV: Grain given to others (quintal/household/year), GSE: Grain used for social events (quintal/household/year).

Analysis of climate variability

A two decade of meteorological data of rainfall and temperature was gathered from the RMSA. From these data, PCI, CV of the rainfall values at different timescales and annual temperature were calculated and presented in tables and graphs. According to [29], precipitation concentration index is calculated using formula (2).

$$\mathrm{PCI} = 100 \times \frac{\left(\sum \mathrm{pi}^2\right)}{\left(\sum \mathrm{pi}\right)^2} \quad (2)$$

where p_i = is the rainfall amount of the ith month of a year and $\sum$ = summation over the 12 months. According to Oliver, PCI values of less than 10 indicate uniform monthly distribution of rainfall, values between 11 and 20 indicate high concentration and above 21 indicate very high temporal rainfall concentration as in indicated in Woldeamlak [29]. Rainfall variability over a period of time was also analyzed by calculating the CV of the rainfall values at different timescales using formula (3).

$$\mathrm{CV} = \frac{\sqrt{\sum f(xi - x)^2}}{n/x} \quad (3)$$

where x = mean annual rainfall in mm, n = number of years for which the rainfall data are available for a given station, xi = annual rainfall (mm) of the year ith of a given station.

Result and discussion

Food security status of the households

Household Food Balance Model was used to measure food security though there are many approaches to measure food security. The result showed that 15.7% were food secure and the rest 84.3% were food insecure. This indicates that more than three-fourth of the households was exposed to critical shortage of food, which falls to meet the minimum calorie requirement. For the purpose of lucid understanding of food security and to have good picture of households' food gap, households were grouped into three categories based on their calorie availability/adult equivalent/day. Food secure ($\geq$ 2100 kcal/adult equivalent/day), moderately food insecure ($\geq$ 1050 to < 2100 kcal/adult equivalent/day) and severely food insecure (< 1050 kcal/adult equivalent/day). The mean daily calorie intake of the households was 1181.3 cal, which only meets 56.2% of the minimum daily requirement (2100 kcal). The minimum and maximum calorie intake was 103 and 3000 kcal, respectively, which shows a great variation in the availability of food across households. So, the mean value falls under the category of moderately food insecure with incredible minimum and maximum daily calorie intake value. Table 2 indicates female-headed households were heavenly exposed to food insecurity than their counterpart male ship. Studies conducted [30] in Zimbabwe, Malawi and Zambia founds that female-headed households were more vulnerable to food insecurity than male-headed households where their means of livelihood is casual agricultural labor which complements with this finding. Similarly, anthropometrics measurement also verified that children from female-headed households show more stunting (low height-for-age) as compared to those from male-headed households [31]. In addition, findings of [32–34] in Pakistan, South Africa and Ethiopia showed female-headed households are more food insecure than male-headed households. This is mainly related to lack of access to financial, social, human, natural and physical assets.

Honey was not included directly in computing total annual grain availability and was not converted into its respective calorie equivalent, because it was not incorporated in the food composition table and contributes insignificant amount in their daily consumption. Generally, all animal products were not included when calculating the total grain availability and not converted to their respective calories as they contribute insignificant in the daily household food consumption. Most of the time households sell life animals and buy grains to fill the gap in grain requirements of their families as the focus group discussants implied.

Moreover, postharvest losses were considered as negligible and are not included in calculating the total grain availability of households. This is due to the introduction of different types of pesticides, good awareness in postharvest losses prevention, control and traditional knowledge of the community; therefore, postharvest losses are considered as insignificant. Except the aforementioned

Table 2 Food availability in calorie

Calorie availability/adult equivalent/day	Household head			
	Male		Female	
	Number	%	Number	%
FS ($\geq$ 2100 kcal)	26	24.4	8	18.6
MFI (1050 to < 2100 kcal)	33	30.7	17	39.5
SFI (< 1050 kcal)	48	44.9	18	41.9
Total	107	100	43	100

variables which are not included in computing total grain availability, the rest of all variables are computed using the HFBM and converted into their respective calorie equivalent using the EHNRI and FAO food composition table.

Coping strategies of the households

Coping mechanisms used by different rural households are many depending on the availability of resources they have. Respondents replied. It is not uncommon to experience different coping strategies during food shortage to minimize or avert the prevalence of asset depletion and other catastrophic impact. Result of the analysis discloses (Table 2) that daily laborer is a dominant means of coping mechanism. The key informants and focus group discussants said, although there is job opportunity in their residential area, most of the capable bodies are fled to western Tigray during the weeding and harvesting season since daily wage is better. Following daily laborer, eating less preferred foods and a combination of one or two coping strategies (selling wood, work as daily laborer, selling alcoholic beverages in village towns, mining and guarding schools, clinics, etc.), with a percent of 22 and 14.7%, respectively. Likewise, 3.3% of the respondents said reducing quantity of meals they eat, 2.7% borrowing or purchasing on credit, 2% selling productive assets, 1.3% reducing number of meals and selling woods but consuming seed stocks and getting free aid (1%) are barely practiced as listed in Table 3.

Findings of [33, 35] in Ghana, Bangladesh, also showed that eating less preferred foods commonly practiced coping mechanism, which has quite close similarity. However, study conducted [36] in Malaysia indicated that selling valuable materials and borrowing money are the most practiced coping strategies, which is less practiced in the study area. In general a review [37] in Ethiopia, household employ seasonal labor migration, consume less preferred foods, borrow grain to be returned double at the next harvesting season, send their children to relatives, rent their economic assets like land, and children are forced to drop from their school.

Analysis of climate variability

The objective of this analysis is not to show sophisticated model analysis and correlations using rigorous analytical model, rather to focus on general implication of temperature and rainfall, and their implication to food availability/production. If there is increasing/decreasing trend so what will be the fate of agricultural production or food availability. Food production in a given years is mostly related to optimum level of temperature, precipitation and radiation. Therefore, extreme events are a cause to failure of food availability as a result of inter-annual climate variability. Profound investigation of climate variability is crucial in Ethiopia, as large portion of the population means of living is rain-fed agriculture.

Table 3 Coping strategies

Coping strategies	Frequency	%
Eating less preferred foods	33	22.0
Borrowing or purchasing on credit	4	2.7
Consuming stock seeds	1	0.7
Reducing number of meals	2	1.3
Reducing quantity of meals	5	3.3
Selling household assets	3	2.0
Selling woods and charcoal	2	1.3
Daily laborer	77	51.3
Aid	1	0.7
A combination of different coping strategies	22	14.7
Total	150	100.0

Temperature trend

Generally in tropical countries and subtropical countries, crop and crop productivity decrease as temperature increases from 2 to 3 °C and the forecast indicates that a significant reduction in crop productivity will be observed in 2020 [38]. Studies [29] implied differences in precipitation have a higher explanatory power than temperature in describing changes in crop yields: correlation between precipitation and yields is higher than temperature. This suggests that wetter years are associated with higher crop production. While higher temperature could be generally associated with lower yields.

The result showed that temperature is increasing from time to time; it increases from the year 1992–2003 and falls in the consecutive years after 2003. This fall in temperature might be due to some conservation practices. Basically, a little rise in temperature leads to the dwindling of plants and animals performance directly or indirectly. Figure 2 shows that there is fluctuation of temperature every year. This has impacts on their production according to the focus discussants. The variance analysis shows every year there is a variation of 0.65 °C. Even this variation is not undermined; its impact on the biological system is not overstated.

Rainfall trend

Report [39, 40] indicates that annual rainfall is significantly changed across Ethiopia and suggesting high inter-annual variability with a slight but statistically significant negative trend. This is in line with previous analyses, which highlight the importance of inter- and intra-seasonal rainfall variability over total annual rainfall in determining livelihood and food security

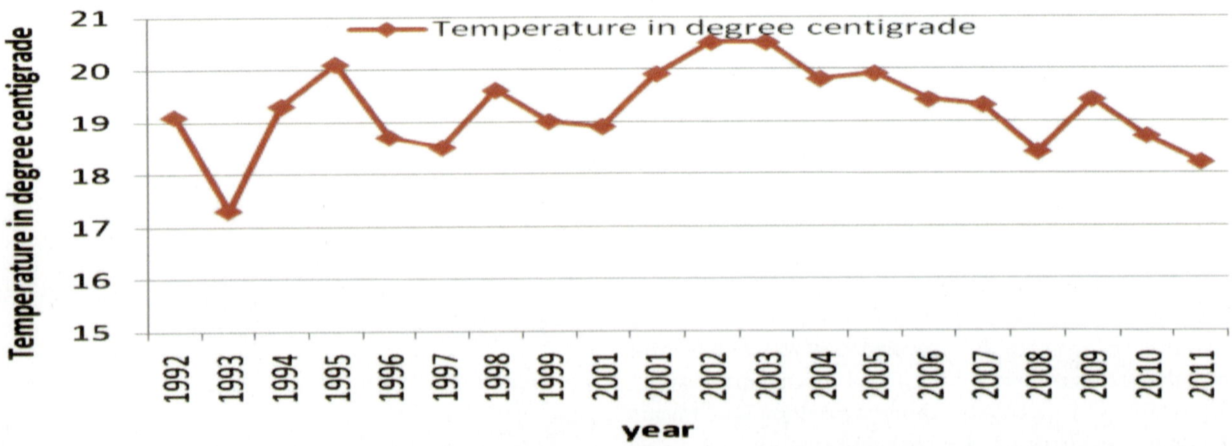

Fig. 2 Temperature trend. *Source*: Own compilation based on RMSA (2013)

outcomes [41]. The result indicates the minimum and maximum annual rainfall is 354.5 and 1037 ml, respectively, with mean annual rainfall of 725 ml. Besides, the minimum and maximum PCI is 15.6 and 41.7 ml, respectively, with average PCI of 24.4 ml (Table 4). It is clear that optimum rainfall is necessary for increased crop productivity and food security. However, high concentration of rainfall has a serious consequence on crop production and food security. According to [29], PCI is classified into three: < 10 uniform monthly distribution, 11–20 high concentration and > 21 very high concentration of rainfall. Therefore, the result showed high concentration rainfall variability, which is similar to the finding of [42]; this implies that food availability is in question, because high concentration of rainfall exposes arable and non-arable land to severe soil erosion and environmental degradation. According to [43] PCI greater than 21 ml is a challenge for water and soil conservation. So this implies that food availability of rural people who dominantly depend on rain-fed agriculture is extremely suffered by food shortage. Generally, the PCI indicates heavy rain mixed with hails every year in one or more than one of the kebeles will be happened. As a result crop, production or availability will be declined; then, this leads to mass poverty and food insecurity. However, the inter-annual rainfall variation is medium with rainfall variability of 0.16, which has not significant effect in crop production (Fig. 3).

Table 4 PCI and CV. *Source*: Own compilation based on RMSA (2013)

Station	PCI %	CV
Aksum	24.4	0.16

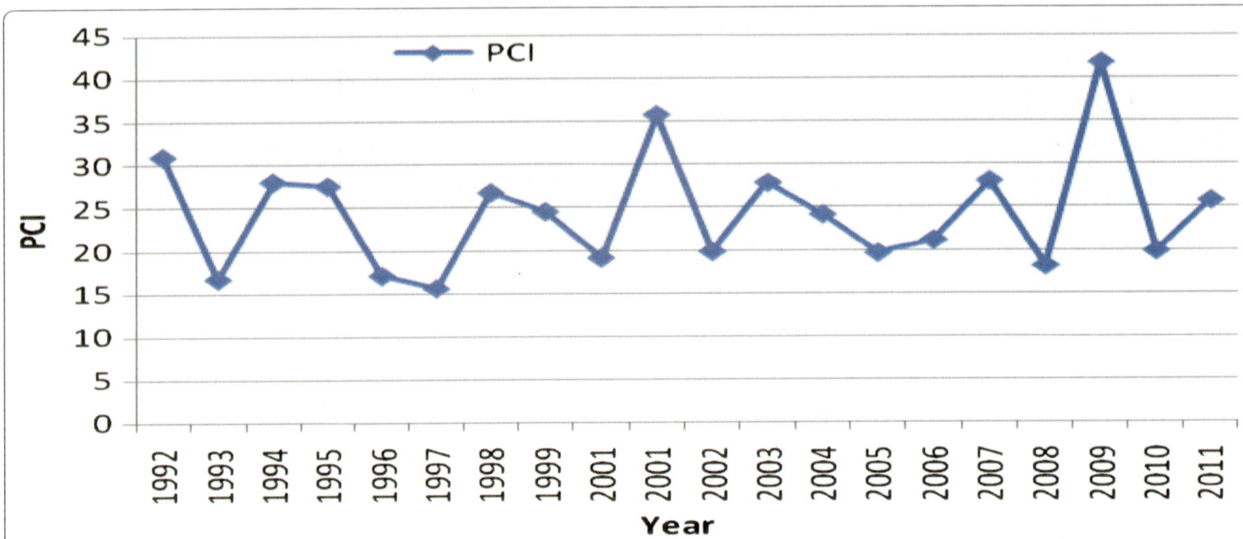

Fig. 3 Precipitation concentration index (PCI). *Source*: Own compilation based on RMSA (2013)

Conclusions

Different efforts are in place to minimize the prevalence of food insecurity and poverty. However, from the sampled households more than 75% are food insecure. Besides, the prevalence of food insecurity is higher in female-headed households. This report is a bell for the government to invest aggressively in modern agriculture. Since the population is getting higher and higher, the production is lower. Therefore, appropriate development strategies should be devised, and booming agricultural production and productivity using improved seeds, fertilizers and investing on irrigation rigorously will be paramount. In addition, the community should also engaged in honey bee production, poultry production and fattening to diversify their income and to solve the problem of food shortage.

The commonly practiced coping mechanism is working as daily laborer, eating less preferred foods and combination of different strategies while consuming stock seed and getting aid are rarely practiced. Therefore, heavy investment on industries is mandatory to increase the employment opportunity, since most of the rural households are flew more than 300 km from their residential to work. The annual rainfall concentration is very high; this implies that there is a great variation in rainfall between rainy months. It may be also followed by heavy hails which damages the crop production and productivity. Therefore, building strong early warning system and rehabilitation program is crucial.

In general, food insecurity is pervasive in female-headed households, daily laborer is the mostly practiced coping mechanism and intra-month rainfall concentration is very high. Therefore, devising packages that matches with the physical capacity of female-headed households to minimize the risk of food insecurity in particular and different adapting mechanism should also developed to halt the progressive impact of rainfall variability.

In conclusion, the government and the community should give due value to a forestation, reforestation and soil and water conservations. As these practices are very important, to minimize the risk of climate variability and to boost agricultural production and to generate extra income from the conserved ecosystem.

Abbreviations

UN: United Nation; IFPRI: International Food Policy Research Institute; WBoA: Woreda Bureau of Agriculture; WPF: Woreda Plan and Finance; FAO: Food and Agriculture organization; PSNP: Productive Safety Net Program; FDRE: Federal Democratic Republic of Ethiopia; MoFED: Ministry of Finance and Economic Development; HFBM: Household Food Balance Model; RMSA: Regional Meteorological Service Agency; PCI: Precipitation Concentration Index; CV: coefficient of variability; ENHRI: Ethiopian Nutrition and Health Research Institute.

Authors' contributions

SK carried out reviewing relevant materials, data gathering, analyzes the data and preparing the tables and figures. DG carried out editing, formatting and organizing the references. Both authors read and approved the final manuscript.

Acknowledgements

We are grateful to Addis Ababa University College of development studies for its support.

Competing interests

This is our original work; there is no any competing interests in regard to this paper.

Funding

Partial fund from Addis Ababa University.

References

1. IPCC. Fourth assessment report. Cambridge: Cambridge University Press; 2007.
2. Schmidhuber J, Tubiello FN. Global food security under climate change. Proc Natl Acad Sci. 2007;4(50):19703–8.
3. Shishay S, Messay M. Determinant of rural household food insecurity in Laelay Maichew woreda Tigray, Ethiopia. Afr J Agric Food Secur. 2014;2(1):106–12.
4. Devereux S, Maxwell S. Food insecurity in Sub-Saharan. Pietermaritzburg: University of Natal Press; 2003.
5. IFPRI. Reaching sustainable food security for all by 2020: getting the priorities and responsibilities right. Washington, DC; 2002.
6. Hart T. Exploring definitions of food insecurity and vulnerability: time to refocus assessments. Agrekon. 2009;48(4):1–22.
7. Arora VPS, Bargali SS, Rawat JS. Climate change: challenges, impacts, and role of biotechnology in mitigation and adaptation. Progress Agric. 2011;11:8–15.
8. Morede T. Vulnerability, land, livelihoods and migration nexus in rural Ethiopia: a case study in South Gondar Zone of Amhara Regional State; 2012.
9. Devereux S. Food insecurity in Sub-Saharan Africa. London: Institute of Development Studies; 2000.
10. Nath KP, Behera B. A critical review of impact of and adaptation to climate change in developed and developing economies. Environ Dev Sustain. 2011;13:141–62.
11. Mwaniki A. Achieving food security in Africa: challenges and issues. Cornell University, U.S. Plant, Soil and Nutrition Laboratory; 2005.
12. Kotir JS. Climate change and variability in Sub-Saharan Africa: a review of current and future trends and impacts on agriculture and food security. Environ Dev Sustain. 2010;13(3):587–605.
13. FAO. Food insecurity in the horn Africa, sub-Regional emergency and rehabilitation, office for Eastern and Central Africa (REOA); 2010.
14. Kotir JS. Climate change and variability in Sub-Saharan Africa: a review of current and future trends and impacts on agriculture and food security. Environ Dev Sustain. 2010;13(3):587–605.
15. Regassa N, Stoecker BJ. Household food insecurity and hunger among households in Sidama district, Southern Ethiopia. Public Health Nutr. 2010;15(7):1276–83.
16. Feleke ST, Kilmer RL, Gladwin CH. Determinants of food security in Southern Ethiopia at the household level. Agric Econ. 2005;33:351–63.
17. Degefa T. Household Seasonal Food Insecurity: causes. Social Science Research Report Series No. 26. Addis Ababa: Organization for Science Research in Eastern and Southern Africa; 2002.
18. Frehiwot FS. Food insecurity and its determinants in rural households: In Amhara region. MA thesis, Addis Ababa University: www.edri.eth.org. Publisher; 2007.
19. Messay M. Determinants of agricultural productivity and household food security: case studies from Kuyu district, central Ethiopia: LAP Lambert

Academic Publishing; 2011. ISBN-13:978-3-8465-2442-1,.
20. Misgna AR. Rural household food security status and its determinants: the case of Laelay Maichew *woreda* central zone of Tigra, Ethiopia. MA thesis, Haramaya University: LAp LAMBERT Academic Publishing; 2010.
21. *Woreda* Bureau of Agriculture. Unpublished report; 2012.
22. *Woreda* plan and Finance. Unpublished report; 2012.
23. Degefa T. An assessment of the role of local institutions and social capital in household food security: a case study at two rural communities in Oromia Zone, Amhara Region. In: Proceedings of the 16th international conference of Ethiopian Studies; 2005.
24. Welch S, Commer J. Quantitative methods for public administration. 2nd ed. Belmont: The Dorsey Press; 1988.
25. Federal Democratic Republic of Ethiopia. Food security strategy. An update. Addis Ababa, Ethiopia; 2002.
26. MoFED. Ethiopia: sustainable development and poverty reduction program. Addis Ababa: Ministry of Finance and Economic Development; 2002. p. 2002.
27. Degefa T. Belg crop production as a strategy of households' food security: a comparative study of Belg Grower and Non-Belg Farmers in Munessa Wereda, Arsi Zone (Unpublished MA Thesis), Addis Ababa University; 1996.
28. Messay MT. Challenges and opportunities of resettlers in Jiru Gamachu resettlement villages, West Shewa Zone of Oromiya National Regional State. J Sustain Dev Africa. 2009;11:4.
29. Woldeamlak B. Rainfall variability and crop production in Ethiopia: case study in the Amhara region. In: Ege S, editor. Proceedings of the 16th international conference of Ethiopian Studies. Trondheim International Fund for Agricultural; 2009.
30. IRIN. Integrated Regional Information Networks, Southern-Africa: female-headed households most vulnerable to food insecurity. (http://www.irinnews.org/print.asp?ReportID=42410 (2006).
31. Babatunde RO, Omotesho OA, Olorunsanya EO, Owotokit GM. Determinants of vulnerability to food insecurity: a gender-based analysis of farming households in rural India. Indian J Agric Econ. 2008;63(1):116.
32. Tibesigwa B, Visser M. Assessing gender inequality in food security among small-holder farm households in urban and rural South Africa. World Dev. 2016;88:33–49.
33. Farzana DF, Rahman AS, Sultan S, Raihan MJ, Hague MdA, Waid JL, Choudhury N, Ahmed T. Coping strategies related to food insecurity at the household level in Bangladesh. PLoS ONE. 2017;12(4):e0171411.
34. Ahmed UJ, Ying L, Bashir MK, Zulfigar F. Status and determinants of small farming households' food security and role of market access in enhancing food security in rural Pakistan. PLoS ONE. 2017;12(10):e0185466.
35. Kuwornu JKM, Suleyman DM, Amegashie DPK. Analysis of food security status of farming households in the forest belt of the central region of Ghana. Russ J Agric Socio-Econ Sci. 2013;1(13):26–42.
36. Shariff ZM, Khor LG. Household food insecurity and coping strategies in a poor rural community in Malaysia. Nutr Res Pract. 2008;2(1):26–34.
37. Mohamed AA. Food security situation in Ethiopia: a review study. Int J Health Econ Policy. 2017;2(3):86–96. https://doi.org/10.11648/j.hep.20170203.11.
38. Mariara JK, Drb E. Climate change and food security in Kenya does climate change affect food insecurity in Kenya? 2015.
39. NMA [National Meteorological Agency of Ethiopia]. Weather station data; 2012.
40. Livelihoods, Early Assessment and Protection (LEAP). 2013.
41. WFP. Climate risk and food security in Ethiopia: analysis of climate impacts on food security and livelihoods.
42. Hadgu G, Tesfaye K, Mamo G, Kassa B. Trend and variability of rainfall in Tigray, Northern Ethiopia: analysis of meteorological data and farmers' perception. Acad J Environ Sci. 2013;1(8):159–71.
43. Shamsuddin S, Osman SK. Spatio-temporal variability of rain fall over Bangladesh during the time period 1969–2003. Department of Geography, University Malaya 50603 Kuala Lumpur, Malaysia; 2009.

17

Predictors of food insecurity and coping strategies of women asylum seekers and refugees in Durban, South Africa

Carin Napier[1*], Wilna Oldewage-Theron[2] and Beryl Makhaye[1]

Abstract

Background: This cross-sectional study assessed the prevalence of household food insecurity and the association between food insecurity and coping strategies among women asylum seekers and refugees in the city of Durban in South Africa.

Methods: A coping strategy questionnaire was used to determine food security as this gives a good indication of what the women have to go through in order to provide safe and nutritionally sound meals to family members. The questionnaire also presents an indicator of food security, with a higher score indicating a higher level of food insecurity. Participants were recruited in the Durban city centre from church and community meetings and in their workplace. Data were collected from 200 women. Households were classified as food-secure and food-insecure groups for comparison purposes. Data were analysed on SPSS, version 23.0, for descriptive statistics, correlations, and logistic and bivariate linear regression.

Results: The results showed that 96.2% of the women were living in poverty (< US$1.9 per person per day). Only 8.0% of the women were food secure. Regarding the coping strategies, except for 'sending children to the refugee preschools' the number of coping strategies was significantly higher in the food-insecure group ($p = 0.000$). The number of people in the household, the unemployment status of the women and household income were predictors of poverty ($R^2 = 0.859$, $p = 0.000$, SEE = 0.189). Larger household size, households spending between ZAR700 and ZAR900 on food monthly (OR 1.01; CI 0.26–4.56) and households purchasing food from street vendors (OR 4.02; CI 01.35–46.23) or informal community shops (OR 4.1; CI 0.42–40.22) were indicators of food insecurity.

Conclusion: This study points to a serious problem of household food insecurity among black women asylum seekers and refugees in the city of Durban, South Africa. The food security status of this group of women can possibly impact on the nutritional status of the family unit and make the family vulnerable to disease and illness. This will place an additional burden on the government in providing healthcare services to refugees and asylum seekers.

Keywords: Food insecurity, Refugee, Households, Women, Coping strategies

Background

Since 1994, South Africa has become a popular destination for asylum seeker and refugees, especially those from other African countries [14]. According to the United Nations High Commissioner for Refugees UNHCR [29], South Africa (SA) currently hosts 1,096,063 asylum seekers and 121,645 refugees. The Geneva Convention defines an asylum seeker as a person who crosses international borders out of fear of being persecuted for belonging to a specific group or ideology, race, religion or gender. These people seek protection in a foreign country because the country of origin is not willing to protect them or is the cause of violation of human rights. They need to apply to the host country to be registered as refugees. A refugee is defined as a person who has well-founded fears of being persecuted for a variety of reasons

*Correspondence: carinn@dut.ac.za
[1] Department of Food and Nutrition Consumer Sciences, Durban University of Technology, P. O. Box 1334, Durban 4000, South Africa
Full list of author information is available at the end of the article

including their political or religious affiliation, having a specific racial or social profile, or who has experienced persecution or been unfairly prejudiced because of their gender or sexual orientation. The term 'refugee' refers to someone who has been processed and registered with the (South African) Department of Home Affairs as a refugee UNHCR [28].

Annual reports of UNHCR in South Africa indicate that asylum seekers and refugees are mainly urban city dwellers; therefore, promotion of self-reliance is crucial to prevent dependency. As of 2012, when this study was undertaken, the UNHCR regional office through its implementing partners in various provinces such as Refugee Social Services in KwaZulu-Natal, Jesuit Refugee Services in Gauteng and the Cape Town Refugee Centre has assisted 6000 refugee children with basic education and 20,000 refugee learners with education support material. Furthermore, vocational training was offered to 10,000 adult refugees and English classes were made available to 20,000 recipients to empower refugees and asylum seekers with skills to access the labour market UNHCR [27].

The most vulnerable groups which include children, women with children, the elderly, and disabled asylum seekers and refugees are provided with material assistance such as food, shelter and medical referral for a period of 3 months. Thereafter, implementing partners, such as the Consortium for Refugees and Migrants in South Africa (CoRMSA) and the Department of Social Development, need to facilitate access to local charitable organisations and state structures for grants. The majority of asylum seekers and refugees in Africa are women and children, and they are the most vulnerable groups in a population. These women have to cope with a variety of challenges while settling into the new camp or host country [10]. Not only do they have to adjust to a new culture and language, and more often than not xenophobia and crime situations, but they also have to provide for their children and often have to cope with unfamiliar food choices and limited resources [6, 10, 14]. Access to healthy foods, specifically in urban areas where most asylum seeker and refugees are based, is often difficult and depends on many factors such as distance to shops and access to transport, availability of affordable quality foods and lack of nutrition knowledge [4, 7, 8]. A large number of asylum seekers and refugees in South Africa thus survive through personal networks and family support. Women with children often approach faith-based organisations (FBO) and local non-governmental organisations (NGOs) for food and other material assistance (African Centre for Migration and Society (ACMS) [1]. There is a paucity of information about the food security status of refugee families; however, the few studies available showed a high prevalence of food insecurity, hunger and malnutrition among asylum seeker and refugees. Poverty and food insecurity, leading to low food accessibility and poor dietary quality, can thus have a detrimental effect on their health and quality of life [6, 14].

Although a study of this limited literature has demonstrated the high prevalence of food insecurity and its effect on asylum seeker and refugees, little is known about how women asylum seekers and refugees cope with food insecurity and their living conditions. Furthermore, socio-economic factors are important indicators of health, food and nutrition security [2]. In an effort to address this gap in the literature, this study had several aims, namely to (a) determine the prevalence and severity of food insecurity; (b) examine the use of various coping strategies used by the food-insecure refugees, (c) describe the relationship between socio-economic factors and food insecurity and (d) describe the socio-economic predictors of food insecurity among women asylum seekers and refugees in Wards 26 and 28 in the city of Durban.

Methods

A mixed-methods design was used. A sample size calculator (The Survey System) was used to calculate a statistically representative sample. Female asylum seekers and refugees were recruited after church and community meetings as well as during or after their shift work as car guards in Durban CBD and were selected using targeted sampling until the sample size of 200 women was obtained. Women were selected to participate in the study if they met the following criteria: (a) being in possession of Section 22 or 24 documents from the Department of Home Affairs for asylum seekers or refugees, respectively, (b) aged 19–50 years of age, (c) living in Ward 26 area (North and South Beach) and Ward 28 area (Central Business District—CBD) in the city of Durban, and (d) conversant in English and Swahili. Each participant was informed about the study before implementation. Participation was voluntary, and informed consent was obtained from all participants included in the study. The women were informed that they could withdraw from the study at any time without being penalised. Furthermore, the participants received food parcels on completion of the interview, but had not been informed that they would receive the food parcels on recruitment. Receiving the food parcels would thus not have affected their decision to participate in the study.

Trained final-year Food and Nutrition students from Durban University of Technology (DUT) acted as fieldworkers to collect data from the study participants. Fieldwork was conducted between February and August 2013. All the questionnaires were translated into Swahili and either administered in English or Swahili, based on the participant's preference, by trained fieldworkers in a one-on-one interview situation. The Swahili questionnaire was translated back into English by another person to ensure conceptual agreement. Each interview lasted for approximately 30 min. A valid and reliable socio-demographic questionnaire was used [18] to collect socio-demographic data and determine the socio-economic profile of the women. A number of independent variables such as age, marital status, employment status, accommodation arrangements, household composition, education level, household income and food procurement and consumption patterns were included in the questionnaire.

The validated coping strategy questionnaire developed by Maxwell et al. [16] was adjusted in a focus group discussion with 10 women from the refugee community, who were not included in the main study, to determine which coping strategies, as listed in the questionnaire but supplemented with others as suggested by the women, should be used to address food and nutrition insecurity in the refugee community. These coping strategies were listed, and the same women ranked the severity of each of the coping strategies during the focus group discussion from one (least severe) to four (most severe) as described by Maxwell et al. [16]. The additional coping strategies identified by the women that were included were: 'Rely on food aid from Non-Governmental Organizations and Faith Based Organizations' and 'Send young children in the household to be fed at the local refugee pre-school' and these were both ranked as a four severity weight. This list of coping strategies with their severity rankings became the coping strategy questionnaire administered to the women. This was not pre-tested and internal validity was thus not determined. The women had to report their experience of food security for the preceding one-month period [17]. The questionnaire included the following nine coping strategies: skipping meals for an entire day, reducing the number of meals per day, relying on aid from a local NGO/FBO, relying on the community or friends for food, buying less preferred and less expensive foods, restricting the amount of food consumed by adult women, feeding the employed man of the household first, sending children to eat at a local refugee pre-school and limiting portion sizes at meal times.

Statistical analysis

The socio-demographic questionnaires were captured on a Microsoft Excel spreadsheet and analysed on the Statistical Package for the Social Sciences (SPSS) software version 23 for descriptive statistics (frequencies for categorical data, means and standard deviations [SDs] for numerical data) presented in graphs and tables. The coping strategy questionnaires were captured on a Microsoft Excel spreadsheet and analysed. Frequencies for each of the coping strategies were calculated. The mean ± SD coping strategies index (CSI) was calculated. The CSI was chosen for this study as it measures food consumption behaviour when the respondents cannot access enough food and was adopted by many agencies such as the World Food Programme Vulnerability Analysis Mapping unit, the United Nations Food and Agriculture Organization and the Global Integrated Phase Classification team, among others, to use as an indicator to measure food security [31]. It was also found that the CSI and the Household Food Insecurity Access Scale (HFIAS) corroborated the full set of food security measures focusing on consumption behaviours including CSI, reduced coping strategy index (rCSI), HFIAS, household dietary diversity score, food consumption score (FCS), self-assessed measure of food security (SAFS) more than the others [17]. The CSI does not have universal cut-off points for different categories of insecurity [17], and the researchers used the classification of food security based on other studies undertaken to classify the women as food secure or food insecure as follows: food secure (CSI 0–2), mildly food insecure (CSI 3–12), moderately food insecure (13–40) and severely food insecure (CSI > 40) [17]. The sample was further classified as food secure (including mildly food secure) and food insecure (including moderately and severely food insecure) for comparison between food-secure and food-insecure groups. Analysis of variance (ANOVA) with the correction of Bartlett's test for equal variances at a significance level of $p \leq 0.05$ was used to determine significant differences for the numerical data between the groups. Nonparametric hypothesis test summaries were performed for categorical socio-economic data to determine significant differences between the groups.

Pearson and Spearman correlations were used to determine significant relationships between numerical and categorical variables, respectively. Only significant relationships are reported in the results. Bivariate linear regression analysis for predictors of poverty (< US$1.9 per person per day cut-off—The World Bank [25]) was also employed to identify socio-economic factors that could explain the existing poverty and food insecurity status among this group of women asylum seekers and refugees. A logistic regression analysis was used to estimate socio-economic and demographic variables contributing to food insecurity. Those variables with significant correlations as determined by the Pearson and Spearman correlations were included in the logistic regression analyses. The results from the logistic regression model are stated as odd ratios with 95% confidence intervals.

Results

Respondent characteristics

The mean age ± SD of the participants was 38.5 ± 7.8 years. The sample was exclusively black women who spoke mainly Burundi (31.0%) and Swahili (28.0%). The average household size was 6.9 ± 2.8 persons and ranged from 2 (4.0%) to 11 (26.5%) people. The majority (82.5%) of the women had been living in South Africa for less than 5 years. The majority of women lived in a rented apartment (71.5%), followed by 25.0% who lived in hostel accommodation. The women had a relatively high literacy rate as 49.0% had completed secondary school, and 23.0% had a college qualification and 4.5% had a post-school qualification. Although 36.5% of the women were employed, the mean ± SD monthly household income was ZAR1656.99 ± 956.97. A large percentage (96.2%) of the women were living in poverty according to the < US$1.9 per person per day cut-off (The World Bank [25]), and 91.3% and 96.0%, respectively, were living in poverty according to the KwaZulu-Natal lower-bound poverty line (LBPL) of ZAR539 and upper-bound poverty line (UBPL) of ZAR757 (Statistics South Africa (Stats SA) [24]). In 39.0% of the households, the woman was a mother and the head of the household. In the majority of the households, the women were responsible for food preparation (98.5%) and feeding the children (90.8%), as well as making decisions about the amount of money spent on food (65.8%) and the type of food to buy (86.5%). The mean ± SD spent on food per month was ZAR789.53 ± 432.13 and contributed 60.3% of the total income. Food was mainly bought from the informal food vendors in the residential area (93.0%) weekly (56.0%) by the majority of households, followed by daily (25.5%) and monthly (18.5%). Most of the meals (92.9%) are consumed at home and the majority of the women consumed only one meal per day (55.5%), followed by 34.5% and 1.5% consuming two or three meals, respectively, with 8.5% who had not consumed any meals the day before data collection. The majority of the women (68.3%) indicated that they sometimes experienced a food shortage, whereas 18.6% and 9.5% indicated that they often and always experienced not having enough money to buy food, respectively. All the women had potable water and sanitation facilities available in the house; however, they had a problem with cockroaches (86.0%) and mice (44.0%) and ants (37.5%) in the house. Although they all had electricity available, only 3.5% used electricity for food preparation. Wood fires (93.5%) were mainly used for food preparation.

The results in Table 1 show a comparison between the socio-demographic characteristics of the food-secure and food-insecure households. No significant differences were observed in socio-economic variables between the food-secure and food-insecure groups. However, the food-insecure group experienced a shortage of money to buy food often or always in 36.3% of the households compared to 22.0% in the food-secure households ($p < 0.05$). The latter indicates that although the women were classified as food secure by the CSI, 4.6% and 17.4% of the women in the food-secure group indicated that they always and often did not have enough money to buy food, respectively, thus having to employ coping strategies such as relying on the community or family to provide them with food.

Prevalence of food insecurity

The mean ± SD household CSI was 50.6 ± 34.8 and ranged from 0.0 to 171.0. Only 8.0% of the women were food secure compared to 46.5%, 45.0% and 0.5% who were mildly, moderately and severely food insecure, respectively [16] (Fig. 1).

The results in Fig. 2 show that all the coping strategies identified by the respondents were used by the majority of the households, except for sending young children to eat at the local refugee pre-school (45.0%). The three most used coping strategies were: reducing the number of meals of all the adult women in the house (92.5.0%), buying less preferred and expensive foods (90.0%) and limiting portion sizes at meals (83.5%) (Fig. 2).

All the coping strategies were applied significantly ($p = 0.000$) more in the food-insecure groups than in the food-secure group, except for sending out the children to eat at the refugee pre-school for which no significant difference ($p = 0.083$) between the groups was observed. More than 90% of the women in the moderately and severely insecure group employed coping strategies with 98.9% skipping meals for an entire day, 100.0% reduced the number of meals eaten in a day, 100.0% relied on less preferred and more affordable foods, 93.4% restricted food consumption for themselves and 98.9% limited portion sizes at meals (Fig. 3). The coping strategies most employed by the food-secure group included a reduced number of meals per day (86.2%) and relying on less preferred and expensive foods (81.7%).

A significant positive relationship was observed between CSI (food security/insecurity indicator) and being unemployed ($r = 0.157$; $p = 0.012$) at the 0.001 level (99% confidence level), and a significant adverse relationship with money spent on food ($r = -0.130$; $p = 0.021$) and not enough food in the household ($r = -0.162$; $p = 0.004$) was also observed. Furthermore, significant relationships were found between the available household US$ per person per day (proxy for poverty) and the variables measured in this study (Table 2).

The logistic regression analysis carried out on the socio-economic and demographic predictors of poverty ($ per person per day) showed that the number of people

Table 1 Association between food security status and socio-economic factors

Variable	Food secure, $n = 109$ (54.5%)	Food insecure, $n = 91$ (45.5%)
Number of household members		
1–3	12 (11.0)	9 (9.9)
4–6	45 (41.3)	37 (40.7)
7–11	52 (47.7)	45 (49.4)
Current employment status		
Employed	34 (31.2)	39 (42.9)
Unemployed	75 (68.8)	52 (57.1)
Education status		
No schooling or up to primary school	9 (8.2)	7 (7.7)
Standard 8	18 (16.5)	13 (14.3)
Standard 10	57 (52.3)	41 (45.0)
College/FET or completed post-school	25 (23.0)	30 (33.0)
Do not have enough money to buy food*		
Always	5 (4.6)	14 (15.4)
Often	19 (17.4)	19 (20.9)
Sometimes	81 (74.3)	55 (60.4)
Seldom or never	4 (3.7)	3 (3.3)
Types of places or shops for purchasing food		
Supermarket	100 (91.7)	81 (89.0)
Wholesalers	3 (2.8)	2 (2.2)
Street vendor	2 (1.8)	3 (3.3)
Informal community shop	4 (3.7)	5 (5.5)
Variable	Food secure, $n = 72$ (%)	Food insecure, $n = 60$ (%)
Household monthly income		
500–1000	16 (22.2)	9 (15.0)
1001–1500	22 (30.6)	16 (26.7)
1501–2000	12 (16.7)	11 (18.3)
2001–2500	10 (13.9)	9 (15.0)
2501–≥3000	12 (16.7)	15 (25.0)
Money spent on food on monthly basis		
100–300	8 (11.1)	6 (10.0)
301–500	10 (13.9)	8 (13.3)
501–700	27 (37.5)	18 (30.0)
701–900	16 (22.2)	15 (25.0)
901–≥1500	11 (15.3)	13 (21.7)

$^*p < 0.05$

in the house, unemployment status of the woman and household income were predictors of poverty ($R^2 = 0.843$, $p = 0.000$, SEE $= 0.189$). The binomial and multinomial regression analyses showed no significant socio-economic and demographic predictors of food insecurity.

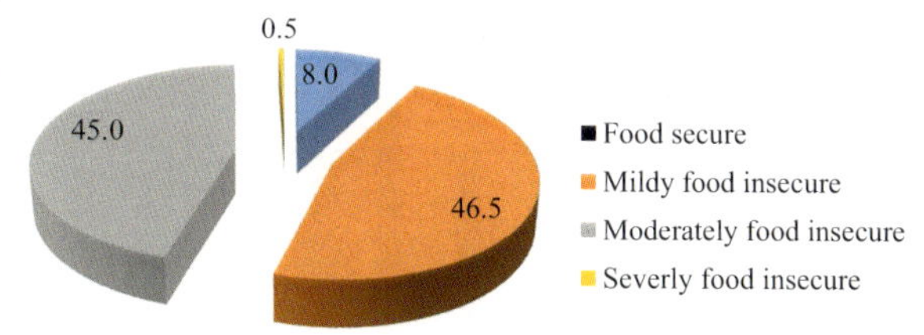

Fig. 1 Household food insecurity status ($n = 200$)

Discussion

South Africa has become a popular destination for refugees from other African nations [14]. Although South Africa is viewed as food secure at the country level [14], it is estimated that 26.6% of households are food insecure [23]. Little information is available about the food insecurity situation among refugees in South Africa, and the objective of this research was thus to contribute an understanding of the extent of food insecurity and to investigate the coping strategies and socio-economic predictors of food insecurity of women asylum seekers and refugees in Durban.

Despite the considerable progress in the political and economic status since 1994, poverty and unemployment are still major challenges faced by many people in the country [13], especially refugees who have to compete with South African citizens for the already scarce resources, and are often faced with difficult circumstances. The monthly household income of these women was ZAR1657, which was slightly more than the estimated average income of the poor in South Africa [13]. However, almost all the women in this study lived in poverty (< 1.9US$ per person per day—The World Bank [25]) and only about a third were employed despite having a relatively high literacy rate. It was thus not surprising that there was no statistically significant difference between the incomes of the food-secure and food-insecure households.

As expected, the prevalence of food insecurity among these women asylum seekers and refugees was very high (92%), which is much higher than the national prevalence of 26.6% food insecurity and 28.3% at risk of hunger [23]. These results are consistent with studies conducted among refugees from Liberia, West Africa and Somali in the USA [6, 11, 12]. Of the food insecure in this study, 46.5% of the women were categorised as mildly food insecure, 45.0% as moderately food insecure and 0.5% as severely food insecure. Food insecurity usually occurs when people do not have access or money to purchase

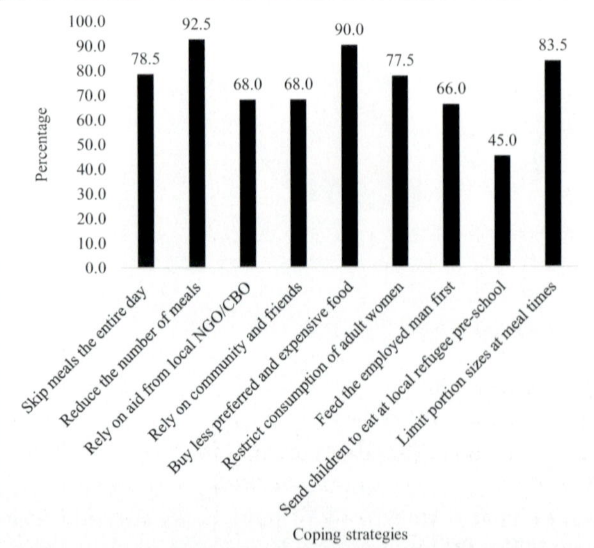

Fig. 2 Coping strategies at the household level ($n=200$)

food at all times [5, 21]. Food-insecure households are often at risk of food shortages [15, 19, 22]. In this study, 60.3% of the total household income was spent on food and the majority of the women indicated that they sometimes (68.3%), often (18.6%) or always (9.5%) experienced a food shortage, resulting in the majority of the women only consuming one meal per day. Furthermore, food shortages in the household led to the adoption of coping strategies that could adversely impact the household's food security situation in the short and long term [15, 19, 22]. Coping strategies were also adopted by this refugee community. Common household responses to food shortages include reduced food intakes, changes in types of foods served and adjusting the food budget [21]. These responses were also found in this study as the two coping strategies used by more than 90% of the women included reducing the number of meals and buying less preferred and less expensive foods. The aforementioned was also consistent with the finding that the majority of women indicated that they consumed only one meal a day. More than 75% indicated that they would limit portion sizes, restrict the consumption of the adults and go without food a whole day. The latter response was also found in this study in that 8.5% of the women indicated that they had not eaten the whole day before the survey was undertaken. More than 60% of the women relied on the community, friends and local NGOs for food. This is consistent with the findings from another study where refugees in Kenya reported that the informal social support from family, friends and others in the community was an important coping strategy [10]. All the coping strategies

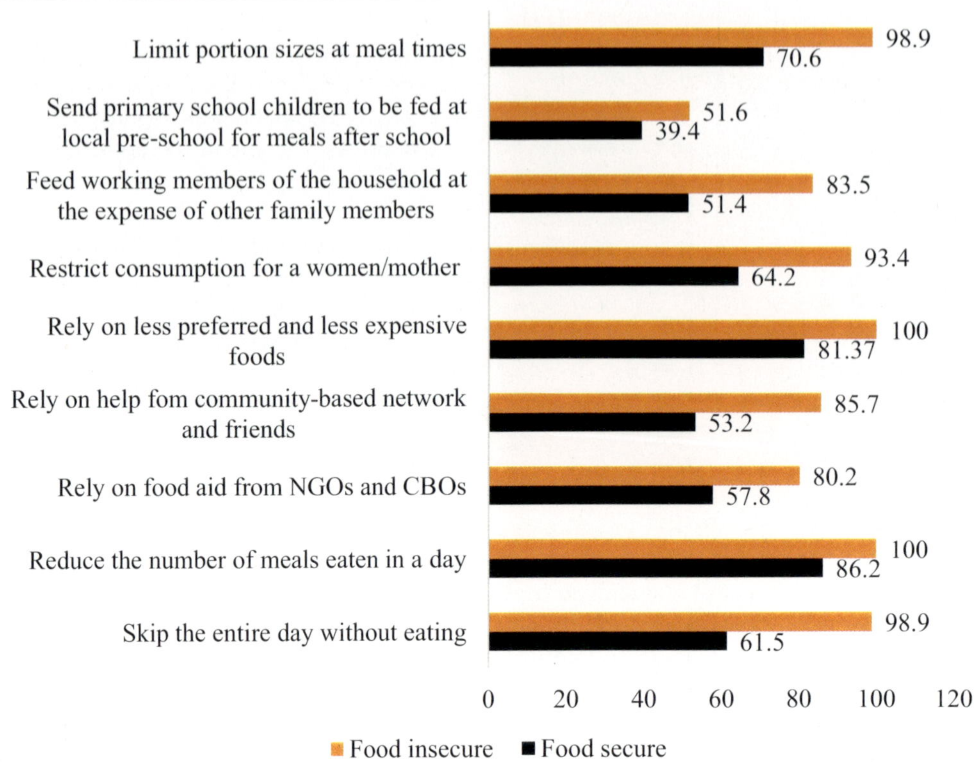

Fig. 3 Comparison of coping strategies between the food-secure and food-insecure groups ($n=200$)

Table 2 Significant correlations between the CSI and the amount of money available per person per day and various variables

Variables	Correlation *r*	Significance *p*
US$ per person per day (poverty indicator) with		
Number of people in house	−0.525	0.000
Total income	0.605	0.000
Money spent on food per month	0.321	0.000
Unemployment	−0.185	0.007
Frequency of purchasing food	0.168	0.010
Not enough food in the house	0.281	0.000
Education	0.159	0.012
Mother as head of the household	−0.143	0.031
Number of meals per day	0.172	0.004
Food insecurity (CSI) with		
Not enough food in the house	−0.162	0.004
Money spent on food	−0.130	0.021
Unemployment	0.157	0.012

were applied significantly ($p=0.000$) more in the food-insecure group than in the food-secure group, except for sending the children out to eat at the refugee pre-schools for which no significant difference ($p=0.083$) between the groups was observed. This finding suggests that the food-insecure women managed their food insecurity by significantly utilising coping strategies more frequently than their food-secure counterparts. A study in Kenya has found that the primary concerns of women refugees were mainly focused on their daily needs for food and shelter [10].

In examining the relationship of food insecurity to socio-economic variables, a significant positive relationship was observed between the CSI (as a proxy for food insecurity) and being unemployed, as well as a significant adverse relationship with available money to spend on food. Food access is thus severely compromised by unemployment and a lack of financial resources in this refugee community. Although these two factors were not found to be predictors of food insecurity, both unemployment and household income were predictors of poverty in this community. Although poverty has been identified as a risk factor for food insecurity [6], no significant relationship between poverty and food insecurity was found in this study.

A limitation of this study is the small sample size. The targeted sampling strategy may have introduced a selection bias, and it is thus not representative of all women asylum seekers and refugees in Durban. Another limitation may be that different measures were used to determine food security (10-item Radimer/Cornell hunger scale, HFIAS and CSI) in the various studies used in comparing our results. However, all these measures included coping strategies and, as previously mentioned, the CSI corroborates well with HFIAS among others [17, 20]. The coping strategies questionnaire used for this study was based on focus group results and was not pre-tested; thus, the internal consistency of the scale was not tested. Our findings can thus not be generalised to the larger refugee population in South Africa and may reflect on only those women asylum seekers and refugees who attend church and community meetings, and those that work as car guards in the Durban CBD. The results, however, represent a good first attempt at understanding household food insecurity and the coping strategies adopted to cope with this among women refugees and asylum seekers in South Africa. The results further highlight the need for more research among women refugees and asylum seekers, especially on the impact of food insecurity on health and nutritional status.

Conclusions

It was difficult to compare our results across other similar studies because of the paucity of data on the extent of food insecurity and coping strategies among refugees in South Africa. However, this study points to a serious problem of household food insecurity among black women asylum seekers and refugees. This is also true for the USA [6]. Larger households were associated with both poverty and food insecurity in this group of women. Poverty and food insecurity have both been identified as predictors of nutritional outcomes among adults [3]. Specifically, food insecurity has a detrimental effect on nutritional and health status as a number of studies have found that food insecurity is associated with poor dietary intake because of limited economic resources [5, 26, 30]. Food security refers to both adequate nourishment (intake) and adequate nutrition (quality of diet and utilisation) [9]. In this study, a number of coping strategies were employed that could adversely affect dietary intake and nutritional status. More research is thus needed to investigate how food insecurity affects the food choices and dietary intake as well as nutritional status of these women refugees and asylum seekers. More longitudinal research in South Africa and/or sub-Saharan Africa is also needed to explore the various socio-demographic variables that contribute to food and nutrition insecurity among refugees in order to plan and implement appropriate strategies to address food and nutrition insecurity effectively and efficiently among this group of vulnerable people.

Abbreviations

UNHCR: United Nations High Commission for Refugees; SA: South Africa; CoRMSA: Consortium for Refugees and Migrants in South Africa; FBO:

Faith-based organisations; NGOs: Non-governmental organisations; ACMS: African Centre for Migration and Society; CBD: Central Business District; DUT: Durban University of Technology; SPSS: Statistical Package for Social Sciences; SDs: Standard deviations; CSI: Coping strategy index; HFIS: Household Food Insecurity Scale; RCSI: Reduced coping strategy index; FCS: Food consumption score; SAFS: Self-assessed measure of food security; ANOVA: Analysis of variance; US$: US Dollar; ZAR: South African Rand; LBPL: Lower-bound poverty line; UBPL: Upper-bound poverty line; Stats SA: Statistics South Africa.

Authors' contributions

BM collected and captured the data; CN and WOT analysed and interpreted the data, and both contributed to writing the manuscript. All authors read and approved the final manuscript.

Authors' information

Professor Carin Napier is an Associate Professor in the Department of Food and Nutrition at the Durban University of Technology in South Africa (SA). She started her work career as a food service manager in community hospitals. Working in communities, and dealing with food and nutrition issues, convinced her to pursue a career in this field of nutrition. Working at an academic institution for the last 18 years has given her the ideal opportunity to practice food and nutrition in the context of educating students, doing research as well as uplifting communities served by the University she worked in. She has successfully supervised 23 Masters and 2 PhD students and is currently supervising 16 Masters and 4 PhD students. She has published 23 articles in accredited journals and presented papers at numerous international and national conferences. She is a registered Nutritionist with the Health Professions Council of South Africa. Her beliefs are that Nutritionists affiliated with academic institutions can impact on the nutritional health of the communities that they serve and sharing this information through publications and presenting it to policy makers. Planned interventions through postgraduate and independent studies can impact on the nutritional health of the communities. Her field of research is in line with national and international policies. Food and Nutrition security research can contribute towards achieving the Sustainable Development Goals, the SA National Development Plan, Food and Nutrition Security policy as well as the Integrated Nutrition Programme of the South African Department of Health. Dr. Wilna Oldewage-Theron is a Professor of Nutrition in the College of Human Sciences at Texas Tech University. She has eighteen years Community Nutrition research experience in Africa and is a National Research Foundation (NRF)-rated researcher in South Africa (SA). Her research interests include the factors contributing to household food insecurity and malnutrition in resource-poor communities where she has investigated the effect of various interventions, such as food fortification, supplementation, nutrition education, food product development and implementation, as well as school feeding programs on food insecurity, dietary diversity and nutritional status of women and children as well as the elderly. Her community research and development programme at present is mainly focused on the double burden of disease (micronutrient deficiencies and lifestyle diseases) as well as soy applications and the nutritional benefits of soy for human health in SA. She has also been involved in impact studies for the Council for Scientific and Industrial Research (SA), Joint Aid Management, the US Potato Board, and the World Initiative for Soy in Human Health (USA). Mrs Beryl Mbalenhle Makhaye was a postgraduate student in the Department of Food and Nutrition Consumer Science at the Durban University at the time of the study; she studied towards her Masters of Applied Science in Food and Nutrition. She collected the data in the refugee community in the Durban City Centre. She is an entrepreneur and manages her own business.

Author details

[1] Department of Food and Nutrition Consumer Sciences, Durban University of Technology, P. O. Box 1334, Durban 4000, South Africa. [2] Department of Nutritional Sciences, College of Human Sciences, Texas Tech University, P. O. Box 41270, Lubbock, TX 79409-1270, USA.

Acknowledgements

The authors would like to thank the Kwazulu-Natal Refugee Council, the participants and the fieldworkers who assisted with data collection. Sanjoy Saha at Texas Tech University assisted with the statistical analyses.

Competing interests

The authors declare that they have no competing interests.

Funding

No funding was received towards this study.

References

1. African Centre for Migration and Society (ACMS), University of Johannesburg. 2006. African Cities Study 2006. http://www.migration.org.za/projects. Accessed 03 Oct 2016.
2. Baiphethi MN, Jacobs PT. The contribution of subsistence farming to food security in South Africa. Agrekon. 2009;48(4):459–82.
3. Bhattacharya J, Currie J, Haider S. Poverty, food insecurity and nutritional outcomes in children and adults. J Health Econ. 2004;23(4):839–62.
4. Caraher M. Food policy and regeneration: food for thought. Local Work: Voice. Centre Local Econ Strateg. 2005;64:1–4.
5. Dharod JM, Croom J, Sady CG, Morrell D. Dietary intake, food security, and acculturation among Somali refugees in the United States: results of a pilot study. J Immigr Refug Stud. 2011;9(1):82–97.
6. Dharod JM, Croom JE, Sady CG. Food insecurity: its relationship to dietary intake and body weight among Somali refugee women in the United States. J Nutr Educ Behav. 2013;45(1):47–53.
7. Dibsdall LA, Lambert N, Frewer LJ. Using interpretative phenomenology to understand the food-related experiences and beliefs of a select group of low-income UK women. J Nutr Educ Behav. 2002;34(6):298–309.
8. Dibsdall LA, Lambert N, Bobbin RF, Frewer LJ. Low-income consumers' attitudes and behaviour towards access, availability and motivation to eat fruit and vegetables. Public Health Nutr. 2003;6(02):159–68.
9. Fanzo J. Strengthening the engagement of food and health systems to improve nutrition security: synthesis and overview of approaches to address malnutrition. Global Food Secur. 2014;3(3):183–92.
10. Gladden J. Coping strategies of Sudanese refugee women in Kakuma Refugee Camp, Kenya. Refug Surv Q. 2013;32(4):66–89.
11. Hadley C, Sellen D. Food security and child hunger among recently resettled Liberian refugees and asylum seekers: a pilot study. J Immigr Minor Health. 2006;8(4):369–75.
12. Hadley C, Zodhiates A, Sellen DW. Acculturation, economics and food insecurity among refugees resettled in the USA: a case study of West African refugees. Public Health Nutr. 2007;10(04):405–12.
13. Labadarios D, Mchiza ZJR, Steyn NP, Gericke G, Maunder EMW, Davids YD, Parker WA. Food security in South Africa: a review of national surveys. Bull World Health Organ. 2011;89(12):891–9.
14. Maharaj V, Tomita A, Thela L, Mhlongo M, Burns JK. Food insecurity and risk of depression among refugees and immigrants in South Africa. J Immigr Minor Health. 2016;19:1–7.
15. Maxwell D, Ahiadeke C, Levin C, Armar-Klemesu M, Zakariah S, Lamptey GM. Alternative food-security indicators: revisiting the frequency and severity of coping strategies. Food Policy. 1999;24(4):411–29.
16. Maxwell D, Watkins B, Wheeler R, Collins G. 2003. The coping strategies index: A tool for rapidly measuring food security and the impact of food aid programs in emergencies. Nairobi: CARE Eastern and Central Africa Regional Management Unit and the World Food Programme Vulnerability Assessment and Mapping Unit. http://www.fao.org/3/a-ae513e.pdf. Accessed 20 Mar 2017.
17. Maxwell D, Vaitla B, Coates J. How do indicators of household food insecurity measure up? An empirical comparison from Ethiopia. Food Policy. 2014;47:107–16.
18. Napier C. 2006. Evaluation of a feeding programme in addressing malnutrition in a primary school. DTech thesis. Vaal University of Technology.
19. Ngidi MS, Hendriks SL. Coping with food insecurity in rural South Africa: the case of Jozini, KwaZulu-Natal. Mediterr J Soc Sci. 2014;5(25):278–89.
20. Radimer CL, Olson CM, Greene JC, Campbell CC, Habicht JP. Understanding hunger and developing indicators to assess it in women and children. J Nutr Educ. 1992;24(suppl 1):36S–44S.
21. Seligman HK, Laraia BA, Kushel MB. Food insecurity is associated with chronic disease among low-income NHANES participants. J Nutr. 2009;140(2):304–10.
22. Shariff ZM, Khor GL. Household food insecurity and coping strategies in a poor rural community in Malaysia. Nutr Res Pract. 2008;2(1):26–34.
23. Shisana O, Labadarios D, Rehle T, Simbayi L, Zuma K. South African National Health and Nutrition Examination Survey (SANHANES-1). Cape Town: HSRC Press; 2013. p. 135–44.

24. Statistics South Africa (StatsSA). (2015). Methodological report on rebasing of national poverty lines and development of pilot provincial poverty lines. Technical report 03-10-11. Pretoria: Statistics South Africa.
25. The World Bank. 2016. Poverty & Equity Data. http://povertydata.worldbank.org/poverty/home. Accessed 08 Dec 2016.
26. Townsend MS, Peerson J, Love B, Achterberg C, Murphy SP. Food insecurity is positively related to overweight in women. J Nutr. 2001;131(6):1738–45.
27. United Nations Refugee Agency (UNHCR). (2011). Global report 2011. http://www.unhcr.org/gr11/index.xml. Accessed 01 Mar 2017.
28. United Nations Refugee Agency (UNHCR). (2012). South Africa. http://www.unhcr.org/51b1d637b.html. Accessed 01 Mar 2017.
29. United Nations Refugee Agency (UNHCR). (2015). UNHCR Statistics. The world in numbers. http://popstats.unhcr.org/en/overview#_ga=1.30122495.1963645507.1481038969. Accessed 01 March 2017.
30. Vozoris NT, Tarasuk VS. Household food insufficiency is associated with poorer health. J Nutr. 2003;133(1):120–6.
31. Vhurumuku, E. 2014. Food security indicators. http://www.fao.org/fileadmin/user_upload/food-security-capacity-building/docs/Nutrition/NairobiWorkshop/5.WFP_IndicatorsFSandNutIntegration.pdf. Accessed 15 Sep 2017.

Determinants of seasonal food security among smallholder farmers in south-central Uganda

Nataliya Apanovich[1*] and Robert E. Mazur[2]

Abstract

Background: Food security is threatened by deteriorating soil quality and declining crop yields in Uganda and throughout Africa where a majority of the population relies on small-scale farming for its livelihood. To understand these dynamics, this study investigates the relationships among soil quality, crop yields, socioeconomic factors (independent variables), and food security (dependent variable), measured as household meals per day, during the seasons of scarcity and plenty from interviews with 150 farmers in south-central Uganda.

Results: Using multivariate logistic regression, total acreage, and banana (*Musa × paradisiaca* L.) and bean (*Phaseolus vulgaris* L.) yields were positively associated with food security during the season of plenty (OR = 1.44, 1.86, and 2.21, respectively). During the period of scarcity, total acreage, and maize (*Zea mays* L.) and bean yields were associated with food security (OR = 1.70, 2.57, and 3.18, respectively), while off-farm income was associated with less food security (OR = 0.13).

Conclusions: The results indicate that land size and crop yields are more important for smallholder food security than soil quality in terms of direct relationships. We recommend that land fragmentation and low crop yields should be addressed to improve food security in south-central Uganda.

Keywords: Uganda, Crop yields, Soil quality, Meal consumption, Multivariate logistic regression

Background

'Food security exists when all people, at all times, have physical and economic access to sufficient, safe, and nutritious food that meets their dietary needs and food preferences for an active and healthy life' [1]. Food insecurity remains one of the greatest global development challenges with 815 million people (11% of the global population) being food insecure in 2016 [2], despite declining steadily since 1990–1992. Food insecurity is manifest through hunger and, consequently, malnourishment.

In sub-Saharan Africa (SSA), food insecurity constitutes a humanitarian crisis in which approximately 23% of the population is undernourished [2]. This situation reflects high annual population growth (2.5%) without a corresponding growth in food supply. Food insecurity in SSA is the outcome of multiple causal factors: socioeconomic (high population growth, poor infrastructure, slow economic development) [3], political (government instability, corruption, lack of political transparency) [4], and biophysical (poor or deteriorating soils, deforestation, erratic rainfall patterns, crop pests, and diseases) [5]. Agriculture plays a central role in food security, especially in SSA where most of the population depends on subsistence farming. However, agriculture's impact is dependent on the availability of good quality soils and household socioeconomic status. It is important, therefore, to incorporate natural and human resources in the analysis of food security.

This paper utilizes the sustainable livelihoods approach as an analytical tool to understand the association between capitals (assets) and food insecurity. Previous

*Correspondence: apanovichn1@gmail.com
[1] Graduate Program in Sustainable Agriculture, Iowa State University, Ames, IA 50011, USA
Full list of author information is available at the end of the article

studies on food security demonstrate the importance of some of these assets [6]. For example, human and financial assets (farm size, age, education, and income) are among the most common predictors of food security in SSA, as discussed below.

A study in Kwara State, Nigeria, conducted by [7] utilizing logistic regression found farm size, gross farm income, and total non-farm income to be significant predictors of rural household food security measured as daily per capita calorie consumption. Frongillo and Nanama [8] found economic status to be negatively correlated with food insecurity in northern Burkina Faso. Abafita and Kim [9] used two measures of household food security, a self-reported food security status and a multidimensional index generated through principal component analysis, to identify important food security determinants in Ethiopia. Their ordinary least squares (OLS) regression analysis found age, education, rainfall, livestock, off-farm activities, and soil conservation to be significantly and positively related to household food security.

Natural assets (land and soil), on the other hand, are less studied in the context of food security. Good quality soil, for example, is a key factor in agricultural productivity but is rarely analyzed as a food security predictor [10]. Depletion of soil fertility is a major biophysical cause of low per capita food production in Africa. Small-scale farmers often remove nutrients from the soil when harvesting crops and the associated biomass without adding inputs, thus creating a negative soil nutrient balance [11]. Mining of soil nutrients in Africa is estimated at annual depletion rates of 22 kg N, 2.5 kg P, and 15 kg K per hectare of cultivated land over the 30 years since 1975 [11, 12]. These practices contribute to declining productivity.

A series of experiments conducted on tropical soils determined that yield decline follows a curvilinear, negative exponential pattern with different soils showing different degrees of impact [13]. For example, Ferralsols (35% of the tropics and subtropics) have low resilience to soil nutrient loss, meaning that once vegetation is removed they degrade quickly [14]. However, research on cause–effect relationships linking soil quality degradation to food production is scarce, especially for soils of SSA [15]. According to [16], during the last 50 years the population of Uganda increased by a factor of five while food production increased by a factor of three; the latter involved doubling the food cropping area but only 50% increase in yields.

This paper examines household food security in relation to socioeconomic factors, soil quality, and crop yields. In this paper, soil quality refers to the ability of soil to sustain crop productivity. Food security, measured as the number of meals eaten per day at household level, has been assessed by other researchers [17–19]. Predictors of food security are determined using a multivariate logistic regression model [7–9]. The following hypothesis was developed based on the presented evidence: Socioeconomic status, soil quality, and crop yields are associated with increased meal consumption per day with seasonal variability.

Methods and data

Farmer interviews were conducted during September–October 2015 in Masaka District covering six sub-counties (Bukakata, Mukungwe, Buwunga, Kabonera, Kyanamukaka, and Kyesiga) and one division (Katwe-Butego). In total, 150 smallholder farmers were interviewed representing 92 villages and 30 parishes. One objective of this survey was to collect data on demographics, soil type and quality, soil management practices, farming practices, and off-farm income. Figure 1 shows the location of the households surveyed in Masaka.

Additional file 1: Table S1 shows the conversion of local units to their equivalent in kg as well as the time it takes for each crop to reach physiological maturity.

Study location

Masaka District is one of the most densely populated (245 people per square kilometer) in Uganda due to its favorable climate and proximity to Lake Victoria. Annual rainfall ranges from 1100 to 1200 mm during two rainy seasons, which allows for two growing seasons per year. Temperatures range between 10 and 30 °C [20]. Topography of the region is characterized by hills and ridges with an average altitude of 1150 m above the sea level. Masaka District has a total area of 1603 km^2, half of which is land and the other half is wetlands [20]. The dominant coffee–banana crop production system is supplemented with cassava (*Manihot esculenta* Crantz), sweet potatoes (*Ipomoea batatas* [L.] Lam), beans, groundnuts (*Arachis hypogaea* L.), and maize.

High population density and growth have led to agricultural land fragmentation, cropping intensification, deforestation, and decreased soil fertility. Despite the region's good climate and overall fertile soils, the quality of those soils has been deteriorating at a high rate. In Masaka District, soil degradation was estimated to affect 50% of the area [21]. The situation is exacerbated by insufficient soil conservation practices due to lack of knowledge, interest or incentives. This makes Masaka District susceptible to food insecurity and a good site to study the dynamics of soil quality and food production. An estimated 73% of all households in the district depend on farming for their livelihoods [20].

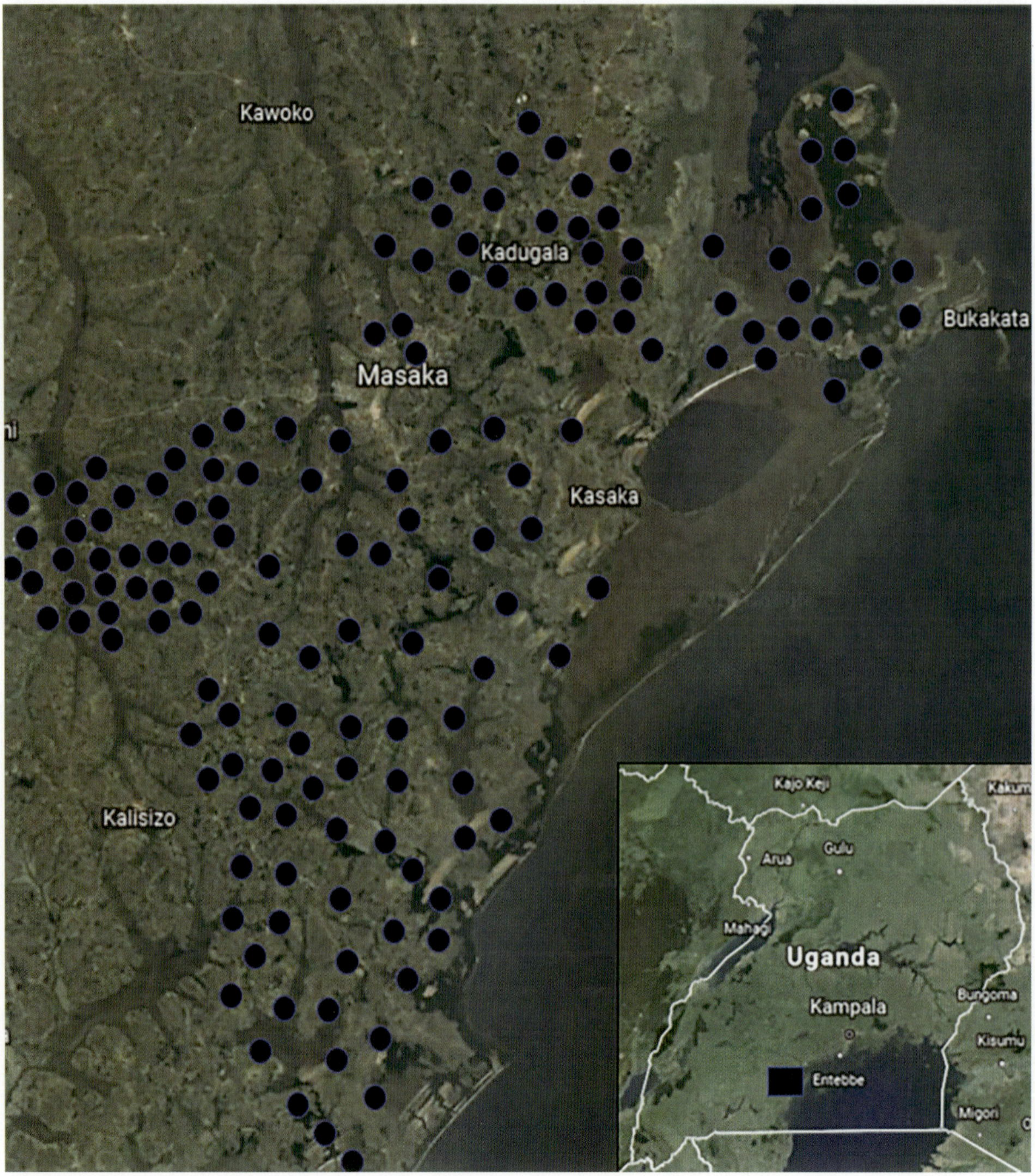

Fig. 1 Map of 150 sampled households in Masaka District, Uganda, 2015

Sampling

District-level administrative units consist of sub-counties, parishes, and villages. To choose 150 farmers, five villages were randomly chosen per parish. Then, one to six farmers were identified in each village with an average of two households per village. Farmers were identified from either farmer training records kept by local extension services or by village leaders. A list of all potential farmers (approximately 500 farmers) was created and then used to randomly select 150. In order to capture cross-village variation, the number of farmers surveyed per village was limited to a maximum of six.

Farmers interviewed represented a range of locations and farming systems. Interviews consisted of 22 questions covering agricultural production and food security topics. Interviews were conducted by two researchers. The majority (66%) of those interviewed were women age of 30 or older. Most farming households had four to seven members. Most farm sizes were between one to two hectares, with the average size of 1.61 ha. Seven or eight years of primary schooling was the most common educational level achieved. Only 32% of farmers surveyed had access to electricity.

Description of explanatory variables

The concept of food security is complex and includes availability, access, affordability, use, and stability of food [1, 22]. This study focuses primarily on food availability, considering a household 'food secure' when household members eat more than two full meals a day. The study captured meal consumption during the season of scarcity and plenty. The same measure was used by [23] to examine food availability on a seasonal basis, and [24] who found that inter-annual and intra-annual rainfall variability has a negative impact on food security. The 'season of plenty' refers to the period right after harvest, and it usually occurs from June to August, and December to February in Uganda. The 'season of scarcity' coincides with the periods of intensive agricultural production activities, which are from March to May, and September to November in Uganda.

Socioeconomic factors are commonly considered to explain variation in food security among poor households [25]. Bashir et al. [26] used household size, education, livestock, and income to understand variation in food security in rural Pakistan. Mengistu et al. [27] used marital status, household size, livestock, and income in their study. Omotesho et al. [7] used household size and farm size. The current study used four continuous variables (total amount of land to which a farmer has use rights, years lived in village, household size, and education) and two binary variables (off-farm income and access to electricity) as socioeconomic indicators. Crop production variables include soil quality and crop yields (coffee [*Coffea robusta* (L.) Linden], banana (*Musa × paradisiaca* L.), beans (*Phaseolus vulgaris* L.), and maize (*Zea mays* L.)).

Measurement and analysis

Eleven variables (household size, education, years lived in a village, total acreage, soil quality, banana yield, coffee yield, maize yield, beans yield, electricity, and off-farm income) are used in a multivariate logistic regression model for each season to determine which variables are associated with having three meals per day (compared to two or less). The soil quality variable has ordinal categories (responses: poor, fair, good, excellent) which reflect farmers' knowledge and terminology used when describing their soil. Farmers were asked to characterize their soil according to those ordinal categories and explain their characterization which was confirmed through observations of the soil and crops by the researcher and extension agent. This reduced subjectivity among farmers' responses. Soil type is a categorical variable that is based on farmers' indigenous nomenclature taking into account soil color and texture. Seven soil types are most common—*liddugavu* (black fertile soil), *luyinjayinja* (stony soil), *limyufumyufu* (red soil), *lusenyusenyu* (sandy soil), *lubumbabumba* (rich in clay soil), *kiwugankofu* (silty and dusty soil), *and kikusikusi* (brown soil). Crop yields were calculated based on farmers' reported acreage and harvested amount for each crop, which produced values that are not normally distributed, and hence, the natural logarithm of yields was used [28, 29]. Household size, education, years lived in a village, and total acreage are continuous measures.

Analyses were performed in R 3.4.0 using the stats package by the model fitting functions glm () [30]. The analysis employs multivariate logistic regressions to identify relationships among variables. Multicollinearity was not a problem since the variance inflation factor did not exceed the cutoff value of 2 for any of the independent variables. Only 84 observations were included in the first regression analysis because not all households grew all four crops chosen for investigation. All 150 observations were included in the second regression that analyzed soil quality and number of meals. The results are reported in odds ratios (OR), which are calculated by exponentiating the variables' coefficients. This measure tells the expected change in the odds ratio of meals per day for each unit increase in each variable. A logistic model provides a better fit to the data if it demonstrates an improvement over the intercept-only model (null model). An improvement over this baseline is examined by using the likelihood ratio test using lrtest () function from the lmtest package in R [30]. The H_0 holds that the null model is true and a P value for the overall model fit that is less than 0.05 provides evidence against the null model. Goodness-of-fit statistics assessed the fit of logistic models against actual outcomes by using the hoslem.test () function from the ResourceSelection package in R [30]. The Hosmer–Lemeshow test, an inferential goodness-of-fit test, yielded χ^2 of 6.748 and 10.516 for the plenty and scarcity models (Table 1), respectively, and was insignificant for both models ($P \geq 0.05$), suggesting that the models were fit to the data well.

Table 1 Coefficients and statistical significance for odds ratio (OR) of having three versus two or less meals per day during seasons of plenty and scarcity ($n = 84$)

Variables	Model	Coefficients	*P* value	OR
Soil quality (fair)	Plenty > 2	−0.081	0.901	0.923
	Scarcity > 2	−0.587	0.507	0.908
Soil quality (good)	Plenty > 2	−0.002	0.998	0.998
	Scarcity > 2	−0.899	0.536	1.704
Off-farm income**	Plenty > 2	−0.067	0.909	0.935
	Scarcity > 2	−2.053	0.033	0.128
Total acreage**	Plenty > 2	0.368	0.015	1.444
	Scarcity > 2	0.533	0.007	1.704
ln (banana yield)**	Plenty > 2	0.622	0.043	1.863
	Scarcity > 2	0.497	0.197	1.644
ln (coffee yield)	Plenty > 2	−0.077	0.636	0.926
	Scarcity > 2	−0.112	0.568	0.894
ln (maize yield)*	Plenty > 2	0.247	0.508	1.281
	Scarcity > 2	0.946	0.097	2.574
ln (bean yield)*	Plenty > 2	0.793	0.064	2.210
	Scarcity > 2	1.157	0.060	3.180
Years lived in village	Plenty > 2	−0.017	0.380	0.983
	Scarcity > 2	−0.007	0.792	0.993
Household size	Plenty > 2	−0.094	0.420	0.910
	Scarcity > 2	−0.048	0.759	0.953
Electricity	Plenty > 2	0.172	0.787	1.188
	Scarcity > 2	0.879	0.293	2.408
Education	Plenty > 2	−0.025	0.816	0.976
	Scarcity > 2	0.131	0.358	1.138
Test		χ^2	***df***	***P* value**
Overall model evaluation				
Likelihood ratio test	Plenty	21.462	1	0.029
	Scarcity	31.566	1	0.001
Goodness-of-fit test				
Hosmer and Lemeshow	Plenty	6.748	8	0.564
	Scarcity	13.763	8	0.088

**Significant at $P \leq 0.05$; *significant at 0.10

Results and discussion

The results indicate that total acreage, and banana, maize, and bean yields are positively associated with meals per day, while off-farm income is negatively associated. Even though soil quality was not significant, we chose to analyze it further. Soil quality represents a natural asset, and, due to limited research on the role of natural assets in food security, we decided it was appropriate to continue investigating this determinant. Additionally, soil quality is expected to play a bigger role in food security due to climate change, agricultural intensification, and increasing population; by continuing to analyze soil quality, we can assess its significance for food security in Uganda.

Food security

In both seasons, farmers' modal response was two meals per day. As Fig. 2 shows, during the period of scarcity, 20, 61, and 19% had one, two, and three meals, respectively. During the period of plenty, corresponding values were 7, 64, and 29%, respectively. Thus, the 'period of scarcity' is very aptly named for a significant proportion of farming households.

When considering both seasons, 13.5% of households have one meal per day. This number differs from the six percent reported by the World Food Programme in 2013 for Uganda's rural and urban populations [31]. This difference is likely due to higher meal consumption rates in urban areas. Indeed, according to the same

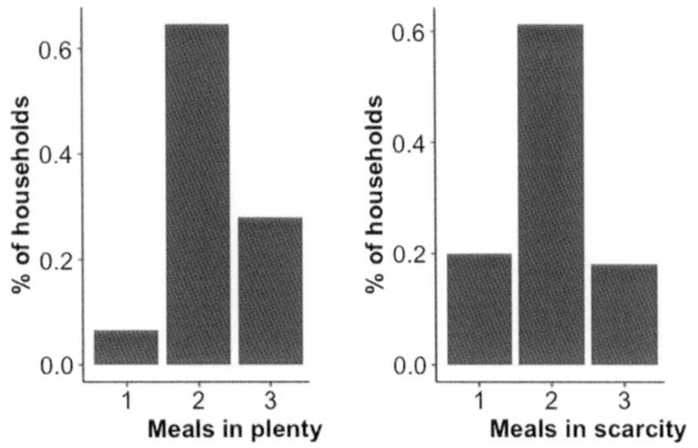

Fig. 2 Household number of meals per day in season of plenty and scarcity (pairwise comparison using Welch two sample t-test showed a significant difference in meal consumption between two seasons with the mean number of meals in plenty and scarcity of 2.23 and 1.99, respectively, $P \geq 0.001$)

study, in the country's capital city, Kampala, only five percent consumed one meal per day. Our results show that eating only one meal per day is much more common during the season of scarcity when food availability is less and costs are higher.

In the season of plenty, the odds ratio of having more than two meals increased with total acreage, and banana and bean yields. The significance of banana and bean yields in the season of plenty could be attributed to banana's importance as a staple energy food and to beans' role as the main source of protein. During the period of scarcity, the odds ratio of having more than two meals was negatively associated with off-farm income and positively associated with total acreage and maize and bean yields. The negative association might be explained by households diverting their food growing time to non-food related income earning activities, or by households with limited agricultural resources being forced to rely on non-agricultural activities yet earning income too low to ensure food security.

Five variables (off-farm income, total acreage, and banana, maize, and bean yields) are all associated with our measure of food security. In the season of plenty, each additional acre increases the probability of food security by 44%, and each unit increase in ln banana and ln bean yields is associated with greater probability (86% and 121% higher, respectively) of food security.

During the period of scarcity, each additional acre is associated with 70% higher probability of having more than two meals. During the same period, off-farm income lowers the probability by 87% while ln maize and ln bean yields are associated with 157% and 218% higher probability, respectively. Table 1 contains data regarding factors associated with the odds ratio of household members eating more than two meals per day during seasons of scarcity and plenty.

Total acreage and bean yield are the only variables that are significant in both seasons. Both variables are especially important in the season of scarcity when the probability of having more meals is greater than in the season of plenty. Land size was also important for food security in Kenya because larger and higher quality land endowments allow accumulation of cash and livestock, both of which are related to improved food security [32]. The significance of total acreage could also be related to total crop production. It was previously demonstrated that households with insufficient food availability obtained the majority of their energy from crops produced on-farm [33]. Beans are important in food security because it is a staple crop for dietary protein in East Africa [34]. In Uganda, common bean ranks third as the most widely grown crop and is the most important source of protein [35].

In Uganda, banana, maize, beans, and cassava are the staple food crops grown primarily for home consumption, and they are commonly intercropped. Coffee is the main cash crop for rural households and is most often intercropped with banana; beans and maize constitute another common intercropping system. Cassava is often sparingly grown in bean–maize systems. These farming practices obfuscate crop planting areas and yields. As many farmers operate under subsistence conditions, they tend to consume their crops during the growing season and sell them incrementally before, during, and after the harvest. These factors could contribute to the low or non-significance of crop yields data by making them spatially and temporally dispersed [36].

Banana, however, is the most important staple food of the central region where 'matooke,' mashed steamed green bananas, is eaten as often as three times a day. Banana 'plantations' are grown close to the house and often receive more household organic food scraps in addition to ash, which might contribute to improved soil quality and yields [37]. However, farmers have been reducing the size of their banana 'plantations' in the region due to increased pest and disease problems. Banana weevil, banana bacterial wilt, sigatoka and nematodes are the most common problems [38]. The role of banana in the diet of the central region, therefore, is expected to change in the future as farmers are slowly replacing banana plantations with new crops. These changes will influence farmers' food consumption patterns.

Off-farm income was another significant variable in the analysis. Off-farm income was negatively associated with the number of meals. This can be due to more time being diverted to non-food production activities.

According to the survey, 27% of the farmers spent their off-farm income on school fees, which further contributes to the idea of non-farm income earning activities competing with food production efforts. Similarly, some East African studies on the effects of off-farm income found decreased agricultural production with increased off-farm income [39]. Off-farm income was also found to be more important in households with higher food availability [33]. It may be that the households in this study are too food insecure and/or poor for off-farm income to make a difference in their food security status.

Another important characteristic in crop production is acreage prioritization according to crops. Figure 3 shows crop acreage distribution for coffee, banana, beans, maize, and cassava. As shown, smaller fields are associated with a greater variety of crops while coffee and banana start to dominate from 0.75 ha upwards. Figure 4 shows the crop acreage relationships for coffee and the four staple crops (banana, beans, maize, and cassava), with banana and cassava acreages being positively associated with coffee acreage. Coffee is favored relative to beans in large farms as shown by the smaller size of bean fields relative to coffee in farms with 1.5 ha or more of land. This can be due to either widespread pests and diseases or general preference for coffee over beans. A study by [40] found that farmers derived more income from coffee than from other crops such as beans and maize. Preference for coffee as a source of income could potentially lead farmers to allocate land to cash crops, leaving them with less land for growing staple crops. It also indicates that farmers grow a great diversity of crops on smaller acreages, which supports the claim that beyond a certain acreage coffee is grown as a monocrop. On the other hand, having large coffee acreage generates extra income which improves farmers' livelihoods. However, with the current trends of climate change and the rapid spread of pests and diseases in central Uganda, having large acreages under any one crop creates risk for farmers.

Relationship between soil type and quality

The majority of soils in Masaka District are Ferralsols, which are old and highly weathered [41]. These soils used to be quite fertile, but due to land use change and mismanagement they have become among of the poorest [42]. In the study area, the most common soil types identified by the farmers are black (Phaeozem) and red (Ferralsol). Figure 5 shows seven local soil types classified by the criteria that were identified by farmers and their perceived quality based on crop yields. The black soil was most often described as fair in quality, while the rest of the soils were perceived as fair or poor. Only a few

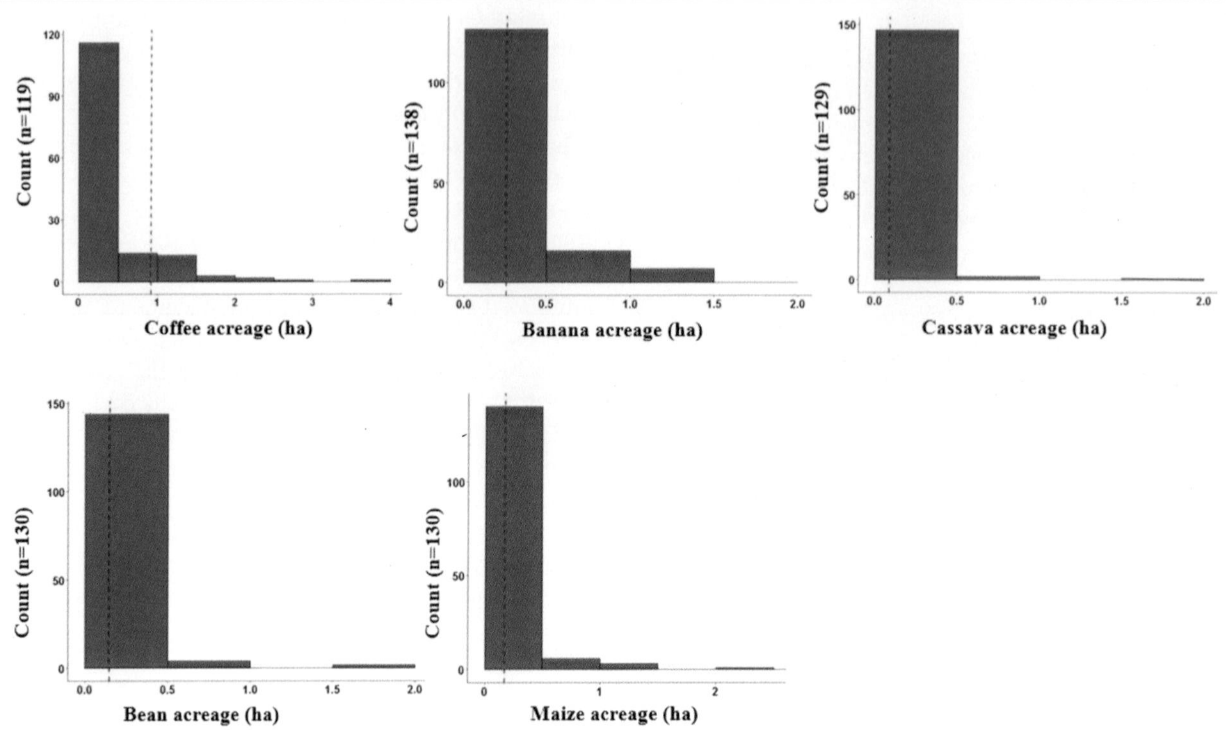

Fig. 3 Coffee, banana, beans, maize, and cassava crop acreage distributions shown as histograms with a mean (denoted by the red line)

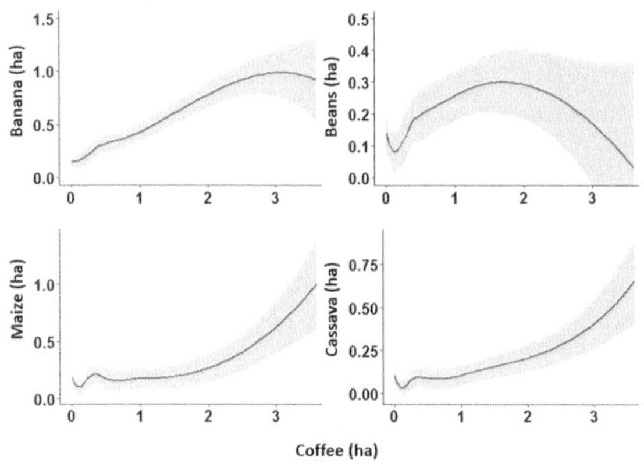

Fig. 4 Relationships between hectares of coffee and beans, maize, banana, and cassava expressed as a quadratic function with 0.90 confidence interval

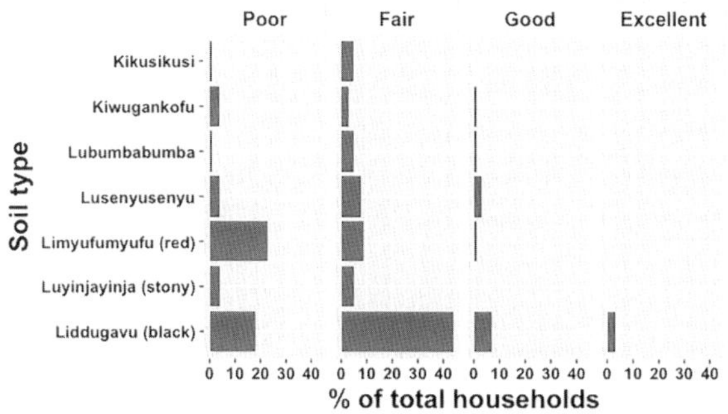

Fig. 5 Soil types according to local classification and their perceived quality

farmers reported their soil as excellent; therefore, this soil quality was excluded from the regression analysis.

Only 40.6% of farmers used a soil conservation practice, with no farmer using more than one. The most common soil management practice was digging of trenches (14.0%) followed by burying of weeds (6.3%), compost manure (4.9%), fallowing (4.2%), and green manure (2.8%).

Relationship between meals and soil quality

The primary analysis did not find soil quality to be significant, but we chose to look at it because it is often ignored in food security studies. Additionally, soil quality is expected to have a bigger impact on food security in SSA with climate change and growing populations. When the soil quality variable is analyzed in isolation, the regression analysis shows a direct relationship between soil quality and the number of meals consumed by a household. Table 2 presents regression coefficients and odds ratios for soil quality and three meals a day versus two meals or less. Soil quality reported as good is statistically significant in both seasons, thus indicating that there is

Table 2 Coefficients and statistical significance for odds ratio (OR) of having three versus two or less meals per day during seasons of plenty and scarcity based on soil quality ($n = 150$)

Variable	Season	Coefficient	*P* value	OR
Soil quality (fair)	Plenty	0.363	0.377	1.33
	Scarcity	0.369	0.461	1.27
Soil quality (good)**	Plenty	1.407	0.029	3.77
	Scarcity	1.750	0.017	5.04

**Significant at $P \leq 0.05$

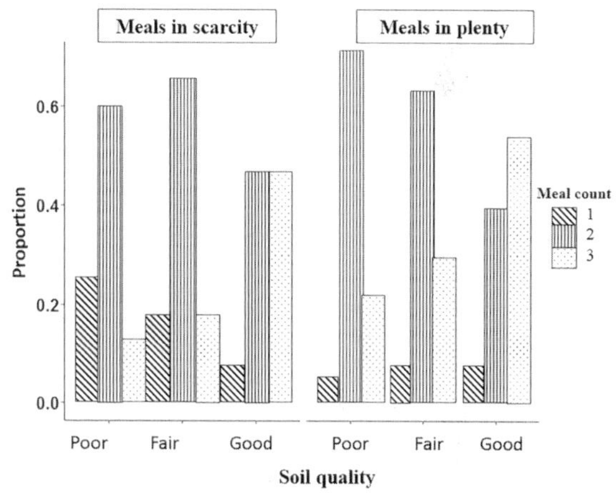

Fig. 6 Proportion of farmer household meals per day in the plenty and scarcity seasons grouped by soil quality types

a strong positive relationship between good soil quality and having three meals a day. However, because the primary analysis did not find soil quality to be significant and the second model is limited in scope, these results can only imply a positive association between soil quality and number of meals.

Most (80%) farmers who characterized their soil as either poor or fair were having two or fewer meals per day. Half (50%) of farmers with good soil quality had three meals per day. The odds ratio of having three meals per day is more than five times as high for soil characterized as good compared to fair during the season of scarcity, and nearly four times as high during the season of plenty, respectively. This indicates that households with poor soil quality experience greater food insecurity in the season of scarcity. Figure 6 shows the proportion of farmer household meals per day in seasons of plenty and of scarcity grouped by soil quality types.

Results only partially support the hypothesis because only socioeconomic and crop production determinants were found to be associated with increased meal consumption.

Limitations

The study acknowledges the limitations that are associated with measuring food security only by the number of meals eaten per day. The central Ugandan diet is rich in calories but generally lacks amino acids [31]. Additionally, by classifying a household as food insecure only because its members eat two or fewer meals a day could be an inaccurate representation of the reality. Some households eat only two meals a day but in large quantity which eliminates a need for more meals. Therefore, more indicators such as caloric and nutrient intake should ideally be measured when analyzing determinants of household food security status.

Crop yields were calculated from crop production and area planted data that were based on farmer recall. In some instances, farmers were not sure about the size of their land or their crop production. In these cases, the interviewer (with the help of a skilled extension agent) probed farmers to develop an estimate. However, this method produced data that are not normally distributed, and hence, yield transformation with natural logarithm was made. Direct field measurements would have produced more accurate values among farmers.

Soil quality was recorded based on farmer recall. Even though farmers generally know the quality of their soil, the rating of soil quality in a given field could vary among farmers for non-objective reasons. We operated under the assumption that all farmers used the same criteria to characterize their soil.

The survey was designed to gather relatively simple indicators that are comparable across villages and farming systems. Therefore, the information gathered on any one topic, such as food security, was not as in-depth as is possible in location-specific surveys.

Conclusions

This paper examined household food security in south-central Uganda by analyzing soil quality, crop production, and socioeconomic factors. Five factors were found significant: total acreage, off-farm income, and banana, maize, and bean yields. Total acreage and bean yield affect food security in both seasons unlike the other factors. The study also found that the majority of crops are grown on either poor or fair soils and that the number of meals is higher in both seasons for households with good quality soils compared to those with fair soil quality. The study also found that smaller farm size is associated with a greater variety of crops, while larger landholdings permit growing coffee as a monocrop.

The results indicate that land size and crop yields determine rural household meal consumption in south-central Uganda. However, the importance of soil quality in food security is expected to increase with the fast growing population. In the near future, we recommend that land fragmentation and low crop yields should be addressed to improve food security in rural Uganda. In the long term, we recommend further research on food security using the sustainable livelihoods approach which includes natural assets such as soil quality.

Abbreviations

FAO: Food and Agriculture Organization; SSA: sub-Saharan Africa; OR: odds ratio; OLS: ordinary least squares.

Authors' contributions

Both authors conceptualized the survey. NA designed the survey, analyzed data, and wrote the paper. RM edited and proofread the paper. Both authors read and approved the final manuscript.

Author details

[1] Graduate Program in Sustainable Agriculture, Iowa State University, Ames, IA 50011, USA. [2] Department of Sociology, Iowa State University, Ames, IA 50011, USA.

Acknowledgements

We thank Dr. Moses Tenywa from Makerere University for his assistance with data collection in Uganda. We also thank all farmers and extension officers from Masaka District who partook in this study.

Competing interests

The authors declare that they have no competing interests.

Funding

This work was supported by U.S. Agency for International Development (USAID) through the Feed the Future 21st Century Borlaug Leadership Program (8000066664 GS 129805/400-77-11).

References

1. FAO. Rome Declaration on World Food Security. World Food Summit, 13–17 of November 1996, Rome, Italy. 1996. http://www.fao.org/wfs/index_en.htm. Accessed from FAO Corporate Document Repository, 13 January 2015.
2. Food and Agriculture Organization of the United Nations (FAO). The state of food security and nutrition in the world: building resilience for peace and food security. 2017. http://www.fao.org/3/a-I7695e.pdf. Accessed 25 Dec 2017.
3. Babatunde RO, Omotesho OA, Sholotan OS. Socio-economic characteristics and food security status of farming households in Kwara State, North-Central Nigeria. Pak J Nutr. 2007;6(1):49–58.
4. Allouche J. The sustainability and resilience of global water and food systems: political analysis of the interplay between security, resource scarcity, political systems and global trade. Food Policy. 2011;36:S3–8.
5. Gregory PJ, Ingram JS, Brklacich M. Climate change and food security. Philos Trans R Soc B Biol Sci. 2005;360(1463):2139–48.
6. Alinovi L, D'errico M, Mane E, Romano D. Livelihoods strategies and household resilience to food insecurity: an empirical analysis to Kenya. In conference organized by the European Report of Development, Dakar, Senegal. 2010; p. 28–30.
7. Omotesho OA, Adewumi MO, Muhammad-Lawal A, Ayinde OE. Determinants of food security among the rural farming households in Kwara State, Nigeria. Afr J Gen Agric. 2006;2:1.

8. Frongillo EA, Nanama S. Development and validation of an experience-based measure of household food insecurity within and across seasons in northern Burkina Faso. J Nutr. 2006;136(5):1409S–19S.
9. Abafita J, Kim K. Determinants of household food security in rural Ethiopia: an empirical analysis. J Rural Dev. 2013;37(2):129–57.
10. D'Hose T, Cougnon M, De Vliegher A, Vandecasteele B, Viaene N, Cornelis W, et al. The positive relationship between soil quality and crop production: a case study on the effect of farm compost application. Appl Soil Ecol. 2014;75:189–98.
11. Sanchez PA. Soil fertility and hunger in Africa. Science. 2002;295(5562):2019–20.
12. Henao J, Baanante C. Agricultural production and soil nutrient mining in Africa: implications for resource conservation and policy development. Muscle Shoals: International Center for Soil Fertility and Agricultural Development; 2006.
13. Mermut AR, Ortiz Silla R, Faz Cano A. International Symposium on Sustainable Use and Management of Soils in Arid and Semiarid Regions. Sustainable use and management of soils: arid and semiarid regions. Reiskirchen: Catena; 2005.
14. Stocking MA. Tropical soils and food security: the next 50 years. Science. 2003;302(5649):1356–9.
15. Lal R. Soil degradation as a reason for inadequate human nutrition. Food Secur. 2009;1(1):45–57.
16. Leliveld AHM, Dietz AJ, Foeken DWJ, Klaver W, Akinyoade A, Smits HJ, et al. Agricultural dynamics and food security trends in Uganda. ASC Research Report; 2013.
17. Wiesmann D, Hoddinott J, Aberman NL, Ruel M. Review and validation of dietary diversity, food frequency, and other proxy indicators of household food security. Report submitted to the World Food Programme. Rome: International Food Policy Research Institute; 2006.
18. Hoddinott J, Yohannes Y. Dietary diversity as a food security indicator. Food Consumption and Nutrition Division of the International Food Policy Research Institute. Discussion paper no. 136. Washington, DC; 2002.
19. Ruel MT. Is dietary diversity an indicator of food security or dietary quality? A review of measurement issues and research needs. Food Nutr Bull. 2003;24(2):231–2.
20. Masaka District. Masaka District local government statistical abstract. Masaka, Uganda. 2012.
21. Berry L, Olson J, Campbell D. Assessing the extent, cost and impact of land degradation at the national level: findings and lessons learned from seven pilots. World Bank. 2003.
22. Ericksen P, Thornton PK, Notenbaert AMO, Cramer L, Jones PG, Herrero M. Mapping hotspots of climate change and food insecurity in the global tropics. CCAFS Report 5. Copenhagen: CCAFS; 2011.
23. Moreno-Black G, Somnasang P. In times of plenty and times of scarcity: nondomesticated food in northeastern Thailand. Ecol Food Nutr. 2000;38(6):563–86.
24. Generoso R. How do rainfall variability, food security and remittances interact? The case of rural Mali. Ecol Econ. 2015;114:188–98.
25. Carletto C, Zezza A, Banerjee R. Towards better measurement of household food security: harmonizing indicators and the role of household surveys. Global Food Secur. 2013;2(1):30–40.
26. Bashir MK, Schilizzi S, Pandit R. Impact of socio-economic characteristics of rural households on food security: the case of the Punjab, Pakistan. J Anim Plant Sci. 2013;23(2):611–8.
27. Mengistu E, Regassa N, Yusufe A. The levels, determinants and coping mechanisms of food insecure households in Southern Ethiopia: case study of Sidama, Wolaita and Guraghe Zones. Dry Lands Coordination Group (DLCG); 2009.
28. Lobell DB, Bänziger M, Magorokosho C, Vivek B. Nonlinear heat effects on African maize as evidenced by historical yield trials. Nat Clim Change. 2011;1(1):42.
29. Evenson RE, Mwabu G. The effect of agricultural extension on farm yields in Kenya. Afr Dev Rev. 2001;13(1):1–23.
30. R Core Team. R: a language and environment for statistical computing. R Foundation for Statistical Computing, Vienna, Austria. 2013. URL http://www.R-project.org/. Accessed 10 Nov 2016.
31. McKinney P. Comprehensive food security and vulnerability analysis: Uganda. Kampala: WFP; 2009. https://doi.org/10.1002/yd.20093.
32. Stephens EC, Nicholson CF, Brown DR, Parsons D, Barrett CB, Lehmann J, et al. Modeling the impact of natural resource-based poverty traps on food security in Kenya: The Crops, Livestock and Soils in Smallholder Economic Systems (CLASSES) model. Food Secur. 2012;4(3):423–39.
33. Frelat R, Lopez-Ridaura S, Giller KE, Herrero M, Douxchamps S, Djurfeldt AA, et al. Drivers of household food availability in sub-Saharan Africa based on big data from small farms. Proc Natl Acad Sci. 2016;113(2):458–63.
34. Kweka SO, Ndakidemi PA, Mushi CS, Nkonya E, David S. Adoption of Lyamungu 85 bean variety in the medium-altitude zone of Tanzania. In: PABRA millennium workshop. 2005;147.
35. Haggblade S, Dewina R. Staple food prices in Uganda. In: Prepared for the Comesa policy seminar on "Variation in staple food prices: causes, consequence, and policy options," Maputo, Mozambique. 2010; p. 25–26.
36. Fermont A, Benson T. Estimating yield of food crops grown by smallholder farmers. Washington, DC: International Food Policy Research Institute; 2011. p. 1–68.
37. Van Asten PJA, Gold CS, Okech SHO, Gaidashova SV, Tushemereirwe WK, De Waele D. Soil quality problems in east African banana systems and their relation with other yield loss factors. InfoMusa. 2004;13:20–5.
38. Ochola D, Jogo W, Ocimati W, Rietveld A, Tinzaara W, Karamura DA, Karamura EB. Farmers' awareness and perceived benefits of agro-ecological intensification practices in banana systems in Uganda. Afr J Biotechnol. 2013;12(29):4603–13.
39. Holden S, Shiferaw B, Pender J. Non-farm income, household welfare, and sustainable land management in a less-favoured area in the Ethiopian highlands. Food Policy. 2004;29(4):369–92.
40. Bagamba F, Senyonga JW, Tushamereirwe WK, Gold CS. Performance and profitability of the banana subsector in Uganda farming systems. Banana Food Secur. 1998;1998:729–39.
41. Esilaba AO, Nyende P, Nalukenge G, Byalebeka JB, Delve RJ, Ssali H. Resource flows and nutrient balances for crop and animal production in smallholder farming systems in eastern Uganda. Agric Ecosyst Environ. 2005;109(3–4):192–201.
42. Pender J, Nkonya E, Jagger P, Sserunkuuma D, Ssali H. Strategies to increase agricultural productivity and reduce land degradation: evidence from Uganda. Agric Econ. 2004;31(2–3):181–95.

Achieving sustainable development goals on no poverty and zero hunger: Does technical efficiency of Ghana's maize farmers matter?

Camillus Abawiera Wongnaa* and Dadson Awunyo-Vitor

Abstract

Background: Whether or not maize farmers are technically efficient determines their choice of productivity improvement strategy and very important for achieving the sustainable development goals on no poverty and zero hunger. This study examined technical efficiency and its determinants of Ghana's maize farmers. Analysis of the determinants is crucial to devising strategies aimed at improving technical efficiencies of the farmers as well as ensuring no poverty and zero hunger.

Methods: The study used cross-sectional data collected from 576 maize farmers in the four main agro-ecological zones of Ghana with the aid of structured questionnaire. Descriptive statistics and the stochastic frontier production function were the methods of analysis used.

Results: The results showed that the mean technical efficiency estimate for maize farmers in Ghana was 58.1%. They also revealed that an increase in educational level, maize farming experience, extension contact as well as uses of fertilizer and improved seeds would increase the technical efficiency of maize producers in Ghana. Similarly, male Ghanaian maize farmers were more technically efficient than female farmers. Furthermore, membership of a farmer association will increase their technical efficiencies. Finally, an increase in farm size and land fragmentation will decrease technical efficiency of the maize farmers.

Conclusion: Efficiency improvement programmes by stakeholders in the maize industry could target literate maize farmers since education enhances the managerial and technical skills of farmers and also increase farmers' ability to utilize existing technologies. Ministry of Food and Agriculture (MOFA) could also liaise with Ministry of Education to provide maize farmers with no formal education with special training prior to introducing new efficiency enhancing programmes to them. MOFA should analyse the problems extension officers face in the discharge of their duties to pave the way for the provision of appropriate incentives to extension officers by stakeholders in the maize industry.

Keywords: Determinants, Maize, Technical efficiency, Stochastic frontier analysis

Background

The importance of the contribution of maize production to economic growth, poverty alleviation, food security, gross domestic product, employment and incomes in Ghana cannot be overemphasized. For most rural households in Ghana, maize is the main staple food and an ideal crop whose increased productivity has the potential to support Ghana's strive to meeting the sustainable development goals on no poverty and zero hunger. However, there has been an unstable trend in the country's maize production over the last two decades, which threatens household food security and income sources [27]. Because maize production is critical for the overall process of social and economic development of Ghana, it is crucial that the country takes steps, especially through research, to help grow and develop the maize sector. A very important factor that will help sustain an increase in maize production is an increase in its productivity driven

*Correspondence: wongnaaa@yahoo.com
Department of Agricultural Economics, Agribusiness and Extension, Kwame Nkrumah University of Science and Technology, Private Mail Bag, University Post Office, Kumasi, Ghana

by increase in economic efficiency of its producers [6]. Economic efficiency is incomplete if technical efficiency is not studied [19]. The farmer's ability to achieve maximum output with available resources is referred to as technical efficiency. Whether or not maize farmers are technically efficient determines their choice of productivity improvement strategy and very important for achieving the sustainable development goals on no poverty and zero hunger. This is because, if they are technically efficient, then new inputs or production technologies will be required to shift the production curve or frontier upward. Conversely, if their productivities are below potential yields, then their technical efficiencies could be improved by dealing with the causes of technical inefficiency [32].

Various studies have analysed technical efficiency of Ghana's maize farmers even though the scopes of these studies did not cover the entire nation like the current study and in most of these studies, the stochastic frontier analysis procedure was employed as the method of analysis. In spite of the drivers of technical efficiency identified in previous studies, maize farmers continue to experience low yields [27] depicting the failure of recommendations of previous studies to help address the productivity and efficiency challenges farmers have faced over the years. By way of literature review, [2] employed the stochastic frontier analysis procedure and estimated the technical efficiency of maize farmers in Northern Ghana to be 74%. The study concluded that their technical efficiencies were positively influenced by agricultural mechanization, experience and gender. Sienso et al. [38], studying the technical efficiency of maize production in Nkoranza district of Brong-Ahafo Region of Ghana using the stochastic frontier analysis procedure, found the technical efficiency of the farmers to be 91%. In this study, variety of maize cultivated, male gender, farming experience and number of extension visits were found to influence technical efficiency of the farmers. Also, using the stochastic frontier analysis procedure, [24] reported a technical efficiency of 51% for maize farmers in Ghana's eastern region. The findings from the inefficiency model of this study also showed that technical efficiencies of the farmers were determined by contact with agricultural extension workers, membership of a farmer association, number of meetings held by association, formal education and use of credit. Bempomaa and Acquah [13] also employed the stochastic frontier analysis procedure and reported a mean technical efficiency of 67% for maize farmers in the Ejura/Sekyedumase district of Ashanti region of Ghana. According to the study, the technical efficiencies of the farmers were influenced by their ages, male gender as well as off-farm activities. In addition, [3] estimated the technical efficiency of maize farmers across three agro-ecological zones of Ghana using the stochastic frontier analysis procedure to be 64.1%. The results of this study also revealed that their technical efficiencies were positively influenced by male gender, contact with agricultural extension officers, farmer's age, access to credit, whether or not farmer owned land and mono-cropping. Furthermore, in analysing the technical efficiency and production risk in Ghana's maize production, [31] estimated the technical efficiency of the farmers to be 62% with the combined farm specific factors explaining technical efficiency variations. Finally, [20] obtained a mean technical efficiency of 58% for maize farmers in the Mfantseman Municipality of Ghana by the stochastic frontier analysis approach. According to the study, the mean technical efficiency was influenced by size of household, male gender, years of schooling, off-farm income as well as farmer's age.

In addition to the variables considered in the aforementioned studies, other variables including land fragmentation, access to ready market, use of fertilizer, pesticides and improved seeds as well as farming in particular agro-ecological zones may also influence technical efficiency. This study analyses Ghana's maize farm level technical efficiency variations caused by the aforementioned factors. The study provides a nationwide analysis of technical efficiency of Ghana's maize farmers by using data from the Northern Savannah, Transitional, Forest and Coastal Savannah zones. This is important because existing similar studies [2, 13, 20, 24, 31, 38] used data that were restricted to only certain districts, municipalities or agro-ecological zones which are not very representative of the nation. Also, it has been argued that technical efficiency of maize farmers might differ depending on their respective agro-ecological zones as they might be using different technologies. The current study provides evidence of same or similar technologies used by maize farmers across Ghana's agro-ecological zones and therefore employs the stochastic production frontier instead of the metafrontier.

Methods

Study area and data collection

The study was conducted in Ghana's four main agro-ecological zones, namely Northern Savannah, Transitional, Forest and Coastal Savannah zones. The Northern Savannah zone comprises the three northern regions, namely Northern, Upper West and Upper East regions. The Transitional zone is made up of the Brong-Ahafo region and the northern part of Ashanti region. The Forest zone consists of the middle and central parts of Ashanti region as well as the Eastern and Western regions. Finally, the

Coastal Savannah zone comprises the coastal parts of Greater Accra, Central and Volta regions of Ghana.

The study used primary cross-sectional data which was collected from 576 maize farmers with the aid of a structured questionnaire. Generally, the questionnaire comprised four sections. The first section consisted of identification of the enumerator, the respondent as well as the district and operational area of the respondent. The second section included questions on maize producers' personal and household characteristics. The third section dealt with questions on inputs or resources employed in maize production, while the fourth section posed questions on the output of maize and its marketing. Basically, the study employed multi-stage sampling technique. In stage one (1) of the sampling design, two districts/municipalities were purposively selected from each zone based on the level of maize production [27]. The selected districts/municipalities were East Gonja and West Mamprusi (Northern Savannah zone), Nkoranza and Ejura Sekyedumase (Transitional zone), Fanteakwa and Sekyere South (Forest zone) and Gomoa and Ketu (Coastal Savannah zone). In the second stage, nine (9) villages or communities were randomly selected from each district/municipality. Stage three involved random sampling of eight (8) maize farmers from lists of maize farmers obtained from agricultural extension agents operating in the various villages/communities.

Analytical framework

The study employed descriptive statistics in analysis of socio-economic characteristics of maize farmers in Ghana. The stochastic frontier analysis approach was also employed in analysing the technical efficiency of the farmers. The function was independently proposed and defined by [5, 26] as;

$$y_i = f(x_i;\, \beta)\exp^{v_i - u_i} \text{ where } i = 1, 2, \ldots, N \quad (1)$$

where y_i is output of the ith farmer; $f(x_i;\, \beta)$ is a suitable production function, x_i is a vector of farm inputs, β is a vector of parameters to be estimated, and N is the sample size. v_i, which represents a random error, has zero mean and is assumed to be symmetric and independently distributed as $N(0;\, \sigma^2 v,)$ and is particularly related to random factors which cannot be controlled by the farmer such as measurement errors and weather factors. u_i, which is *non-negative* truncated half normal, random variable, $N(0;\, \sigma^2 v,)$, is linked to the technical inefficiency of the farmer.

According to [37], technical efficiency of a farm firm is calculated using the ratio of observed output to the potential or frontier output. That is,

$$\text{Technical efficiency} = \text{TE}_i = \frac{y_i}{y_i^*} = \frac{f(x_i;\, \beta)\exp^{v_i - u_i}}{f(x_i;\, \beta)\exp^{v_i}} = \exp^{-u_i} \quad (2)$$

where y_i is the highest or maximum predicted output for the ith farm and y_i^* is the potential or frontier output. In employing the stochastic frontier analysis approach, [16] present two estimation procedures. In the first procedure, in the absence of a specified clear distribution of the efficiency component, stochastic corrected ordinary least squares (COLS) is appropriate. On the other hand, maximum likelihood estimation (MLE) technique is employed as the estimation procedure if a clear distribution is specified. MLE is more efficient than COLS because it uses the specific distribution of the error term [22]. The stochastic frontier and the inefficiency model are jointly estimated using frontier computing packages such as FRONTIER, STATA and Limdep which apply MLE [11, 21]. This study used STATA.

The selection of functional form is essential in employing the stochastic frontier analysis approach. The Cobb–Douglas production function introduces severe restriction on the technology employed in the farm by limiting the elasticities of production to be fixed and the input substitution elasticities to unity [43]. This is, however, not the case with the translog functional form. Diagnostically, to determine which functional form is more appropriate for the data collected, the generalized likelihood ratio test was employed as follows:

$$\lambda = -2\left[\ln \frac{L(H_0)}{L(H_1)}\right] = -2[\ln L(H_0) - \ln L(H_1)] \quad (3)$$

where $L(H_0)$ = value of the log-likelihood of the adopted model, $L(H_1)$ = value of the log-likelihood of the restricted model.

λ has a Chi-square distribution with degree of freedom equal to number of restrictions. The main hypothesis tested is whether or not the Cobb–Douglas functional form fits the data collected in the study, given the translog functional form. Cobb–Douglas functional form was rejected in favour of the translog functional form. After settling on the ideal functional form, separate stochastic production frontier (SPF) models were defined for maize farmers in different agro-ecological zones of Ghana. After the estimation of the separate stochastic production frontier (SPF) models, it was necessary to verify if maize farmers in the various agro-ecological zones shared the same technology. This was also done using a likelihood ratio (LR) test, where $L(H_0)$ is the value of the log-likelihood function for a stochastic production frontier estimated by pooling the data for all agro-ecological zones and $L(H_A)$ is the sum of the values of the log-likelihood functions from the individual stochastic production frontiers. The degree of freedom for the Chi-square statistic is the difference between the number of parameters estimated under H_A and H_0. If the null hypothesis that the stochastic frontier for the pooled data exists is rejected in favour of the individual frontiers, then the data should not be pooled and in such a case, the metafrontier is the appropriate framework to estimate and compare technical efficiency across agro-ecological zones [12]. Empirically, the stochastic frontier translog production function is specified as:

$$\ln y_i = \beta_0 + \sum_{k=1}^{m} \beta_k \ln x_{ki} + \frac{1}{2}\sum_{k=1}^{m}\sum_{j=1}^{m} \beta_{kj} \ln x_{ki} \ln x_{ji} + v_i - u_i \quad (4)$$

where y_i=total quantity of output (kg), x_i=vector of inputs which includes *SED*=quantity of seed (kg), *LANDSZ*=farm size (ha), *LAB*=quantity of labour (Man-days), *CAP*=capital (depreciated charges on farm tools and implements), *FET*=quantity of fertilizer (kg), *MAN*=quantity of manure (kg), *PET*=quantity of pesticides (litres), *HEB*=quantity of herbicides (litres), m=number of production inputs, ij=positive integers $(i \neq j)$, $\beta's$=vector of parameters to be estimated, v_i and u_i have their usual meanings. Empirically, the inefficiency model is also specified as:

$$u_i = \delta_0 + \sum_{m=1}^{N} \delta_m z_i \quad (5)$$

where z_i=vector of farmer characteristics thought to influence technical efficiency which includes *ROAD*=access to good roads, measured as a dummy (1 for access to good road and 0 otherwise), *SEX*=gender of maize farmer, measured as a dummy (1 for male and 0 for female), *AGE*=age of maize farmer, measured in years, *EDU*=educational level of maize farmer, measured in years of schooling, *HOSIZE*=household size, measured as number of family members living with maize farmer. *EXP*=maize farming experience, measured in number of years in maize farming, *MGROUP*=membership of a farmer association, measured as a dummy (1 for membership of an association and 0 otherwise), *CREDIT*=access to credit, measured as a dummy (1 for access to credit and 0 otherwise), *INCOME*=previous year's maize income, measured in Ghana Cedis, *NPLOTS*=land fragmentation, measured as a dummy (1 for owning more than one plot and 0 otherwise), *NOEXTVI*=extension contact, measured in number of meetings of maize farmer with agricultural extension agents, *REDYMKT*=access to ready maize market, measured as a dummy (1 for available maize market and 0 otherwise), *FERTus*=use of inorganic fertilizer, measured as a dummy (1 for use of inorganic fertilizer and 0 otherwise), *PESTus*=use of pesticides, measured as a dummy (1 for use of pesticides and 0 otherwise), *SEDtyp*=seed variety planted by maize farmer, measured as a dummy (1 for improved variety and 0 for traditional variety), *NOSAV*=living in the Northern Savannah zone, measured as a dummy (1 for living in Northern Savannah zone and 0 for living in the Coastal Savannah zone), *TRASIT*=living in the Transitional zone, measured as a dummy (1 for living in Transitional zone and 0 for living in the Coastal Savannah zone), *FOREST*=living in the Forest zone, measured as a dummy (1 for living in Forest zone and 0 for living in the Coastal Savannah zone), δ=vector of parameters to be estimated. The parameters in the stochastic frontier production function and those of the inefficiency model can be jointly estimated in STATA [21]. The joint model generates variance parameters, i.e., $\lambda = \sigma_u/\sigma_v$, variance of the model, σ, variance of the stochastic model, σ_v^2, and variance of the inefficiency model, σ_u^2. Breusch Pagan statistic and variance inflation factor (VIF) were employed in testing for heteroscedasticity and multicollinearity, respectively. Also, the partial input elasticities can be calculated by taking the partial derivative of y_i with respect to x_{ki} in Eq. (4) and substituting the sample input means. This, according to [1, 45], is specified as:

$$e_j = \frac{\partial \ln y_i}{\partial x_{ki}} = \beta_k + \sum_{j=1}^{m-1} \beta_{kj} \ln x_j + 2\beta_{kk} \ln x_k \quad (6)$$

Results and discussion

Socio-economic characteristics of the respondents

The results presented in Table 1 showed that most of the respondents (77.4%) were males, an indication of the popularity of males in Ghana's maize production than females. It could also be inferred from this study that both men and women could take maize production as a business and a source of employment. With majority of the farmers with ages within 18–45 years' bracket (56.9%), the mean age was 45.15 years (Table 2). The implication was that the farmers were old. According to studies conducted by [20], [13] as well as [3], old age of farmers could adversely affect their technical efficiencies. The results in Tables 1 and 2 also showed that generally, majority of the farmers who took part in the survey (64.1%) received at least six (6) years of formal education and this could help them operate close to the production frontier [20, 24, 31]

Table 1 Socio-economic characteristics of respondents. *Source*: Survey, 2015

Variable	Frequency	%
Ready market last year		
No	70	12.2
Yes	506	87.8
Total	576	100
Gender		
Male	446	77.4
Female	130	22.6
Total	576	100
Age group of farmers (years)		
18–45	328	56.9
46–60	180	31.2
Greater than 60	68	11.8
Total	576	100
Educational level of maize farmer		
No formal education	207	35.9
Primary school	84	14.6
JHS/JSS/Middle school	200	34.7
SHS/SSS	69	12
Tertiary	16	2.8
Total	576	100
Number of plots		
One plot	454	78.8
More than one plot	122	21.2
Total	576	100
Group membership		
No	436	75.7
Yes	140	24.3
Total	576	100
Access to extension		
No	331	57.5
Yes	245	42.5
Total	576	100
Access to credit		
No	475	82.5
Yes	101	17.5
Total	576	100
Access to good roads		
Bad (tarred with potholes/rough and marshy)	292	50.7
Good (asphalt/tarred but not asphalt/rough and smooth)	284	49.3
Fertilizer use		
No	207	35.9
Yes	369	64.1
Total	576	100
Pesticides use		
No	560	97.2
Yes	16	2.8
Seed variety planted		
Traditional	346	60.1
Improved	230	39.9
Total	576	100
Land fragmentation		
Owned more than one plot	223	38.7
Owned only one plot	353	61.3
Total	576	100

Table 2 Descriptive statistics of quantities of inputs and output of respondents. *Source*: Survey, 2015

Variable	Min	Max	Mean	Std Dev
Size of farm (ha)	0.22	69.81	2.92	12.7
Number of farm plots	1.00	7.000	1.53	0.78
Quantity of fertilizer (kg)	0.01	1114	211	143
Quantity of herbicide (L)	0.01	88.14	11.4	13.1
Quantity of pesticide (L)	0.02	36.21	9.13	5.21
Number of extension visits/year	0.00	7.000	2.74	4.81
Quantity of labour (man-days)	2.18	8107	104.0	94.5
Quantity of manure (kg)	0.01	1019	99.3	154
Capital (depreciation)	40.0	2499.9	558	766
Quantity of output (Mt)	0.01	10.414	4.874	5.88

With a high mean household size of 7.61 (Table 2), the labour available to the farmers could be enhanced, thereby making farmers carry out husbandry practices on time which could help improve their technical efficiencies [20]. Generally, the respondent maize farmers had 14 years of experience in maize farming. With high experience levels, it will not be surprising if farmers' technical efficiencies are high given that such farmers are supplied with the required productivity enhancing technologies [2, 38]. Also presented in Table 1 is that 57.5% of the farmers in the sample had no access to extension service, and for those who received extension service, extension officers visited them thrice (Table 2), indicating poor services received by the farmers. This could cause little or no use of productivity enhancing technologies since farmers will likely be unaware of these technologies because those who are supposed to disseminate the technologies to them (i.e., extension agents) are far from them. The results could be that the technical efficiency of the farmers would be relatively low [24, 38]. Also worthy of note is the fact that as high as 75.7% of the respondents were not members of any farmer group (Table 1). This could increase technical inefficiency of the farmers since most extension agents disseminate agricultural technologies through farmer groups [7, 14, 28]. In fact, most of the respondents (82.5%) too received no credit from any source and this could stop or delay application of key production inputs to crops since farmers will have no money to purchase the inputs, thereby making them technically inefficient [3, 24].

The findings of the current study show that Ghana's maize production is basically on a small scale as the average farm size was estimated to be 2.86 ha (Table 2). Almost half of the sampled maize farmers (49.3%) remarked they have access to good roads (asphalt/tarred but not asphalt/rough and smooth) as against 50.7% of them that reported dilapidated nature of their roads (tarred with potholes/rough and marshy) (Table 1). Okoboi [30] emphasized the importance of good roads in the agricultural production and marketing process. According to the study, roads are needed for regular access to inputs and output markets. The average yield of maize grain produced in Ghana according to the results is 1800 kg/ha (1.8 metric tonnes/ha) (Table 2). This is in line with MOFA's that the average yield of maize was 1.9 metric tonnes/ha instead of a potential of 6.0 metric tonnes/ha [27], indicating that maize yield is still low.

The study revealed that 60.1% of the respondents used traditional maize seeds, while 39.9% used improved seeds (Table 1). Most improved seeds are relatively high yielding so most farmers using traditional seeds could pose a negative effect on their technical efficiencies [38]. The average quantity of seed used per hectare for all the sampled maize farmers is 18 kg (Table 2). This is quite low, even though most farmers use traditional varieties and seeds are not difficult to come by since they can get it from their previous season. The results also show a 64.1% use of fertilizer (Table 1). Fertilizer supplies the crops with the required plant nutrients, and therefore, its use in recommended quantities is expected to improve the farmers' technical efficiencies [44]. The number of maize farmers using fertilizer is therefore quite high in the study only that the quantities applied by the farmers were lower than recommended rates. This corroborates the results obtained by [34] that use of fertilizer greatly exceeds earlier reports, even though the extent of use was 50% of the recommended rate. Use of pesticides is not very common among maize farmers in the study area as 97.2% of the farmers did not apply pesticides (Table 1). Farmers could therefore be technically inefficient if steps are not taken to improve pesticide usage. The situation could be worse today, especially with the recent invasion of the fall army worms which are reported to have destroyed vast maize farms which has threatened Ghana's ability to meet the sustainable development goals on no poverty and zero hunger.

Stochastic frontier production function analysis of maize production

Two likelihood ratio tests were conducted in the study. The first test was to ascertain the appropriateness of either Cobb–Douglas or translog functional form for the data, and the second was to confirm whether or not maize farmers in the four agro-ecological zones shared same or similar technologies. The results of the generalized likelihood ratio test rejected the stochastic frontier Cobb–Douglas production function in favour of the stochastic frontier translog production function as a fitting functional form for the data collected in all agro-ecological zones (Table 3). Table 5 presents the estimates of the stochastic production frontiers for the pooled and each agro-ecological zone from whose log-likelihoods the second likelihood ratio test was conducted. The results of the second likelihood ratio test examining whether or not the agro-ecological zones shared the same technology revealed that they shared the same technology as is not significant (Table 3). The implication is that there would be no reason for estimating a metafrontier production model. This suggests that the stochastic frontiers for maize farms in the four agro-ecological zones are not different and that any efficiency comparison across these four subsamples should be undertaken with respect to the pooled stochastic frontier instead of the metafrontier. The variance parameters for the chosen pooled

Table 3 Results of test of hypotheses

Restriction	Pooled				Northern Savannah				Transitional				Forest				Coastal Savannah			
	$L(H_0)$	λ	χ^2	D	$L(H_0)$	λ	χ^2	D	$L(H_0)$	λ	χ^2	D	$L(H_0)$	λ	χ^2	D	$L(H_0)$	λ	χ^2	D
$H_0: \beta_{ij} = 0$	−98.2	38.2	23.3	R	−85.3	24.2	23.3	R	−99.4	52.5	23.3	R	−81.3	44.1	23.3	R	−112.2	62.8	23.3	R
$\delta_m = 0$	−148.7	28.4	10.1	R	−178.3	45.8	10.1	R	−83.4	31.2	10.1	R	−94.8	21.7	10.1	R	−138.7	34.4	10.1	R

Hypothesis (restriction)	$L(H_0)$	λ	χ^2	Decision
Pooled sample versus sum of individual log-likelihood		13.4	45.6	Do not reject

Agro-ecological zones share same technology. Critical values are at 5% significance level and are obtained from χ^2 distribution table

$L(H_0)$, Log-likelihood function; λ, test statistic; D, decision on whether hypothesis accepted or rejected; R, hypothesis is rejected; NR, hypothesis is not rejected; β_{ij}, parameters in the square and cross terms; δ_m, parameters in the inefficiency term

frontier model are also presented in Table 4. The gamma (γ) value calculated for the respondents was 0.999. The closeness of this value to one (1) and the high Lambda (λ) value imply inefficiencies in maize production in the study area. This also makes the stochastic frontier model appropriate for the study [33]. The results therefore suggest that about 0.1% of the variations in maize outputs for the maize farmers are as a result of the presence of random shocks outside the control of the farmer. Examples of these random shocks include bad weather, diseases, topology, bushfires as well as statistical errors in measuring data. The values of λ ($p < 0.01$) and σ^2 ($p < 0.1$) indicate a good fit of the model and correctness of the specified distributional assumptions.

Table 4 also presents statistically significant Wald Chi-square statistic of 3.1×10^{10} ($p<0.01$), indicating joint significance of the model. The variables included in the model were tested for multicollinearity using variance inflation factor (VIF). The mean VIF calculated for the model is 1.252. This is small and implies that there was no problem of multicollinearity with the model [18]. In addition, the statistically insignificant Breusch Pagan (BP) statistic of 0.566 indicates homoscedasticity of the model.

The results of the production function analysis showed that whereas fertilizer ($p<0.1$), herbicides ($p<0.01$), pesticides ($p<0.05$) and farm size ($p<0.01$) positively influenced maize output, the effect of labour input ($p<0.01$) was negative (Table 5). The production elasticities for fertilizer, herbicide, pesticide, farm size and labour inputs are 0.49, 0.18, 0.01, 1.15 and −0.25, respectively (Table 6). Also, the return to scale of 3.327 (Table 6) suggests maize production activities in the study area exhibit increasing returns to scale implying that outputs of Ghanaian maize farmers, especially those in the study area, could be increased with increase in intensity of use of herbicide, fertilizer, farm size and pesticides. This finding is in line with the results of similar work done by [31] which reported the influence of herbicide, seed, labour, cost of intermediate inputs and farm size on the output of maize. With maize being the main economic and staple crop for Ghana's rural households, the influence of the above variables on maize output is critical to devising strategies aimed at meeting the sustainable development goals on no poverty and zero hunger.

Table 4 Variance parameters for the production function. *Source*: Survey, 2015

Variable	Parameter	SE
Sigma squared $\sigma^2 = \sigma_u^2 + \sigma_v^2$	0.722*	0.006
Gamma $\gamma = \sigma_u^2/\sigma^2$	0.999***	0.046
Lambda $\lambda = \sigma_u/\sigma_v$	3,764,018***	0.009
Log-likelihood	−246.316	
Mean technical efficiency	58.1	
Sample size	576	
Wald statistic	3.1×10^{10}***	
Mean variance inflation factor (VIF)	1.252	
Breusch Pagan statistic	0.566	

The asterisks indicate levels of significance. *** is significant at 1%, ** is significant at 5%, and * is significant at 10%

Technical efficiency of the farmers

The mean technical efficiency of the maize farmers was estimated to be 58.1%, with a standard deviation of 23.5% and 0.6% and 99.9% as the minimum and maximum, respectively, indicating that the farmers are inefficient with 41.9% of potential maximum output (6.0 metric tonnes per ha) lost to inefficiency. It could therefore be inferred that maize farmers in Ghana are on average 58.1% technically efficient in the use of the technologies available to them. The results imply that the outputs of Ghanaian maize farmers and more importantly those in the study area can be increased by 42% if they are able to use the resources available to them more efficiently. Since technical efficiency is key to enhancing farm incomes and ensuring food security, the finding presents a direction for meeting the sustainable development goals on no poverty and zero hunger. The results corroborate those of previous similar studies in Ghana [2, 3, 13, 20, 24, 31, 38].

The distribution of technical efficiency scores among maize farmers is presented in Fig. 1. The figure shows that majority of maize farmers in the sample (71%) have technical efficiencies ranging from 41 to 60%. The implication is that most maize farmers in Ghana have at least 40% of their potential output lost to inefficiency and this development is worrying as it presents an impression of Ghana struggling to meet the sustainable development goals on no poverty and zero hunger.

Sources of technical efficiency variations in Ghanaian maize farms

From the inefficiency model presented by Table 7, a negative coefficient implies an increase in the variable concerned would increase technical efficiency and productivity and vice versa. The coefficient of use of fertilizer by farmers in the sample has the expected negative sign and is statistically significant at the 10% level. This suggests that maize farmers who use fertilizer produce maize more efficiently. The implication is that fertilizer plays an important role in ensuring technical efficiency of maize farmers in all the maize growing areas of Ghana. This makes the results of [15] that found significant positive relationship between fertilizer use and technical efficiency in developing country

Table 5 Maximum likelihood estimates of stochastic frontier production functions for pooled and separate ecological zones. *Source*: Survey, 2015

Variable	Pooled		Northern Savannah zone		Transition zone		Forest zone		Coastal Savannah zone	
	Coeff	SE	Coeff	SE	Coeff	SE	Coeff	SE	Coeff	SE
Constant	6.141		6.841	0.373	10.280***	0.505	11.176		9.154	
lnFET	0.036*	0.020	0.175***	0.023	0.032***	0.008	0.028**	0.012	0.077***	0.015
lnHEB	0.285***	0.084	0.358	0.046	− 0.385***	0.075	1.549**	0.770	0.190	0.199
lnPET	0.009**	0.004	0.044*	0.004	0.192***	0.048	0.012**	0.011	0.001*	0.002
lnSED	0.025	0.016	− 0.030	0.048	− 0.101***	0.018	2.607***	0.783	0.070***	0.010
lnLAB	− 0.287***	0.076	− 0.363	0.005	− 0.183	0.001	0.123	0.466	− 1.617***	0.200
lnMAN	− 0.005	0.004	0.045***	0.009	0.150***	0.040	0.011	0.011	0.050***	0.003
lnLAD	0.726***	0.143	0.708***	0.161	0.548**	0.244	2.658***	0.339	− 0.507	0.351
lnCAP	− 0.044	0.031	− 0.295***	0.052	− 0.199***	0.063	− 0.461	0.577	− 0.217**	0.091
lnFET × lnFET	− 0.008***	0.002	− 0.013***	0.003	− 0.022***	0.003	− 0.001	0.005	0.050***	0.002
lnPET × lnPET	0.009	0.017	− 0.072	0.063	− 0.001	0.013	− 0.069***	0.019	− 0.085***	0.008
lnHEB × lnHEB	− 0.038***	0.011	− 0.215***	0.023	− 0.147***	0.043	0.063	0.047	0.064***	0.006
lnSED × lnSED	− 0.098***	0.009	− 0.055***	0.021	− 0.082**	0.032	0.314	0.263	− 0.100**	0.046
lnLAB × lnLAB	− 0.016**	0.008	0.010	0.013	− 0.032***	0.007	− 0.062**	0.027	0.221***	0.013
lnMAN × lnMAN	− 0.005**	0.002	− 0.002***	0.005	− 0.053***	0.006	− 0.008**	0.003	− 0.019***	0.001
lnLAD × lnLAD	0.018	0.021	0.099**	0.047	− 0.130	0.109	0.129**	0.064	− 0.581***	0.101
lnCAP × lnCAP	− 0.010***	0.003	0.009**	0.004	− 0.003*	0.002	0.001	0.005	− 0.018***	0.004
lnFET × lnPET	0.036***	0.010	− 0.019**	0.010	0.008	0.011	0.012	0.010	− 0.096***	0.003
lnFET × lnHEB	− 0.002	0.003	− 0.014***	0.004	0.002*	0.001	0.001	0.005	− 0.019***	0.001
lnFET × lnSED	0.007	0.005	0.051**	0.024	− 0.042**	0.013	0.075***	0.012	− 0.084***	0.014
lnFET × lnLAB	− 0.004	0.003	0.008	0.025	0.032**	0.013	− 0.005***	0.002	− 0.038***	0.008
lnFET × lnMAN	− 0.004**	0.002	− 0.053***	0.008	0.016***	0.003	0.017**	0.008	0.006***	0.002
lnFET × lnLAN	0.003**	0.003	− 0.099***	0.022	0.001	0.002	0.006**	0.003	0.014***	0.010
lnPET × lnCAP	0.005***	0.002	0.015***	0.001	0.035***	0.006	− 0.007	0.005	− 0.011***	0.001
lnPET × lnHEB	− 0.032***	0.009	0.026*	0.014	0.110***	0.008	− 0.018	0.012	− 0.019***	0.003
lnSED × lnLAB	− 0.044**	0.020	0.035	0.026	− 0.083***	0.013	0.087	0.172	0.122	0.085
lnSED × lnMAN	− 0.011	0.020	0.109***	0.015	− 0.111***	0.017	− 0.021**	0.010	− 0.012	0.032
lnSED × lnLAD	− 0.164***	0.034	0.028	0.039	0.105***	0.052	− 0.518**	0.227	0.763***	0.090
lnSED × lnCAP	− 0.006	0.007	0.038***	0.011	0.100***	0.028	0.217	0.191	0.205***	0.052
lnSED × lnHEB	0.046***	0.015	0.015	0.019	0.124***	0.011	0.481*	0.250	0.170***	0.011
lnLAB × lnMAN	0.026***	0.008	0.086***	0.026	− 0.055***	0.013	− 0.031	0.023	0.071***	0.010
lnLAB × lnLAD	0.041	0.015	0.042	0.021	0.272***	0.023	− 0.015	0.187	− 0.321***	0.065
lnLAB × lnCAP	− 0.009	0.009	0.041**	0.018	− 0.014	0.009	− 0.006	0.016	− 0.048*	0.029
lnLAB × lnHEB	− 0.104***	0.011	− 0.099***	0.013	− 0.002	0.018	0.057	0.048	0.025	0.050
lnMAN × lnLAD	0.012***	0.023	0.334	0.013	0.309***	0.108	0.002	0.032	− 0.015***	0.003
lnMAN × lnCAP	0.002	0.003	0.079***	0.008	0.016**	0.007	0.002	0.002	0.033***	0.010
lnMAN × lnHEB	0.015**	0.006	0.111***	0.012	− 0.022***	0.004	− 0.003	0.016	− 0.007	0.005
lnLAD × lnCAP	0.013	0.008	− 0.003	0.014	− 0.105***	0.028	− 0.233	0.166	− 0.062	0.040
lnLAD × lnHEB	− 0.006***	0.002	− 0.005***	0.001	− 0.093***	0.013	− 0.533**	0.272	− 0.343***	0.043
lnCAP × lnHEB	− 0.005	0.001	0.002	0.002	− 0.012	0.001	− 0.011**	0.005	− 0.127***	0.011
lnFET × lnCAP	− 0.003	0.002	− 0.005**	0.002	− 0.017***	0.002	− 0.041**	0.012	0.014***	0.003
lnPET × lnSED	− 0.004**	0.001	− 0.012***	0.003	− 0.014***	0.001	− 0.008***	0.001	− 0.001*	0.001
lnPET × lnLAB	0.001	0.0004	0.003***	0.001	0.004***	0.001	− 0.002*	0.001	0.002**	0.001
lnPET × lnMAN	0.006	0.006	0.037	0.057	0.138***	0.007	− 0.042***	0.015	− 0.048***	0.003
lnPET × lnLAD	0.001	0.003	− 0.018	0.043	− 0.021***	0.003	0.013**	0.005	0.007***	0.002
Log-likelihood	− 246.316		17.71978		21.18604		32.206377		64.88732	

The asterisks indicate levels of significance. *** is significant at 1%, ** is significant at 5%, and * is significant at 10%

Table 6 Input elasticities. *Source*: Survey, 2015

Input	Elasticity
Quantity of fertilizer	0.49
Quantity of herbicide	0.18
Quantity of pesticide	0.01
Quantity of seed	0.73
Quantity of labour	−0.25
Quantity of manure	−0.05
Quantity of land	1.15
Quantity of capital	−0.49

Scale elasticity, 3.327

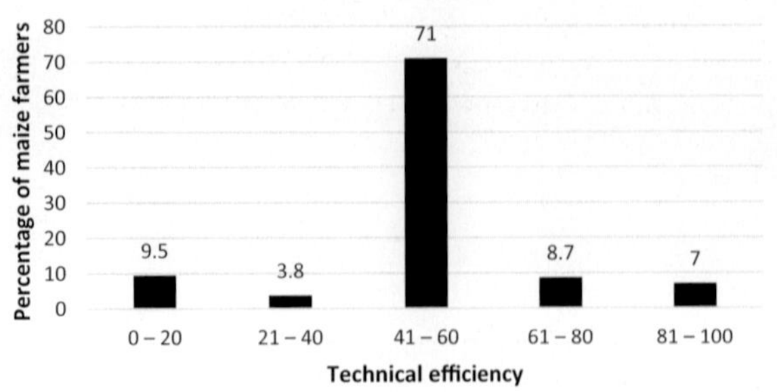

Fig. 1 Distribution of technical efficiency of maize farmers. *Source*: Survey, 2015

Table 7 Sources of technical efficiency among maize farmers. *Source*: Survey, 2015

Variable	Coefficient	SE
Constant	−0.289	0.477
Access to good roads	0.478	0.209
Living in Northern Savannah zone	−0.642***	0.275
Living in Transitional zone	−0.234	0.318
Living in Forest zone	−0.881**	0.399
Gender of maize farmer	−0.222**	0.164
Age of maize farmer	0.007	0.007
Educational level of maize farmer	−0.044***	0.016
Household size of maize farmer	−0.012	0.016
Experience of maize farmer	−0.003**	0.009
Farm size	0.014*	0.007
Land fragmentation	0.007*	0.033
Income	−0.0001	0.00002
Extension contact	−0.302*	0.177
Membership of farmer association	−0.202**	0.200
Access to credit	−0.058	0.197
Access to ready market	−0.294	0.203
Fertilizer use	−0.193*	0.180
Pesticide use	−0.431	0.402
Improved seed use	−0.529***	0.143

The asterisks indicate levels of significance. *** is significant at 1%, ** is significant at 5%, and * is significant at 10%

agriculture still relevant. Therefore, for maize farmers' incomes to be improved as well as being food secure, it is important that mechanisms are put in place to make it more attractive for them to use recommended doses of fertilizer in their farms. The effect of the dummy for improved seeds on technical inefficiency for maize farmers in the sample is negative and is expected. The variable is statistically significant at the 1% level. The implication is that maize farms with improved maize seeds are more technically efficient than farms using traditional seeds. This could be due to the fact that most improved seeds are high yielding. This finding on improved seed use, however, disagrees with the finding of [38] that suggest that technical inefficiency of maize farmers increases with use of improved seeds. According to [38], improved varieties of maize come along with certain agronomic practices which farmers must follow in order to get the maximum level of potential output. However, farmers ranked high cost of inputs as their most pressing problem, and therefore, most of them could not buy the recommended inputs that go with improved varieties. As Ghana strives to achieve the sustainable development goals on no poverty and zero hunger by especially mounting the planting for food and jobs programme, the findings of this study presents a word of caution to forcing farmers to use specified improved seeds which may present other production challenges to them.

The coefficient for maize farming experience with negative sign is expected and is statistically significant at the 5% level. The implication is that experience maize farmers are more technically efficient in the production of maize than inexperienced ones. This is because farmers with many years of maize farming experience will more likely be familiar with the required skills needed for maize production and therefore are more likely to have higher outputs and consequently more technically efficient. This agrees with the findings of [23] that found the number of years in maize farming to have a negative effect on technical inefficiency. The results of the current study also agree with the findings of [2] that concluded that farmers with many years of experience are more technically efficient than those with fewer years.

The coefficient of the dummy representing membership to a farmer association is negatively related to technical inefficiency and statistically significant at 5%. This implies that maize farmers who belong to farmer associations are more technically efficient than those who do not belong to any farmer group. This is because most agricultural technologies and new methods of farming are normally disseminated through farmer-based organizations so it is likely that it is only farmers who are members of such associations that will have access to improved technological packages. This finding is in agreement with [29] that farmers who are members of farmer groups are more efficient and productive because they have more access to extension services even though it contradicts the finding of [24]. Such farmers will likely have higher average incomes and will not struggle having a three square meal on daily basis.

The effect of education on technical efficiency is positive and significant at 1%. The results show that formally educated farmers produce maize more efficiently than those with no formal education. This is true since human capital represented by educational level enhances the managerial and technical skills of farmers. According to [11], education is hypothesized to increase farmers' ability to utilize existing technologies and attain higher efficiency levels. The importance of education of farmers in their use of productivity enhancing technologies is discussed extensively in the literature. According to [25, 40], farmers with many years of education are thought to have greater ability to notice, explain the meaning and respond to new information about productivity enhancing technologies than their counterparts with less number of years of education. Farmers who have received formal education are therefore able to access information from extension agents, which affect their use of production technologies. Furthermore, [42] reported that educational level and the producer's economic status, that influence ability to purchase and use productivity enhancing technologies, are to a larger extent positively related, most especially for farmers in less developed economies. The implication is that educated farmers are expected to be more efficient to understand and obtain new technologies in a shorter period of time than uneducated ones. Also, [36] notes that the complexity of a technology often poses a negative effect on improved input use and that education is thought to reduce the amount of complexity perceived in a technology thereby increasing its use and consequently, making the farmer technically efficient, receiving higher incomes and never going hungry. Djokoto and Gidiglo [17] also suggests that efficiency improving measures including training in farm management (a form of education), are required to make up for the mean technical efficiency difference of 38% experienced by agribusiness firms in Ghana.

The effect of male gender on technical inefficiency is negative for the maize farmers, and this is significant at 5%. The results show that males are more technically efficient in maize production than females. This corroborates the findings of [2, 38, 39] in technical efficiency studies that found negative relationships between gender and technical inefficiency. The negative effect of gender on technical inefficiency can be attributed to the crucial roles women perform in the domestic and economic life of society which negatively affect their technical efficiency. This comprises the unmeasured non-economic activities such as child care, cooking, cleaning, etc., performed by females in the household.

The influence of farm size is positive and significant at the 10% level. This means that an increase in farm size by maize farmers will cause a decline in the technical efficiency of such farmers. The reason is that most farmers are poor and may not have the required resources to meet the production demands of large farms. This, however, disagrees with the findings of [8, 9, 35] that reported a negative correlation between farm size and technical inefficiency.

The influence of land fragmentation on technical inefficiency is also positive and is statistically significant at 10%. This means that owning many farm plots by maize farmers in Ghana causes inefficiencies in maize production. The reasons may include but not limited to losses due to increased travel time, waste of border spaces, ineffective monitoring as well as inability to use farm machinery. This is in line with the results of [35] that reported an inverse correlation between land fragmentation and efficiency of agricultural production. Existence of production inefficiency has the potential to make farmers poor and hungry, and this is a threat to meeting the first two sustainable development goals (no poverty and zero hunger).

Access to extension is positively related to technical efficiency of maize farmers, and this is significant at 10%. The implication is that access to extension service will increase the technical efficiency of maize farmers in Ghana and contribute to achieving the sustainable development goals of no poverty and zero hunger. This is because agricultural production technologies developed by research institutes reach farmers through agricultural extension officers. Extension therefore allows maize farmers to be abreast with the latest recommended farming methods and technologies that are believed will enhance agricultural productivity and efficiency. Several studies including [4, 8, 10], as well as [38, 41] among others have also reported a positive relationship between access to extension service and technical efficiency of farmers.

Finally, the influence of the variables representing maize farms in the Northern Savannah, Transitional and Forest zones on the technical inefficiency of maize production is negative, indicating that improvement in technical efficiency of maize production is independent of agro-ecological zone. The variable representing maize farms in the Northern Savannah zone is significant at the 1% level, while that of maize farms in the Forest zone is significant at the 5% significance level. The results suggest that maize producers in all agro-ecological zones have equal chances of increasing their technical efficiencies and that maize production can take place in every agro-ecological zone. The result is in line with the results of [3] in a study into the technical efficiency of maize farmers across various agro-ecological zones of Ghana that revealed positive relationships between variables representing various agro-ecological zones and the technical efficiency of maize production in Ghana. Similar results were also obtained by [24, 31] whose studies revealed that whether or not a maize farmer is technically efficient is independent of growing location. The implication is that, as Ghana strives to achieve the first two sustainable development goals, steps can be taken to improve technical efficiency of maize farmers in all four agro-ecological zones.

Conclusion

Whether or not maize farmers are technically efficient determines their choice of productivity improvement strategy and very important for achieving the sustainable development goals on no poverty and zero hunger. This study examines technical efficiency and its determinants of Ghana's maize farmers. The study reveals that the mean technical efficiency estimate for maize farmers in Ghana is 58.1%, indicating that maize farmers in Ghana produce below the frontier with 41.9% of potential maximum output lost to technical inefficiency. This development, if unchecked, could threaten Ghana's achievement of the sustainable development goals on no poverty and zero hunger. It also reveals that an increase in educational level, maize farming experience, extension contact as well as uses of fertilizer and improved seeds would increase the technical efficiency of maize producers in Ghana. That is, human capital represented by educational level enhances the managerial and technical skills of farmers. Experience farmers will more likely be familiar with the required skills needed for maize production. Agricultural production technologies developed by research institutes also reach farmers through agricultural extension officers. Fertilizer also enriches the fertility of the soil, and most improved seeds are high yielding. Similarly, male Ghanaian maize farmers are more technically efficient than female farmers. This can be attributed to the crucial roles women perform in the domestic and economic life of society which negatively affect their technical efficiency. Furthermore, membership of a farmer association will increase the technical efficiency of Ghanaian maize farmers because extension officers normally disseminate most efficiency enhancing technologies through farmer groups. Finally, an increase in farm size and land fragmentation will decrease technical efficiency of Ghana's maize farmers. This is because farmers may find it difficult meeting the requirements of maintaining large farms and land fragmentation may lead to losses because of increased travel time, waste of border spaces, ineffective monitoring as well as inability to use farm machinery.

Given that increase in formal education will increase technical efficiency, efficiency improvement programmes by stakeholders in the maize industry could target literate maize farmers. Stakeholders, especially the Ministry of Food and Agriculture (MOFA) could also liaise with the Non-Formal Education Division of the Ministry of Education to provide maize farmers who do not have formal education with special training in at least reading, writing and numeracy prior to introducing new efficiency enhancing programmes to such farmers. The finding on the poor number of maize farmers that had access to agricultural extension service suggests that policy makers through the Ministry of Food and Agriculture should analyse the problems extension officers face in the discharge of their duties. This will pave the way for the provision of appropriate incentives to extension officers by government and other stakeholders in the maize industry that will help improve their commitment to delivery of agricultural extension services to the maize farmers. It is worthy of note that the current efficiency study dwelt on the fact that Ghana's maize farmers use the same technology. However, the importance of the existence of possible technological heterogeneity among farmers in recent times cannot be overemphasized and therefore requires further investigation. Future efficiency and productivity researchers are therefore encouraged to still consider the existence of specific technologies used by maize farmers in different agro-ecological zones to pave the way for analysing the effect of technology gaps across agro-ecological zones. There is no gainsaying the fact that paying attention to the above recommendations will help develop Ghana's maize sector through increased technical efficiency of her maize farmers and consequently contribute to achieving her quest to meeting the sustainable development goals on no poverty and zero hunger.

Abbreviations

BP: Breusch Pagan; JHS: Junior High School; JSS: Junior Secondary School; SSS: Senior Secondary School; SHS: Senior High School; MLE: maximum likelihood estimation; MOFA: Ministry of Food and Agriculture; SRID: Statistics, Research and Information Directorate; VIF: variance inflation factor; COLS: corrected ordinary least squares; NAADS: National Agricultural Advisory Services.

Authors' contributions

CAW conceived the idea and was responsible for analysis of the data as well as preparation of the manuscript. DAV was responsible for designing the study and data collection. Both authors read and approved the final manuscript.

Acknowledgements

We are especially indebted to the staff of the Ministry of Food and Agriculture in the West Mamprusi, East Gonja, Nkoranza, Ejura Sekyedumase, Fanteakwa,

Sekyere South, Gomoa and Ketu districts/municipalities of Ghana for the information they provided about the maize crop and also assisting in the data collection. We are also grateful to the respondent maize farmers in the aforementioned districts/municipalities without whose co-operation the study could not have taken place.

Competing interests

The authors declare that they have no competing interests.

Funding

Not applicable.

References

1. Abdulai A, Eberlin R. Technical efficiency during economic reform in Nicaragua: evidence from farm household survey data. Econ Syst. 2001;25:113–25.
2. Abdulai S, Nkegbe PK, Donkoh SA. Technical efficiency of maize production in Northern Ghana. Afr J Agric Res. 2013;8(43):5251–9.
3. Addai KN, Owusu V. Technical efficiency of maize farmers across various agro ecological zones of Ghana. J Agric Environ Sci. 2014;3(1):149–72.
4. Ahmad MC, Ghulam M, Igbal M. Wheat productivity, efficiency, and sustainability: a stochastic production frontier analysis. Pakistan Dev Rev. 2002;41:4.
5. Aigner DK, Lovell CK, Schmidt P. Formulation and estimation of stochastic frontier production function models. J Econom. 1977;6:21–37.
6. Ajao AO. Empirical analysis of agricultural productivity growth in Sub-Sahara Africa: 1961–2003. Libyan Agric Res Cent J Int. 2011;2(5):224–31.
7. Amaza P, Kwacha A, Kamara A. Farmers' perceptions, profitability, and factors influencing the adoption of improved maize varieties in the Guinea Savannas of Nigeria. Africa: IITA: Research to Nourish Africa; 2007.
8. Amos TT. Analysis of productivity and technical efficiency of smallholder cocoa farmers in Nigeria. J Soc Sci. 2007;15:127–33.
9. Barnes A. Technical efficiency estimation of Scottish agriculture: a note. J Agric Econ. 2008;59:370–6.
10. Basnayake BMJK, Gunaratne LHP. Estimation of technical efficiency and it`s determinants in the tea small-holding sector in the mid country Wet Zone of Sri Lanka. Sri Lanka J Agric Econ. 2002;4:137–50.
11. Battese GE, Coelli TJ. A model for technical efficiency effects in a stochastic frontier production for panel data. Empir Econ. 1995;20:325–32.
12. Battese GE, Prasada Rao DS, O'Donnell CJ. A metafrontier production function for estimation of technical efficiencies and technology gaps for firms operating under different technologies. J Prod Anal. 2004;21:91–103.
13. Bempomaa B, Acquah HDG. Technical efficiency analysis of maize production: evidence from Ghana. APSTRACT: Appl Stud Agribus Commer. 2014;8(2–3):73–9.
14. Bonabana-Wabbi J. Assessing factors affecting adoption of agricultural technologies: the case of integrated pest management (IPM) in Kumi District, Eastern Uganda. Unpublished Masters Dissertation, Virginia Polytechnic Institute and State University, Blacksburg, Virginia; 2002.
15. Bravo-ureta BE, Evenson RE. Efficiency in agricultural production: the case of peasant farmers in eastern Paraguay. J Int Assoc Agric Econ. 1994;10(1):27–8.
16. Bravo-Ureta BE, Robert EE. Efficiency in agricultural production: a case of peasant farms in eastern Paraguay. Agric Econ. 1993;10:27–37.
17. Djokoto JG, Gidiglo KF. Technical efficiency in agribusiness: a meta-analysis on Ghana. Agribusiness. 2016;32(3):397–415.
18. Edriss AK. A passport to research methods. Las Vegas: International Publishers and Press; 2003.
19. Elsamma J, George MV. Technical efficiency in rice production: a frontier production function approach. Agric Econ Res Rev. 2002;15:50–5.
20. Essilfie FL, Asiamah MT, Nimoh F. Estimation of farm level technical efficiency in small scale maize production in the Mfantseman Municipality in the Central Region of Ghana: a stochastic frontier approach. J Dev Agric Econ. 2011;3(14):645–54.
21. Green WH. Limdep version 8.0: econometric modeling guide. New York: Econometric Software, Inc; 2002.
22. Greene WH. Econometric analysis. New York: Pearson Education; 2005.
23. Isaac O. Technical efficiency of maize production in Oyo state. J Econ Int Finance. 2011;3(4):211–6.
24. Kuwornu JKM, Amoah E, Seini W. Technical efficiency analysis of maize farmers in the eastern region of Ghana. J Soc Dev Sci. 2013;4(2):84–99.
25. Langyintuo A, Mekuria M. Modeling agricultural technology adoption using the software STATA. CIMMYT-ALP training manual no. 1/2005 (Part Two). International Maize and Wheat Improvement Center, Harare, Zimbabwe; 2005.
26. Meeusen W, van den Broeck J. Efficiency estimation from Cobb–Douglas production functions with composed error. Int Econ Rev. 1977;18:435–44.
27. MOFA (Ministry of Food and Agriculture). Agriculture in Ghana: facts and figures Statistics, Research and Information Directorate (SRID); 2013.
28. Mohammed I, Wojciech JF, Shashidhara K. The determinants of farmer adoption of improved peanut varieties and their impact on farm income: evidence from northern Ghana. In: Selected paper prepared for presentation at the agricultural and applied economics association annual meeting, Seattle, WA, August 12–14; 2012.
29. NAADS. Proceedings of the mid-term review of the National Agricultural Advisory Services. Kampala, Uganda: NAADS; 2005.
30. Okoboi G. Improved inputs use, productivity and commercialisation in Uganda maize production. Unpublished Doctoral Dissertation, Makerere University; 2011.
31. Oppong BA, Onumah EE, Asuming-Brempong S. Technical efficiency and production risk of maize production: evidence from Ghana. Asian J Agric Ext Econ Soc. 2016;11(3):1–9.
32. Owuor G, Shem OA. What are the key constraints in technical efficiency of smallholder farmers in Africa? Empirical evidence from Kenya. In: A paper presented at 111 EAAE-IAAE seminar 'small farms: decline or persistence'. University of Kent, 26–27th June; 2009.
33. Piesse J, Thirtle C. A stochastic frontier approach to firm level efficiency, technological change and productivity during the early transition in Hungary. J Comp Econ. 2000;28:473–501.
34. Ragasa C, Dankyi A, Acheampong P, Wiredu AN, Chapoto A, Asamoah M, Tripp R. Patterns of adoption of improved maize technologies in Ghana. Ghana strategy support program. International Food Policy Research Institute (IFPRI) working paper 36; 2013.
35. Raghbendra J, Hari KN, Subbarayan DP. Land fragmentation and its implications for productivity: evidence from Southern India. Australia South Asia Research Centre (ASARC) working paper 2005/01; 2005.
36. Rogers EM. Diffusion of innovations. 5th ed. New York: Free Press; 2003.
37. Sherlund SM, Barrett CB, Adesina AA. Smallholder technical efficiency controlling for environmental production conditions. J Dev Econ. 2002;69(1):85–101.
38. Sienso G, Asuming-Brempong S, Amegashie DPK. Estimating the efficiency of maize farmers in Ghana. In: Contributed paper prepared for presentation at the 4th international conference of the AAAE, Cape Town, Hammamet, Tunisia, 22–25 September, 2013.
39. Solís D, Bravo-Ureta B, Quiroga RE. Technical efficiency and adoption of soil conservation in El Salvador and Honduras. In: Research paper presented at the international association of agricultural economists conference, Gold Coast, Australia, August 12–18, 2006.
40. Tabi AJ, Vabi MB, Malaa DK. Adoption of maize and cassava technologies in the forest-savannah zone of Cameroon: implications for poverty reduction. World Appl Sci J. 2010;11(2):196–209.
41. Tchale H, Sauer J. The efficiency of maize farming in Malawi: a bootstrapped translog frontier. Cahiers d`économie et sociologie rurales. 2007;82–83:33–56.
42. UBoS. Uganda national household survey: socio-economic module. Abridged report. Kampala, Uganda: UBoS; 2010.
43. Wilson P, Dave H, Stephen R, Loannis K. Measuring and explaining technical efficiency in uk potato production. J Agric Econ. 1998;48(3):294–305.
44. World Bank. World development report 2008: agriculture for development. Washington, D.C.; 2007.
45. Xu T, Sun F-F, Zhou Y-H. Technical efficiency and its determinants in China's hog production. J Integr Agric. 2015;14:1057–68.

Crop diversification and household food security status: evidence from rural Benin

Gilbert Onionkiton Adjimoti* and George Tsey-Mensah Kwadzo

Abstract

Background: The government of Benin has initiated since 2006 many policies to promote crop diversification. The aim was to help rural household to be more resilient to food insecurity. The objective of this research is to determine how crop diversification has affected the food security status of the rural household.

Methodology: Primary data from 420 rural households were collected in the Collines Region in Benin. We use principal component analysis (PCA) to construct a multidimensional food security indices and a Simpson diversity index has been used to measure the degree of crop diversification. A linear regression model was used to determine the effect of crop diversification on household food security status.

Results: We found that crop diversification has a positive effect on household food security status. The diversity of crops grown through dietary diversity can improve household food security. In rural remote areas where household access to food depends largely on it production, crop diversification provides farmers with the different crops that they cannot access either because of the cost or because of the poor infrastructure constraints (physical access). Beyond, the results also show that some other factors are also affecting the household food security status such as access to extension services and storage facilities.

Keywords: Benin, Crop diversification, Food security, PCA, Policy

Background

Achieving the second objective of the sustainable development goals which is zero hunger requires finding ways to increase the income and assets of rural people living in developing countries. Yet, reducing food and nutrition insecurity in the developing countries continues to be a major public policy challenge. Food insecurity has become a global challenge for policy makers particularly in Sub-Saharan Africa where the rate of population growth far exceeds the quantity and quality of food needed to feed the population. Most of poor population like in Benin to some degree depend on agriculture and especially small-scale farming systems as the primary source of their livelihoods. Agriculture is central to rural development, food insecurity reduction and rural poverty alleviation [29]. Smallholder farmers are fundamental in achieving food and nutrition security and sustainable rural development. According to Antonaci et al. [3], smallholder farmers produce 80% of the food in Sub-Saharan Africa and are the largest providers of work for the local labor force in many developing countries. This is not only because poverty is still concentrated in rural areas, but the agricultural sector also typically accounts for a large share of national income and employment. In Benin for instant, the sector employs 70% of the labor force and contributes about 35% of the GDP [41]. According to Cervantes-Godoy and Dewbre [8], people in developing countries who depend on agriculture for their living are typically much poorer than people who work in other sectors of the economy and that they represent a significant share, often the majority, of the total number of the poor people in the countries where they live. Van Wijk [42] recognized that policy can play a key role in defining the different role of agriculture in food security and sustainable development.

*Correspondence: adjilbert@yahoo.fr
Department of Agricultural Economics and Agribusiness, University of Ghana, P.O. Box 65, Legon, Accra, Ghana
All the authors have participated in the different committee meeting that validated each phase of the study.

Although agriculture-led growth played an important role in reducing food and nutrition insecurity and transforming the economies of many Asian and Latin American countries, most countries in Africa have not yet met these criteria for a successful agricultural revolution. Risk and uncertainty are the major characteristics of agricultural production in developing countries [1, 17]. Smallholder farmers are facing a lot of challenges in many areas including low soil fertility, climate change, water shortage, dysfunctional input–output market, weak extension services and poor policy [36]. As a result, many rural households are facing decreasing agricultural productivity, food and nutrition insecurity, and income variability which affect negatively their livelihoods. Among these issues, the food and nutrition insecurity and income variability are the main component of smallholder farmers' livelihoods. Crop diversification can be used as a tool to increase farm income, generate employment, alleviate poverty, conserve soil and water resources. It is therefore reckoned as an important strategy to overcome many of the emergencies faced by agricultural household in developing countries [20]. In India, Mandal and Bezbaruah [23] found that as flood risk-coping mechanism, crop diversification has positive impact on farmers income. In Guatemala, Von Braun [39] showed that diversification into export vegetable production has increased rural employment by 45%. Similarly, Ali and Abedullah [2] reported that diversification toward high-value commodities has generated substantial employment in seed and seedling production. Recently, Birthal et al. [5] acknowledged that households diversifying toward high-value crops are less likely to be poor.

In rural Benin, household access to food largely depends on what the household grows, either because they consume what they grow, or they purchase food with the income earned from what they grow. Also, due to very high transportation costs, poor infrastructure and to the remoteness of the rural area in Benin, many rural households are forced to rely on their agricultural production or on the locally produced crops to satisfy their nutritional needs. On that condition, crop diversification can be an important mean of reducing food insecurity in rural Benin.

Still, reducing rural food insecurity is one of the biggest challenges of public policies in Benin. According to INSAE [18], about 25% of rural household in Benin are highly food insecure (cannot ensure the basic food needs without help) and about 45% are at the risk of food insecurity and the country remains one of the poorest in the world with an incidence of poverty estimated at 46% in rural areas.

Although the researches on the crop diversification are oft remarked especially in Asia, few attempts have been made to link it to the food security in Africa. Pellegrini and Tasciotti [27] recognized that due to very high transportation costs and to the remoteness of the villages they live in, some rural dwellers are forced to rely on their agricultural production and the locally produced crops to satisfy their nutritional needs.

Herforth [16] examined the relationship between farm diversity and dietary diversity among households in central Kenya and northern Tanzania. In both Kenya and Tanzania, the number of crops grown by a household was positively associated with the dietary variety of the household (i.e., the number of unique foods in the diet) and in Tanzania, crop diversity was associated with the diversity of food groups in household and individual child diets. In both countries, crop diversity was also positively associated with the diversity of home-produced fruits and vegetables consumed. In the rural highlands of Ecuador, on-farm species diversity and family-level dietary diversity were also positively correlated [26]. Families with low agrobiodiverse farms in this setting consumed more off-farm food items. In western Mali, the number of crops cultivated by a household was positively associated with adult nutrient adequacy [38].

Recently, Jones et al. [19] found that the production diversity of farms in Malawi was consistently and positively associated with the diversity of household diets. Farm diversity demonstrated a consistent positive association with household dietary diversity independent of differences in household wealth and social standing. This relationship was significantly greater in wealthier households as well as in households headed by women. Farm diversity was especially strongly associated with consumption of legumes, vegetables and fruits.

Although there is no a single approach to measure the household food security status, accurate measurement indicator is necessary to target resources toward those most in need or at risk of sliding into hunger. Accurate measurement of household food security is also essential for effective research and well-targeted policies and programs. According to Carletto et al. [7], a wide range of indicators are used for food security analysis and the best way is to define clearly the intended scope. The definition of food security is generally understood as a situation whereby *"all people, at all times, have physical, social and economic access to sufficient, safe and nutritious food to meet their dietary needs and food preferences for an active and healthy life"* [11]. There are three important components in this definition: food quantity available in the household, food quality as capture by diet diversity and stability of adequate food supplies.

Most studies try to conceptualize the food security into different way. Two conceptual components have been widely used. The first one is to analyze food security through the set "quantity, quality and stability". The second approach is to follow the conventional components embedded in the definition of food security which are availability, accessibility, utilization and the stability: (1) the availability of food in terms of its physical presence in a given country/household; (2) the access to food as reflected by people's ability to obtain food from own stock/home production, or through market purchases, gifts or borrowing; and (3) the utilization of food, in terms of the ability to derive full biological benefits from food, based on food safety and nutritional/socio-cultural value [37].

Food availability is a combination of domestic food production and food stocks, commercial food imports and food aid. The term 'availability' is usually used in reference to food supplies at the regional or national level [30]. At the household level, the indicators commonly used are household food production and social safety nets (food aid). Some researchers Demeke and Zeller [10] also suggested to use the tropical livestock unit as a determinant of the availability. In the case of Collines Region, the livestock is not developed and its contribution to the household income remains very low.

Accessibility is the ability of the household to get food by having adequate resources, or entitlements, to acquire a sufficient quantity of food, and while consumers' purchasing power in the form of real incomes and food prices is important, entitlements are not necessarily monetary. One of the key determinants of the accessibility is the household income. The income here will be measured with the agricultural production income and the non-farm income. Maxwell et al. [24] add the non-agricultural assets. In the context of Collines Region, household may rely on some physical assets when the food shock occurred.

Utilization is related to health and reflects concerns about whether individuals make good use of the food to which they have access [4]. The variety of food group consumed and the calorie intake are the most indicated determinants of the utilization.

Methodology

Location of the study and sampling techniques

The data used for this study are exclusively from primary source and were collected from five municipalities located in the Collines Region in Benin. A questionnaire was designed and administrated at the household level to the head of household through face-to-face interview.

Location of the study

Benin is a coastal country located in West Africa (see Fig. 2). It is divided into twelve Regions including the Region of Collines. Located in the center of Benin, Collines Region covers an area of 13,931 km^2 which represents 12% of the total area of Benin. According to the last general census of the population in 2013, it has 716,558 inhabitants, representing 7.2% of the total population of Benin. The region is divided into six municipalities: Bante, Dassa-Zoume, Glazoue, Ouessè, Savalou and Save. The economic activities of the area are primarily based on agriculture (71%). Regarding agriculture, the region provides an important part of the national supply as evidenced by production statistic. The region is the largest producer of roots and tubers and vegetables in Benin. In regard to cereals, the region usually ranks second or third position. This is an ideal area for the analysis of livelihood strategies of people given the diversity of crops grown. Yet, this is one of the poorest regions of Benin, with an incidence of poverty estimated at 46% in 2011 [18]. The food situation in the region is also critical despite the amount of agricultural production that comes from this region (Fig. 1).

Sampling techniques

The study used the multistage sampling procedure where it started with a purposive selection of the Collines Region. We chose the region because the crops diversification level is very high and households are engaged in growing on average more than three crops. At the same time, the region is one of the most food insecure in the country. There are six municipalities within the region. Based on the importance of agricultural activities in the municipalities, five municipalities out of six served as the study area: Bante, Dassa-Zoume, Glazoue, Ouessè, Savalou. Firstly we compute a sample size using [9] approach:

$$n = z^2 \times p(1-p)/m^2$$

n required sample size, z confidence level, p estimated prevalence of farm attribute in the area, m margin of error.

In our specific case, in order to minimize the error, we used the confidence level of 95% with the standard value of 1.96; the margin of error (m) is 5%. Our general population is all the crop farm households in the Collines Region, and they represent 65% of the total household in the Region [18]. Therefore, $p=0.65$.

$$n = 1.96^2 * 0.65(1-0.65)/0.05^2 = 350$$

We increased the sample size by 20% to account for contingencies such as non-response or recording error.

$$N* = 350 * 1.20$$
$$N* = 420$$

The next stage was the distribution of the 524 respondents among the five municipalities according to the percentage of farm household in the municipalities. The distribution is shown in Table 1.

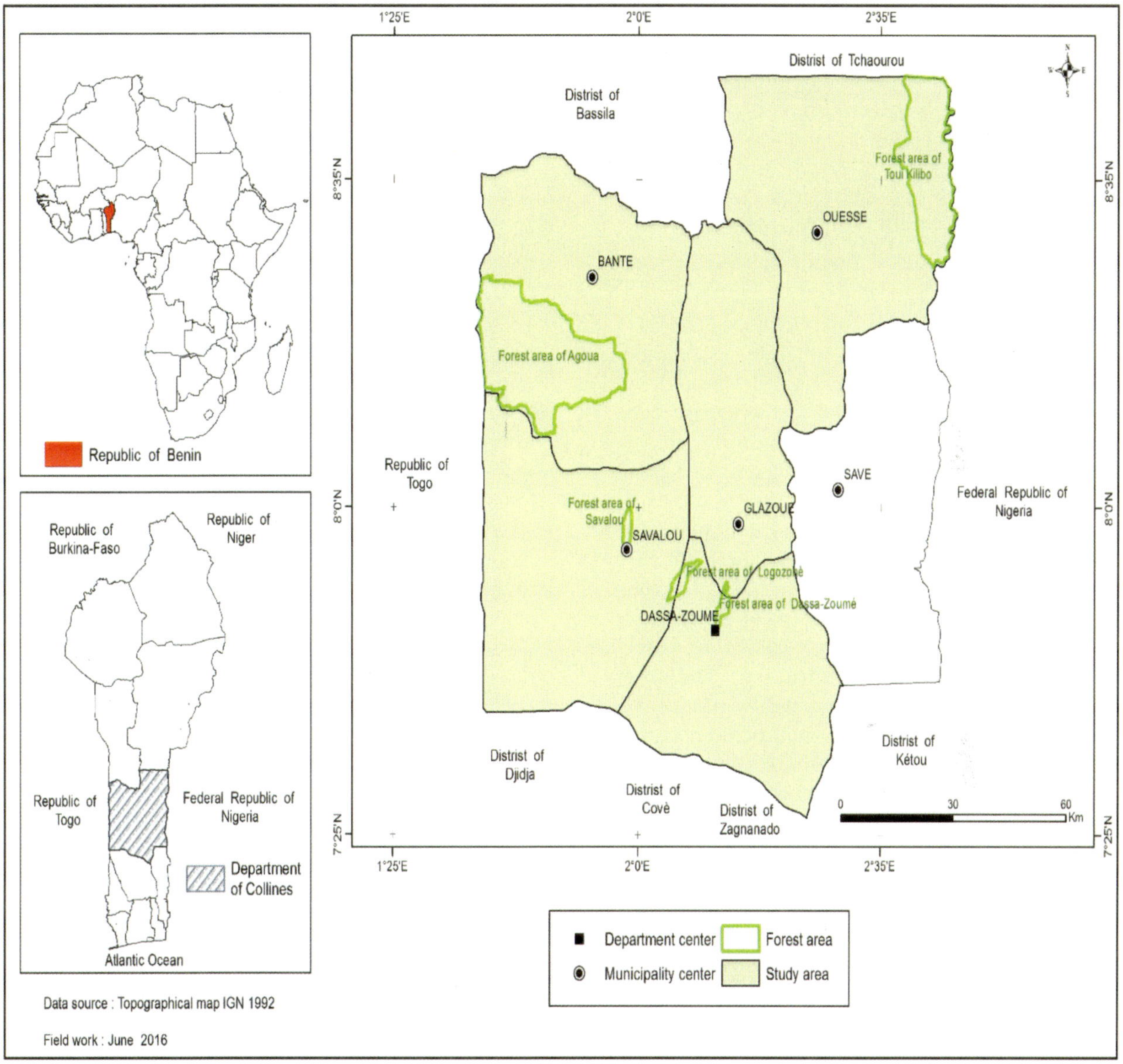

Fig. 1 Location of the study

Each municipality is divided into districts, and each district is divided in villages/communities (Municipalities–Districts–Villages/Communities–Households). First, we randomly selected 50% of the districts in each municipality and where the number of district is odd, we chose the maximum (for example, if the number of the district is 09, we retained 05). Within the districts, we purposively chose one village/community proportionate to the number of the farm household in each village (they are also the main agricultural production villages). The share of the farm household in each municipality and district was collected from the National Statistical Institute of Benin. In each village, we randomly selected 5% of the total farm households based on the information provided by the agricultural extension officers and the farmers-based organizations. The farm households that were enumerated were randomly selected from the total agricultural household list of the villages.

The food security index

As a multidimensional concept, food security study requires the use of multivariate descriptive statistic and a good indicator has to take all the aspects into consideration. To be able to do this, it may be useful to construct a multidimensional index that incorporates the most important indicators from each dimension of food security.

Table 1 Sampling Distribution of household respondents

Municipalities	Bante	Dassa	Glazoue	Ouesse	Savalou	Total
Population[a]	106,945	112,118	123,542	141,760	144,814	629,179
Total household[a]	13,368	16,016	19,006	20,251	18,101	86,742
Crops production households[a]	9438	10,132	11,575	14,062	11,175	56,382
Percentage	16.74	17.97	20.53	24.94	19.82	100
Sample size	70	75	86	105	84	420

[a] Estimation from National Statistical Institute (2014)

The factorial methods have been developed for this purpose and include factor analysis, the principal component analysis and correspondence analysis. However, the major difficulty of the analysis is based on the interpretation of results obtained. Factor analysis is, according to Bry [6], a geometric principle allowing automatic conversion of a large table of data in synthetic images from which will emerge the main structures. By extracting the maximum information, PCA aims to study the linear connections between variables and to identify homogeneous groups of individuals from the correlation matrix or covariance.

The principal component analysis (PCA), therefore, gives a description of the statistical units and observed variables based on the study of the correlation coefficients. It is not based on a probabilistic model and may therefore affect a whole population or a sample. It is not either a regression since we are not explain a variable from other, but rather a summary of the information contained in these variables. Moreover, PCA highlights the similarities and contrasts between the analyzed units. The originally correlated variables are compressed and processed independent variables called principal components or axes. They allow to carry out a geometric representation that best explained the variability in the data. The PCA provides thus a system of orthonormal axes retaining all the distances between the variables, hence the delineation of groups of individuals with similar characteristics. PCA is used for heterogeneous data, while factor correspondence analysis (FCA) is used for contingency tables, but also allows analyzes of qualitative data. The FCA is rather recommended for large tables identical data (expressed in the same unit).

The aim of the PCA is to be able to group all the food security aspect into one indicator and facilitate the interpretation.

Construction of the food security index

We use PCA to construct a composite food security index by incorporating several indicators (food accessibility, food affordability, food utilization and food stability) that are supposed to capture the different dimensions of food security. The PCA extracts the linear combination of these variables which give the maximum variance and transform them into one index [43]. The first principal component is the linear combination capturing the greatest variation among the set of variables. This was converted into factor scores, which served as weights for the creation of an index for each household. In other words, from an initial set of n correlated variables (X_1, X_2 ... X_n), PCA creates m uncorrelated principal components where each is a linear weighted combination of the initial variables as follows:

$$PC_m = a_{m1}X_1 + a_{m2}X_2 + a_{m3}X_3 + \cdots + a_{mn}X_n$$

where a_{mn} represents the weight for the mth principal component and the nth variable. The components are ordered so that the first component explains the largest amount of variance in the data subject to the constraint that the sum of the squared weights ($a_{m1}^2 + a_{m2}^2 + a_{m3}^2 + \cdots + a_{mn}^2$) is equal to one. Each subsequent component explains additional but less proportion of variation of the variables. The higher the degree of correlation among the original variables, the fewer components required to capture common information [40]. Once the first component was identified, we derived the food security index for each household as follows:

$$\mathrm{FSI}_j = \sum F_i \left[\frac{X_{ij} - X_i}{S_i} \right]$$

FSI_j is the food security index for the household j. F_i is the weight for the ith variable. X_{ij} is the jth household's value for the ith variable. X_i and S_i are the mean and standard deviations of the ith variable for overall households.

The results of the PCA were generated through SPSS program. The identification and selection of indicator variable were based on the food security literature as well as the four components of food security.

Choice of the variables

Availability Food availability refers to the physical supply of food for the household. In this study, the availability was measured by the food crop land size, the cereal land size, the cereal production and the cereal production per capita. They all helped to estimate how much food was available in the household. Although cereal land size was already included into food crop land size, we separately considered cereal because of their contribution to the household consumption status.

Accessibility The food access is related to the household ability to acquire food; it depends on the range of food choice open to household given their income, market price and market accessibility. Two indicators have been retained here to measure the food access: income per capita and distance to market. The income per capita was calculated using the total gross household income divided by the household size.

Utilization Utilization is related to health and reflects concerns about whether individuals make good use of the food to which they have access [4]. We use the food consumption score as the measure of utilization for the study because it is a composite score based on dietary diversity and food frequency.

Stability The stability dimension deals with the fact that people's food security situation may change over the time. Therefore, the number of months without food stock was used to measure the stability.

Measures of crop diversification

Crop diversification could be defined as the introduction or development of additional crops to the existing farming system (Makate et al. 2016). The main aspect and the most understood concept of crop diversification is the addition of more crops to the existing cropping system, which could be referred to as horizontal diversification [20]. For example, substitute or add more crops into an existing farming system can be understood as crop diversification. However, this type of crop diversification means the broadening of the base of the system, simply by adding more crops to the existing cropping system utilizing techniques such as multiple cropping techniques coupled with other efficient management practices.

Other words are also used to express crop diversification. For example, we have crop substitution and adjustment which are also related to the primary concept of crops diversification, but they are just a way of optimizing profit and manage land [12]. Diversification at farm level will involve growing of more than one crop in order to achieve self-sufficiency, which is quite different at the national level. At national level, crop diversification will require additional resources and management of a specific or group of crops sold freshly or value added to increase profits [13].

There are several ways to measure crops diversification, and the most important include Herfindahl Index, Simpson Diversity Index, Ogive Index, Margalef Index, Shannon Index, Berger-Parker Index and Entropy Index. Another common method for measuring the crop diversification is to count the number of crop grown by farmer.

Herfindahl index (HI) The Herfindahl index is a concentration index which is often used to determine the concentration of industry. It decreases gradually as the level of diversification increases. It takes the value 1 when there is total concentration and tends to zero as the level of diversification increased [33]. In the context of crops diversification, Herfindahl index is used to study the extent of crops diversification. The formula is given as below.

$$\mathrm{HI} = \sum_{i=1}^{N} P_i$$

where N is the total number of crops cultivated and P_i accounts for the land share of the ith crop in total cropped area.

Simpson diversity index (SDI) SDI was introduced in 1949 to assess the degree of concentration when individuals are grouping into type. The same index is often used to measure the extent of diversification.

The square root of the index has already been introduced by Hirschman. As a result, the same measure is usually known as the SDI in ecology and as the HI or the HHI (Herfindahl–Hirschman Index) in economics. $\mathrm{SDI} = 1 - \sum_{1}^{n} P_i^2$.

P_i is the proportionate area of the ith crop in the gross cropped area; n is the total number of crops grow by the household.

Ogive index (OI) Ogive Index is used to evaluate countries specialization and concentration. It has been also used to estimate the diversification at the farm level. The description of the index is given in the following formula.

$$\mathrm{OI} = \frac{\sum_{i=1}^{n}\left[P_i - \left(\frac{1}{N}\right)\right]^2}{\left(\frac{1}{N}\right)}$$

N represents the number of crops; P_i is the proportionate area of the ith crop in the gross cropped area.

Bhatia's method Bhatia's Index (BI) is inversely proportional to the degree of crops diversification. It means that as the index increased, the degree of crops diversification will reduce and vice versa. After identifying those crops which have more than 10% of cropped areas, we compute the formula below.

$$BI = \frac{\text{Percentage of total cropped area under 'n' crops}}{\text{Number of 'n' crops}}$$

Entropy measures for crop diversification Entropy Index evaluates the shares of farm activity in logarithm term with the inverse measure of the shares. This index has been widely used in the area of crops diversification [31]. The Entropy Index of Diversification (DIE) is computed using the formula DIE = $\Sigma[P^2 it \times \text{In}(1/\text{Pit})]$.

With P_i the proportion of cultivated land for the *i*th crop. DIE is supposed to increase as the level of diversification increases and vice versa. The difference between the first level of diversification and the perfect diversification for a given number of crops is measured by the Berry's index as DIB/$[1-(1/n)]$, while for Entropy Index as DIE/In(n) more the Entropy or Berry's measures value expected is the result or otherwise [33].

All these indices are computed on the basis of proportion of gross cropped area under different crops cultivated in a particular geographical area. Simpson Index is usually known as Simpson Diversity Index in ecology and as the HHI (Herfindahl–Hirschman Index) in economics, and we used it as diversification index for the present study. One of the advantage of using Simpson Diversity Index in the present study is that it does not required farmers to produce all type of crops.

Effect of crop diversification on household food security

The constructed food security index was included as a continuous dependent variable in a regression model [10].

$$Yi = \rho Zi + \varphi Di + \varepsilon i$$

where Yi is the household food security index, Di is the Simpson Diversity Index, Zi is the vector of others explanatory variable for household i, and εi is the error term.

Gender of the household head It is generally shown that female-headed household in the context of developing countries is more food insecure than male-headed households [22]. Recently, Magaña-lemus et al. [21] also showed that due to the lack of resources, female-headed households in Mexico are more likely to be food insecure than male-headed households.

Age of household head Age of the household head can significantly affect their food security status. According to Perz [28], elder household (more than 45 years) heads have more experience in allocating their resources to achieve the household need. On the contrary, Magaña-lemus et al. [21] show that younger household heads are more dynamic in searching for more opportunities and so their households are more likely to be food secure.

Education Education as the main indicator of human capital can improve the opportunities for the household head. Households with more education tend to be more food secure [21]. Lower education level negatively affects the opportunities of getting non-farm employments (Table 2).

Membership of farmers association The new agricultural policy in Benin considers FBOs as a key bridge through which reliable information on agricultural technologies will flow. Farmers associations are also a kind of social capital that can help to overcome some of the food challenges. Being a member of an association may provide reliable information and opportunities.

Access to fertilizer and seed Access to inputs is key components of agricultural production and therefore can enhance the productivity. It is expected that fertilizer use will improve household food security status. All things equal, [32] shows that fertilizer plays an important and critical role in improving yield and therefore food security.

Number of extension visits The number of extension visit as a knowledge will enhance production techniques and productivity. It is expected that the number of extension visit will be positively related to the household food security status.

Location As a dummy variable, the location as a determinant of diversity will capture the difference in culture and environmental conditions in which farm household's work and how this is related to food security.

Total land size the amount of land a household owns plays an important role in the crop production quantity. The more land cultivated by the household, the better the household will be in terms of food security.

Access to formal credit In this study, access to credit is defined as having request for and getting the full amount of credit. Those who get partial amount or who did not request because of lack of collateral, absence of credit institutions or any other reason or did not get any amount at all are considered as not having access to credit. Credit access may facilitate the household capacity to acquire more inputs and food for the household.

Tropical livestock unit As asset accumulation, tropical livestock unit plays an important role in the household especially when crop production fails to provide sufficient food for the household. Households may sell a part of their livestock to buy food.

Storage facilities Storage facilities played a key role in storing and conserving food crops and prevent them

Table 2 Description of Independent variables

Variable	Description	Measurement
Diversity	Simpson Diversity Index (D)	Continuous
Gender	Gender of the household head (SEX)	1 for Male and 0 for Female
Age	Age of the household head (AGE)	Continuous
Family size	Number of person in the household (FSIZE)	Continuous
Education	Whether household head attended school (EDUC)	1 for none and 0 else 1 for primary and 0 else 1 for secondary and 0 else
Credit	Access to credit (CREDIT)	1 for access and 0 else
Association	Membership of farmers association (ASSOC)	1 if yes and 0 else
Location	Location of the household head Dummy variable	1 if Bante and 0 else 1 if Dassa and 0 else 1 if Glazoue and 0 else 1 if Ouesse and 0 else 1 if Savalou and 0 else
Fertilizer	Access to fertilizer (FERTI)	1 if yes and 0 else
Seed	Access to improved seed (SEED)	1 if yes and 0 else
Extension	Number of extension visit per year (EXTEN)	Continuous
Land size	Total land size (LSIZE)	Continuous (in hectares)
Storage	Access to storage facilities (STOR)	1 if yes and 0 else
Livestock	Tropical livestock Unit (TLU)	Continuous

Table 3 Descriptive statistics

	Mean	SD	Analysis *N*
Food crop land	4.088	1.854	420
Cereal land	1.815	1.075	420
Cereal production	3212.57	3425.905	420
Cereal/capita	196.899	269.273	420
FCS	41.181	10.603	420
Income/capita	82,847.684	74,902.324	420
Food stock	1.94	1.542	420
Market	2.593	1.226	420

from being damaged by pest attack. It is measured as a dummy variable.

Result and discussion

Result of the principal component analysis

Table 3 presents the descriptive statistics of the variable loaded for the PCA. There is a disparity for some of the variables such as income per capita, cereal production and cereal production per capita as evidenced by their high their standard deviation.

The suitability of the data for the PCA was checked using the Bartlett's test of sphericity and the KMO measure of sampling adequacy (Table 4). The result shows that the KMO is greater than 0.6 and the Bartlett's test is also significant.

Table 4 Bartlett's test of sphericity and the KMO test

KMO and Bartlett's test	
Kaiser–Meyer–Olkin measure of sampling adequacy	0.615
Bartlett's test of sphericity	
Approx. Chi Square	2116.405
df	28
Sig.	0.000

Table 5 presents the total variance explained in the table to determine how many components to extract considering only those components with eigenvalues of 1 or more. Looking at Table 5, only the first 3 components have an eigenvalues above 1; these three components explained up to 71% of the total variance within this set of data. The components matrix presented in Table 6 shows the contribution of each variable to the principal component which is the food security. The food security index was directly derived from the component matrix. After computation, Fig. 2 shows the distribution of the food security index. About 37.5% of the households were food insecure. According to [25] the calculated food security index can range from 0 (worst case) to 1 (best case) and can be classified into four thresholds:

$0 < FSI < 0.341$ alarming food insecure.
$0.342 < FSI < 0.435$ moderate food insecure.
$0.436 < FSI < 0.558$ moderate food secure.
$0.559 < FSI < 1$ highly food secure.

Table 5 Total variance explained

Component	Initial eigenvalues			Extraction sums of squared loadings			Rotation sums of squared loadings		
	Total	% of variance	Cumulative %	Total	% of variance	Cumulative %	Total	% of variance	Cumulative %
Total variance explained									
1	2.711	33.893	33.893	2.711	33.893	33.893	2.710	33.873	33.873
2	1.897	23.716	57.609	1.897	23.716	57.609	1.891	23.642	57.516
3	1.078	13.479	71.088	1.078	13.479	71.088	1.086	13.572	71.088
4	0.978	12.227	83.315						
5	0.827	10.331	93.646						
6	0.271	3.386	97.032						
7	0.136	1.703	98.735						
8	0.101	1.265	100.000						

Extraction method: principal component analysis

Table 6 Component matrix

Component matrix[a]	Component		
	1	2	3
Income/cap	0.897		
Cereal product	0.884		
Cereal/cap	0.857		
Food cropland		0.961	
Cereal land		0.961	
Market			−0.806
Food stock	0.421		0.527
FCS			0.350

Extraction method: principal component analysis

[a] 3 components extracted

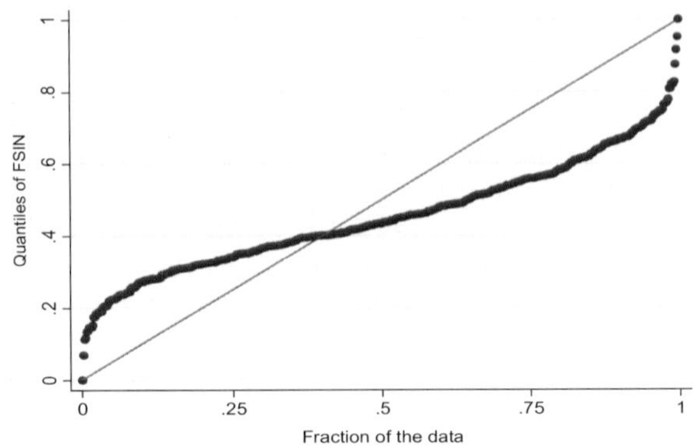

Fig. 2 Distribution of the food security index

Table 7 Estimation results *Source*: Data collected in 2015

Variables	Coefficients	*P* value
Age	−0.001	0.226
Sex	−0.035	0.123
Education		
Primary	0.007	0.567
Secondary	0.119***	0.000
FBOs	0.059	0.200
Extension Services	0.067***	0.000
Location		
Glazoue	−0.015	0.455
Dassa	−0.010	0.631
Ouesse	−0.027	0.175
Savalou	−0.030	0.139
Credit	0.035**	0.023
Simpson Diversity Index	0.198***	0.000
Seed	0.019	0.185
Fertilizer	0.006	0.660
Tropical livestock unit	0.101**	0.023
Storage	0.051***	0.000
Land Size	0.001	0.125
Constant	0.383	0.000
Number of observation = 420		
$F(17, 504) = 8.07$		
Prob > F = 0.000		
R-squared = 0.2139		

*Significant 10%

**Significant 5%

***Significant 1%

Estimation results

Table 7 below presents the results of the effect of crop diversification on household food security. The results indicated that the crops diversification (represented by Simpson diversity index) was significant with positive coefficient. Households growing many crops were more likely to be food secure especially those diversifying into cereals and root and tuber crops. If we assume there is a strong correlation between household dietary diversity and household food security status, we can state that this result are in line with that of Jones et al. [19] who found that more diverse production systems can also lead to more diverse household diets in Malawi. Regarding the remoteness and the poor quality of infrastructure in our study areas, most of the household food access depends on what the household produces mostly because their food diet is linked with their food crops production. Household may choose to diversify crops because of poor infrastructure [14], and this will help them to get the kind of food they need at the local level. Similar results were reported by Herforth [16] who found that in Kenya and Tanzania, the number of crop grown by household is associated with the dietary variety of the household.

The level of education also affected positively the household food security status, and we found that household head who reach secondary school were more likely to have their family food secure. Considering as social capital, education can provide additional opportunity for farmers in term of non-farm employment which may improve household food security status [13]. Also, educated household head are more likely to absorb technologies and new farm technics. Like education, access to extension services was also considered as knowledge acquisition and the result showed a significant and positive relationship between access to extension services and household food security status. This is because contact with extension services tends to enhance the chances of a household having access to better crop production techniques, improved inputs, as well as other production incentives that positively affect farm production and thus household food security.

The results further revealed that access to credit affects positively the household food security status meaning that household with credit access has more chance to be food secure than those without credit access. Access to credit gives the household the opportunity to invest in agricultural business or other non-farm activities which may enhance the household food security profile. The result is in line with the finding of Gecho et al. [15]. According to Tekle and Berhanu [35], credit can also help households to smooth consumption especially during the food shortage period.

Tropical livestock unit also affects positively the household food security status. Livestock plays an important role in household livelihood strategies in the study area and is used mostly as assets accumulation by many households. Livestock contributes to food security status of households in different ways such as by providing cash income, nutrition, draft power, manure, etc. Also livestock serves as savings of assets and used to cope with food insecurity problems during food shortage. Access to storage facilities was also significant and positively associated with household food security status. Storage facilities contribute to reduce food losses and provide households with the opportunity to smooth hunger during lean season. Additionally, farmers are able to improve farm incomes by storing and selling their farm produces at premium prices when demand outstrips supply later in the food shortage period. According to Tefera et al. [34], storage is crucial to improve agricultural incomes and food security for smallholder farmers.

Conclusion and policy recommendations

Food insecurity in rural areas remains a key challenge to be addressed by policies in Benin. This paper investigates the potential effect of crops diversification on household's food security status. Considering the Simpson diversity index as diversification measurement and multidimensional food security index, the study comes out with the findings that crops diversification contributes to the household food security status. In rural remote Benin like the Collines Region, household's access and consumption of food depend mainly on their production and therefore the set of crops to be produced is an important determinant of household food security status. Additionally, many variables affect household's food security status such as: level of education, access to extension services, credit, storage and the tropical livestock unit. The crops diversification policy lunched in 2006 which aim was to help rural households to diversify their livelihood and reduce risk need to be strengthened in order to reduce food insecurity in the country.

Focus on food security ensures that the basic needs of the poorest and most vulnerable groups are not neglected in policy formulation and that requires taking into account all the food security components. Beyond the crops diversification policy, government has to design and implement livelihood diversification policy and design a clear food and nutrition policy document. There is a need to reinforce instruments designed to support crops diversification and strengthen agricultural support services in rural areas.

Authors' contributions

GA is a final year Ph.D. Candidate in Applied Agricultural Economics and Policy. This study is part of his Ph.D. thesis. GA has coordinated the field work for data collection and did the write up of the study under his supervisors. GK is the main Supervisor for this study. He coordinated and supervised the whole process of this research from the proposal to the mains findings. All authors have participated in the different committee meeting that validated each phase of the study. Both authors read and approved the final manuscript.

Authors' information

GA is a final year Ph.D. Candidate in Applied Agricultural Economics and Policy, University of Ghana. This study is part of his Ph.D. thesis. GA areas of study involve Food policy, social protection, gender and rural development. GK is a

Senior Lecturer at the Department of Agricultural Economics and Agribusiness of the University of Ghana. His areas of specialization include Rural Finance, Project management, policy negotiation and agribusiness.

Acknowledgements
Not applicable.

Competing interests
The authors declare that they have no competing interests.

Funding
Not applicable.

References

1. Adesina AA, Ouattara AD. Risk and agricultural systems in northern Cote d'Ivoire. Agric Syst. 2000;66:17–32.
2. Ali M, Abedullah A. Economic and nutritional benefits from enhanced vegetable production and consumption in developing countries. J Crop Prot. 2002;6(1):2002.
3. Antonaci L, Demeke M, Vezzani A. The challenges of managing agricultural price and production risks in sub-Saharan Africa (ESA No 14-09 No. 14–9). Rome: FAO; 2014.
4. Barrett C. Measuring food insecurity. Science. 2010;327:825–8. https://doi.org/10.1126/science.1182768.
5. Birthal PS, Roy D, Negi DS. Assessing the impact of crop diversification on farm poverty in India. World Dev. 2015;72:70–92. https://doi.org/10.1016/j.worlddev.2015.02.015.
6. Bry X. Analyses factorielles simples. (Economica, Ed.) (Economica). Economica; 1995.
7. Carletto C, Zezza A, Banerjee R. Towards better measurement of household food security: harmonizing indicators and the role of household surveys. Glob Food Secur. 2013;2(1):30–40. https://doi.org/10.1016/j.gfs.2012.11.006.
8. Cervantes-Godoy D, Dewbre J. Economic importance of agriculture for poverty reduction. OECD Food, Agriculture and Fisheries Working Papers; 2010. http://www.oecd.org/dataoecd/49/19/46341298.pdf. Accessed 15 Dec 2016.
9. Cochran WG. Sampling techniques. 3rd ed. New York: Willey; 1977.
10. Demeke AB, Zeller M. Using panel data to estimate the effect of rainfall shocks on smallholders food security and vulnerability in rural Ethiopia. Hohenheim; 2009.
11. FAO. Measurement and assessment of food deprivation and undernutrition. In: Proceedings of an international scientific symposium. Rome: FAO; 2002.
12. FAO, IFAD, & WFP. The State of Food Insecurity in the World. Meeting the 2015 International hunger targets: taking stock of uneven progress. ROME; 2015. http://www.fao.org/3/a-i4646e.pdf. Accessed 4 Jan 2017.
13. FAO, WFP. The state of food insecurity in the world: addressing food insecurity in protracted crises. Food and Agriculture Organization of the United Nations. Rome; 2010. http://www.fao.org/docrep/013/i1683e/i1683e.pdf.
14. Frayne B, McCordic C. Planning for food secure cities: measuring the influence of infrastructure and income on household food security in Southern African cities. Geoforum. 2015;65:1–11. https://doi.org/10.1016/j.geoforum.2015.06.025.
15. Gecho Y, Ayele G, Lemma T, Alemu D. Livelihood strategies and food security of rural households in Wolaita Zone, Southern Ethiopia. Dev Ctry Stud. 2014;4(14):123–36.
16. Herforth AW. Promotion of traditional african vegetables in kenya and tanzania : a case study of an intervention representing. Cornell University; 2010. file:///Users/amandinesagnieres/Downloads/Herforth.Anna.pdf.
17. Hua W, Hite D. Assessing the relationship between crop choice and land use change using a Markov model. In: Selected paper prepared for presentation at the American agricultural economics association annual meeting, Providence, Rhode Island; 2005.
18. INSAE. Evaluation de la pauvreté au Bénin. Benin: Cotonou; 2013.
19. Jones AD, Shrinivas A, Bezner-kerr R. Farm production diversity is associated with greater household dietary diversity in Malawi: findings from nationally representative data. Food Policy. 2014;46:1–12. https://doi.org/10.1016/j.foodpol.2014.02.001.
20. Joshi PK, Ashok G, Birthal PS, Tewari L. Agriculture diversification in South Asia: patterns, determinants, and policy implications (No. 57), Washington, D.C. 20006; 2003.
21. Magaña-lemus D, Ishdorj A, Iii CPR, Lara-álvarez J. Determinants of household food insecurity in Mexico. Agric Food Econ. 2016. https://doi.org/10.1186/s40100-016-0054-9.
22. Mallick D, Rafi M. Are female-headed households more food insecure? Evidence from Bangladesh. World Dev. 2010;38(4):593–605. https://doi.org/10.1016/j.worlddev.2009.11.004.
23. Mandal R, Bezbaruah MP. Diversification of cropping pattern: its determinants and role in flood affected agriculture of assam plains. Indian J Agric Econ. 2013;68(2):169–81.
24. Maxwell D, Caldwell R, Langworthy M. Measuring food insecurity: can an indicator based on localized coping behaviors be used to compare across contexts? Food Policy. 2008;33(6):533–40. https://doi.org/10.1016/j.foodpol.2008.02.004.
25. Napoli M. Towards a food insecurity multidimensional index (FIMI). Master Thesis, 2011. http://typo3.fao.org/fileadmin/templates/ERP/uni/FIMI.pdf. Accessed 22 Nov 2016.
26. Oyarzun PJ, Mary Borja R, Sherwood S, Parra V. Making sense of agrobiodiversity, diet, and intensification of smallholder family farming in the highland andes of ecuador. Ecol Food Nutr. 2013;52(6):515–41. https://doi.org/10.1080/03670244.2013.769099.
27. Pellegrini L, Tasciotti L. Crop diversification, dietary diversity and agricultural income: empirical evidence from eight developing countries. Can J Dev Stud. 2014;35(2):211–27.
28. Perz SG. Household demography and land use allocation among small farms in the Brazilian Amazon. Hum Ecol Rev. 2002;9(2):1–16.
29. Ravallion M, Chen S, Sangraula P. New evidence on the urbanization of global poverty (No. WPS4199), Washington; 2008.
30. Riely F, Mock N, Cogill B, Bailey L, Kenefick E. Food security indicators and framework for use in the monitoring and evaluation of food aid programs, Washington, D.C. 20009; 1999
31. Shiyani RL, Pandya HR. Diversification of agriculture in Gujarat: A spacio-temporal analysis. IndianJournal of Agricultural Economics. 1998;53(4):627–39.
32. Stewart WM, Roberts TL. Food security and the role of fertilizer in supporting it. In: 1st international symposium on innovation and technology in the phosphate industry (vol. 46, p. 76–82). Procedia Engineering; 2012. http://doi.org/10.1016/j.proeng.2012.09.448.
33. Swades P, Shyamal K. Implications of the methods of agricultural diversification in reference with Malda district: drawback and rationale. Int J Food Agric Vet Sci. 2012;2(2):97–105. http://www.cibtech.org/jfav.htm
34. Tefera T, Kanampiu F, Groote H De, Hellin J, Mugo S, Kimenju S. The metal silo: an effective grain storage technology for reducing post-harvest insect and pathogen losses in maize while improving smallholder farmers' food security in developing countries. Crop Prot. 2011;30(3):240–5. https://doi.org/10.1016/j.cropro.2010.11.015.
35. Tekle L, Berhanu K. Determinants of rural farm household food security in Boloso Sore District of Wolaita Zone in Ethiopia. Asian J Agric Ext Econ Sociol. 2015;5(2):57–68. https://doi.org/10.9734/AJAEES/2015/14833.
36. Thierfelder C, Mombeyarara T, Mango N, Rusinamhodzi L. Integration of conservation agriculture in smallholder farming systems of southern Africa: identification of key entry points. Int J Agric Sustain. 2013;11:317–30. https://doi.org/10.1080/14735903.2013.764222.
37. Thompson HE, Berrang-ford L, Ford JD. Climate change and food security in Sub-Saharan Africa: a systematic literature review. Sustainability. 2010;2:2719–33. https://doi.org/10.3390/su2082719.
38. Torheim LE, Ouattara F, Diarra MM, Thiam FD, Barikmo I, Hatløy A, Oshaug A. Nutrient adequacy and dietary diversity in rural Mali: association and determinants. Eur J Clin Nutr. 2004;58(4):594–604. https://doi.org/10.1038/sj.ejcn.1601853.
39. Von Braun J. Agricultural commercialization: impacts on income and nutrition and implications for policy. Food Policy. 1995;20(3):1995.

40. Vyas S, Kumaranayake L. Constructing socio-economic status indices: how to use principal components analysis. Health Policy Plan. 2006;21(6):459–68. https://doi.org/10.1093/heapol/czl029.
41. WFP. Analyse Globale de la Vulnérabilité et de la Sécurité Alimentaire (AGVSA). Cotonou: Republique du Benin; 2014.
42. Van Wijk MT. From global economic modelling to household level analyses of food security and sustainability: how big is the gap and can we bridge it? Food Policy. 2014;49:378–88. https://doi.org/10.1016/j.foodpol.2014.10.003.
43. Zeller M, Sharma M, Henry C. An operational method for assessing the poverty outreach performance of development policies and projects: results of case studies in Africa, Asia, and Latin America. World Dev. 2006;34(3):446–64.

Permissions

All chapters in this book were first published in A&FS, by BioMed Central; hereby published with permission under the Creative Commons Attribution License or equivalent. Every chapter published in this book has been scrutinized by our experts. Their significance has been extensively debated. The topics covered herein carry significant findings which will fuel the growth of the discipline. They may even be implemented as practical applications or may be referred to as a beginning point for another development.

The contributors of this book come from diverse backgrounds, making this book a truly international effort. This book will bring forth new frontiers with its revolutionizing research information and detailed analysis of the nascent developments around the world.

We would like to thank all the contributing authors for lending their expertise to make the book truly unique. They have played a crucial role in the development of this book. Without their invaluable contributions this book wouldn't have been possible. They have made vital efforts to compile up to date information on the varied aspects of this subject to make this book a valuable addition to the collection of many professionals and students.

This book was conceptualized with the vision of imparting up-to-date information and advanced data in this field. To ensure the same, a matchless editorial board was set up. Every individual on the board went through rigorous rounds of assessment to prove their worth. After which they invested a large part of their time researching and compiling the most relevant data for our readers.

The editorial board has been involved in producing this book since its inception. They have spent rigorous hours researching and exploring the diverse topics which have resulted in the successful publishing of this book. They have passed on their knowledge of decades through this book. To expedite this challenging task, the publisher supported the team at every step. A small team of assistant editors was also appointed to further simplify the editing procedure and attain best results for the readers.

Apart from the editorial board, the designing team has also invested a significant amount of their time in understanding the subject and creating the most relevant covers. They scrutinized every image to scout for the most suitable representation of the subject and create an appropriate cover for the book.

The publishing team has been an ardent support to the editorial, designing and production team. Their endless efforts to recruit the best for this project, has resulted in the accomplishment of this book. They are a veteran in the field of academics and their pool of knowledge is as vast as their experience in printing. Their expertise and guidance has proved useful at every step. Their uncompromising quality standards have made this book an exceptional effort. Their encouragement from time to time has been an inspiration for everyone.

The publisher and the editorial board hope that this book will prove to be a valuable piece of knowledge for researchers, students, practitioners and scholars across the globe.

List of Contributors

Diriba Welteji and Kerime Mohammed
Department of Rural Development and Agricultural Extension, College of Agriculture and Natural Resources, Madda Walabu University, Bale-Robe, Ethiopia

Kedir Hussein
Department of Public Health, College of Medicine, Madda Walabu University, Bale-Robe, Ethiopia

S. R. Dominick, Nicole J. Olynk Widmar and Jonathan Z. Grennell Weir
Department of Agricultural Economics, Purdue University, West Lafayette, IN, USA

Audrey Ruple
Department of Comparative Pathobiology, Purdue University, 725 Harrison Street, West Lafayette, IN 47907, USA

Lalatendu Acharya
Department of Consumer Sciences, Purdue University, West Lafayette, IN, USA

Benyam Tadesse
College of Agriculture and Natural Resources, Department of Agro Economics, Mizan-Tepi University, Mizan-Aman, Ethiopia

Fayera Bakala
College of Agriculture and Natural Resources, Department of Natural Resource Economics, Mizan-Tepi University, Mizan-Aman, Ethiopia

Lamirot W. Mariam
College of Agriculture and Natural Resources, Department of Horticulture, Mizan-Tepi University, Mizan-Aman, Ethiopia

Farrhin Nowshad, Md. Nazibul Islam and Mohidus Samad Khan
Department of Chemical Engineering, Bangladesh University of Engineering and Technology, Dhaka 1000, Bangladesh

Mohammad Emami, Morteza Almassi and Hossein Bakhoda
Department of Agricultural Mechanization, Science and Research Branch, Islamic Azad University, Tehran, Iran

Issa kalantari
Department of Environment, Tehran, Iran

Sirak Bahta
International Livestock Research Institute (ILRI), Gaborone, Botswana

Francis Wanyoike
International Livestock Research Institute (ILRI), Nairobi, Kenya

Hikuepi Katjiuongua
World Bank Group (WBG), Addis Ababa, Ethiopia

Davis Marumo
Botswana University of Agriculture and Natural Sciences (BUAN), Gaborone, Botswana

Bright Nkrumah
Department of Political Studies, University of the Witwatersrand, Private Bag 3, Wits 2050, South Africa

Md. Taj Uddin and Aurup Ratan Dhar
Department of Agricultural Economics, Bangladesh Agricultural University, Mymensingh 2202, Bangladesh

Belachew Dessalegn
Department of Rural Development and Agricultural Extension, College of Agriculture and Natural Resources, Madda Walabu University, Bale Robe, Ethiopia

Nelson Mango and Clifton Makate
International Centre for Tropical Agriculture (CIAT), Mt. Pleasant, Harare, Zimbabwe

Lawrence Mapemba
Department of Applied Economics, Lilongwe University of Agriculture and Natural Resources, Bunda College Campus, Lilongwe, Malawi

Mathinda Sopo
International Centre for Tropical Agriculture (CIAT), c/o Chitedze Agricultural Research Station, Lilongwe, Malawi

Mahlet Mulugeta
Food and Agricultural Organization, Addis Ababa, Ethiopia

Gebeyaw Tiruneh
GAMBY Medical and Business College, Bahir Dar, Ethiopia

Zewdie Aderaw Alemu
Department of Public Health, College of Health Sciences, Debre Markos University, Debre Marqos, Ethiopia
School of Public Health, Addis Ababa University, Addis Ababa, Ethiopia

Getachew Teferi Moroda, Degefa Tolossa and Negussie Semie
Center for Rural Development, College of Development Studies, Addis Ababa University, Addis Ababa, Ethiopia

Vincent Ezin, Florent Quenum, Roméo Houénagnon Bodjrenou, Chabrolle M. I. Kpanougo and Adam Ahanchede
Department of Crop Production, Faculty of Agricultural Science, University of Abomey-Calavi, Abomey-Calavi, Benin

Emeric M. G. Kochoni
Institut National de la Recherche Scientifique – Eau Terre Enironnement (INRS-ETE), Quebec, Canada

Bienvenu Ifagbémi Chabi
Department of Food Science, Faculty of Agricultural Science, University of Abomey- Calavi, Abomey-Calavi, Benin

Kedir Hussein Abegaz
Biostatistics and Health Informatics, Public Health Department, Goba Referral Hospital, Madda Walabu University, Bale Goba, Ethiopia

Berchie Asiedu
Department of Fisheries and Water Resources, University of Energy and Natural Resources, Sunyani, Ghana

Pierre Failler
Economics and Finance Group, Portsmouth Business School, University of Portsmouth, Richmond Building, Portland Street, Portsmouth PO1 3DE, UK

Yolaine Beygens
Fish Quality Independent Consultant, Dieweg 30A, 1180 Brussels, Belgium

Shishay Kahsay Weldearegay and Dawit Gebremichael Tedla
Food Security, Shire Campus, College of Agriculture, Aksum University, Shire Endaselassie, Tigray, Ethiopia

Carin Napier and Beryl Makhaye
Department of Food and Nutrition Consumer Sciences, Durban University of Technology, Durban 4000, South Africa

Wilna Oldewage-Theron
Department of Nutritional Sciences, College of Human Sciences, Texas Tech University, Lubbock, TX 79409-1270, USA

Nataliya Apanovich
Graduate Program in Sustainable Agriculture, Iowa State University, Ames, IA 50011, USA

Robert E. Mazur
Department of Sociology, Iowa State University, Ames, IA 50011, USA

Camillus Abawiera Wongnaa and Dadson Awunyo-Vitor
Department of Agricultural Economics, Agribusiness and Extension, Kwame Nkrumah University of Science and Technology, Private Mail Bag, University Post Office, Kumasi, Ghana

Gilbert Onionkiton Adjimoti and George Tsey-Mensah Kwadzo
Department of Agricultural Economics and Agribusiness, University of Ghana, Legon, Accra, Ghana

Index